EAI/Springer Innovations in Communication and Computing

Series editor

Imrich Chlamtac, European Alliance for Innovation, Ghent, Belgium

Editor's Note

The impact of information technologies is creating a new world yet not fully understood. The extent and speed of economic, life style and social changes already perceived in everyday life is hard to estimate without understanding the technological driving forces behind it. This series presents contributed volumes featuring the latest research and development in the various information engineering technologies that play a key role in this process.

The range of topics, focusing primarily on communications and computing engineering include, but are not limited to, wireless networks; mobile communication; design and learning; gaming; interaction; e-health and pervasive healthcare; energy management; smart grids; internet of things; cognitive radio networks; computation; cloud computing; ubiquitous connectivity, and in mode general smart living, smart cities, Internet of Things and more. The series publishes a combination of expanded papers selected from hosted and sponsored European Alliance for Innovation (EAI) conferences that present cutting edge, global research as well as provide new perspectives on traditional related engineering fields. This content, complemented with open calls for contribution of book titles and individual chapters, together maintain Springer's and EAI's high standards of academic excellence. The audience for the books consists of researchers, industry professionals, advanced level students as well as practitioners in related fields of activity include information and communication specialists, security experts, economists, urban planners, doctors, and in general representatives in all those walks of life affected ad contributing to the information revolution.

About EAI

EAI is a grassroots member organization initiated through cooperation between businesses, public, private and government organizations to address the global challenges of Europe's future competitiveness and link the European Research community with its counterparts around the globe. EAI reaches out to hundreds of thousands of individual subscribers on all continents and collaborates with an institutional member base including Fortune 500 companies, government organizations, and educational institutions, provide a free research and innovation platform.

Through its open free membership model EAI promotes a new research and innovation culture based on collaboration, connectivity and recognition of excellence by community.

More information about this series at http://www.springer.com/series/15427

Lucia Knapcikova • Michal Balog
Dragan Peraković • Marko Periša
Editors

New Approaches in Management of Smart Manufacturing Systems

Knowledge and Practice

Springer

EAI
RESEARCH MEETS INNOVATION

Editors
Lucia Knapcikova
Faculty of Manufacturing Technologies
Department of Industrial Engineering
and Informatics
Technical University of Kosice
Presov, Slovakia

Michal Balog
Faculty of Manufacturing Technologies
Department of Industrial Engineering
and Informatics
Technical University of Kosice
Presov, Slovakia

Dragan Peraković
Faculty of Transport and Traffic Sciences
University of Zagreb
Zagreb, Croatia

Marko Periša
Faculty of Transport and Traffic Sciences
University of Zagreb
Zagreb, Croatia

ISSN 2522-8595 ISSN 2522-8609 (electronic)
EAI/Springer Innovations in Communication and Computing
ISBN 978-3-030-40175-7 ISBN 978-3-030-40176-4 (eBook)
https://doi.org/10.1007/978-3-030-40176-4

This Springer imprint is published by the registered company Springer Nature Switzerland AG
The registered company address is: Gewerbestrasse 11, 6330 Cham, Switzerland

Preface

New Approaches in Management of Smart Manufacturing Systems provides comprehensive and effective exchange of information on current developments in the management of smart manufacturing systems and Industry 4.0. Manufacturing systems tend to be developed mostly based on lean or agile approaches to reach high level of competitiveness in global environment.

Our priority is to create an environment that rewards excellence transparently and builds recognition objectively regardless of age, economic status, or country of origin, where no membership fees or closed door committees stand in the way of research career.

The world moves further into the digital age, generating vast amounts of data and big digital content. And how will we access this content? We have to systematically develop tools to break down barriers of accessibility for next generations.

In this book, results from different fields of science are presented, but ultimately all have the same goal; they are for people, from people.

My thanks go to all the authors without which this publication would not have been written.

Presov, Slovakia | Lucia Knapcikova

Acknowledgement

I would like to thank all authors and reviewers for their thoughtful comments and efforts towards improving our book.

My thanks is to Dean of Faculty of Manufacturing Technologies with a seat in Prešov of Technical University of Košice Dr.h.c. Prof. Dr. Jozef Zajac, to the Head of the Department of Industrial Engineering and Informatics, to Assoc. Prof. Dr. Michal Balog, and special thanks to my office colleague Dr. Annamária Behúnová for her big support to me.

Lucia Knapcikova

Contents

About the Editors

Lucia Knapcikova works at the Department of Industrial Engineering and Informatics at the Faculty of Manufacturing Technologies (FMT) of Technical University of Košice as researcher. In 2017, she completed her Habilitation for Associate Professor at the FMT in the field "Manufacturing Technologies." In 2011, she completed her PhD study at the above-mentioned Faculty. Assoc. Prof. Knapčíková, PhD, is author for more than 150 scientific papers in Slovakia and abroad, and she is owner of 6 utility models. She participated in several international research projects, mainly in Germany. The most important project is H2020 Less Than Wagon Load (Project duration 2017–2020) in collaboration with Belgium, Germany, Netherland, Italy, and Slovakia, where her position is Faculty project coordinator. The project has the objective to develop a SMART specialized logistics cluster for the chemical industry in the Port of Antwerp in order to shift transport volumes from road to rail freight. Actually, she participates in the project EUROSTARS (BRIDGE2ERA Programm) in collaboration with Technical University of Applied Sciences in Wildau (TH Wildau), Germany. The aim of the project is Industry 4.0 strategic orientation between universities and SME. She lectured at the universities in Germany (TH Wildau), Czech Republic (TU Liberec), and Croatia (J.J. Strossmayer University of Osijek, Mechanical Engineering Faculty in Slavonski Brod). Her area of interest includes SMART technologies, SMART materials, and application possibilities directly in the industry.

In the year 2018, she was awarded "Personality of Science and Technology Under 35 Years." The Ministry of Education, Science, Research and Sport of the Slovak Republic, in support of the development of science and technology and the recognition of the exceptional achievements in this field, awarded the Science and Technology Award—the highest award in science and technology, for the year 2018. She is member of European Alliance for Innovation (EAI), Belgium (www.eai.eu), TEAM Society in Croatia. She is active member in the Editorial Board of Acta Tecnología Journal (Slovakia) and SAR Journal (Serbia). From the beginning of the year 2018, she has been working as Editor-in-Chief of Journal EAI Endorsed Transaction on Future Internet—Ghent, Belgium (https://eudl.eu/journal/ue). She is responsible person for organizing conferences under EAI International Conference on Management of Manufacturing Systems, 2019, and previously MMS2018, MMS2017, and MMS2016.

Michal Balog is Head of Department of Industrial Engineering and Informatics at Faculty of Manufacturing Technologies with a seat in Prešov, TU Košice. His scientific and educational field is oriented on the green logistics, management, quality control, and methods of production systems. He is the organizer of many domestic and international conferences, acts as tutor on PhD study programs, and has participated in numerous Slovak and international projects as team member and as the project head. The most important project is H2020 "Less Than Wagon Load," where he participated as a team member, and in the year 2015 he finished the project entitled "Promotion & Enhancement of Centre for Research on Transportation" with EU Structural Funds. He is author for more than 100 publications in Slovakia and abroad.

Dragan Peraković graduated and completed his Master's degree and PhD (doctoral dissertation title: *The Model of Information Distribution to Traffic System Users*) from the Faculty of Transport and Traffic Sciences (FPZ) at the University of Zagreb. Post graduation, he began his career at the FPZ where he currently holds the positions of Head of Department of Information and Communication Traffic and Head of Chair of Information Communication Systems and Services Management. He has engaged in several international scientific projects and R&D studies as a researcher, leading researcher, and evaluator (CiViTAS ELAN, Harmonized Inland Navigation Transport Through Education and Information Technology—HINT, Green Intermodal Freight Transport—GIFT, etc.). He was the initiator and leading researcher for many important national projects such as the Research of the Context of the use of Smartphones and Related Information and Communication Services, Research of Possibilities on Applying the IoT Concept to Improve the Safety Movement of Blind and Visually Impaired Persons along the Traffic Network, and Information and Communication Services for the Movement along the Traffic Network by the Persons of Reduced and Difficult Mobility. He is an author/co-author of more than 100 science papers (including two CC papers) and studies presented in journals and proceedings of international conferences and five chapters in international scientific books. He also participated in educational activities within the Faculty as a lecturer and mentor for undergraduate and graduate study programs. Until now, he has been mentoring more than 100 undergraduate and graduate students and 3 postgraduate doctoral students. Prof. Dr. Dragan Peraković is a member, board member, and official editor of many journals and conferences of his profession, such as International Telecommunications Society (ITS), International Telecommunication Union (ITU), IEEE Communications Society, European Alliance for Innovation (EAI), DAAAM International, The Society of Digital Information and Wireless Communications (SDIWC), Scientific Council for Traffic of Croatian Academy of Sciences and Arts (HAZU), and the Croatian Chamber of Traffic and Transport Technologies (HKITPT). Prof. Dr. Dragan Peraković is chief editor of the International Journal of Cyber-Security and Digital Forensics (IJCSDF)."

Marko Periša was born on March 2, 1979, in Zagreb where he finished elementary and high school and enrolled in Faculty of Transport and Traffic Sciences, University of Zagreb (postal and telecommunication course). During the academic year 2002/2003, he received the Rector's Award of University of Zagreb for student work titled "Model of multiple-criteria decision making in the function of selecting mobile telephony operatory in Republic of Croatia." On April 11, 2013, he defended doctoral dissertation on Faculty of Transport and Traffic Sciences titled "Dynamic guiding and routing of disabled and visually impaired persons in traffic." He is the Head of the Laboratory of Development and Research of Information and Communication Assistive Technology and Head of Chair of Information Communication Systems and Networks. Since 2005, he has been a part of the Register of Researches at the MZO under the registered number 269095. In his scientific research work, he published more than 40 papers in scientific journals and international scientific conferences and 3 chapters in scientific books. He also participated in several international (EU), national, faculty projects, and supporting programs for research at the University of Zagreb which deal with issues in traffic technologies. He is a member of societies: TM Forum, DAAAM International, International Telecommunications Society (ITS), Institute for Systems and Technologies of Information, Control and Communication (INSTICC), European Alliance for Innovation (EAI). He is also a member of Supervising Board of the Croatian Chamber of Traffic and Transport Technologies (HKITPT), professional class of engineers of postal, information, and communication traffic, and a member of Scientific Council for Traffic of the Croatian Academy of Sciences and Arts (Traffic Medicine and Traffic Safety Section).

RFID System for Effective Managing of Medical Facilities

Michal Balog, Jozef Husár, and Angelina Iakovets

1 Health-Care Industry and Medical Facilities

1.1 Health-Care Institutions in Global Market

The last years of the century are well known due to spread of smart technologies in all sectors of economic activity. The spread of new technologies has also affected one of the most important areas—health care. There are many projects aimed at supporting this idea, like state bills as well as world health organizations such as WHO (World Health Organization), international medical associations, and others.

WHO calls for urgent action target on reducing the harm done to patients under care. Every year, as a result of noncompliance with safety rules, medical care worldwide affects millions of patients and kills 2.6 million people in low- and middle-income countries. Most of these deaths are preventable. Globally, personal, social, and economic damage from harming patients amounts to many trillions of US dollars. The World Health Organization intends to draw public attention around the world to patient safety issues. This information becomes a basis for a great event on September 17, 2019, on the occasion of the first ever World Patient Safety Day that opens a global campaign of solidarity with patients. According to WHO Director-General, Dr. Tedros Adhanom Ghebreyesus, "Medical care should not harm anyone. And yet, all over the world, at least one person dies every minute from unsafe medical care." He encourages to create a patient safety culture within which patients are seen as partners. When providing primary and outpatient care, four out of ten patients are harmed. The most serious consequences are errors in the diagnosis, as

M. Balog (✉) · J. Husár · A. Iakovets
Faculty of Manufacturing Technologies of the Technical University of Kosice
with a seat in Presov, Presov, Slovak Republic
e-mail: Michal.balog@tuke.sk

L. Knapcikova et al. (eds.), *New Approaches in Management of Smart Manufacturing Systems*, EAI/Springer Innovations in Communication and Computing, https://doi.org/10.1007/978-3-030-40176-4_1

well as the appointment and use of drugs. The damage from medication errors alone is about 42 billion US dollars per year [1].

Causing harm in the provision of medical care is unacceptable. WHO calls on countries and partners around the world to take urgent action to reduce the harm done to patients. Patient safety and the quality of health-care delivery are critical to the effectiveness of a health service delivery system and the achievement of universal health coverage [1].

Investing in patient safety can result in significant cost savings. The cost of treating the consequences of harming patients is much higher than the cost of taking preventive measures. For example, only in the USA, targeted measures for improving the safety of patients in hospitals operating under the state medical care program for the elderly allowed saving about 28 billion US dollars in the period 2010–2015 [1].

Based on the information above, it is possible to consider that medical facilities should accept smart solution, which eliminates human failure. As the proposed article is aimed at technical solutions to main activities of health-care facilities, it means that the next part of research will be linked with analysis of activities of health-care institutions.

1.2 Review of Type of Health-Care Facility Activities

Medical institutions have a lot of activities, and for a better understanding of the issue, it is necessary to have an in-depth discussion in this economic field. By studying literary sources, it became possible to identify the main hospital activities: accident and emergency service; anesthetist; asylum activities; children's hospital; convalescent home; ear nose, and throat hospital; ear, nose, and throat specialist; emergency room services; eye hospital; family planning services with accommodation in hospital; general hospital; general hospital psychiatric unit; general medical consultant; genitourinary specialist; geriatric hospital; geriatrician; gynecologist; hospice; hospital and institution for care of the sick; hospital food services; infectious disease hospital; infirmary; isolation hospital; leprosaria activities; maternity hospital; medical and paramedical staff services; medical consultation; medical nursing homes; mental handicap hospital; mental health help; mental hospital; military base hospitals; military hospital; morbid anatomy specialist; nursing home with medical care; operating room services; ophthalmic hospital; orthopedic hospital; pharmacy services; pre-convalescent hospital; prison hospitals; private hospital; psychiatric unit; radiology unit; rehabilitation hospital; sanatoria providing hospital type of care; smallpox hospital; social medicine specialist; special hospital; surgeon; tuberculosis sanatorium or hospital; eye specialized unit; hematological facility; laboratory and technical facilities, including radiologic and anesthesiology services; and urologist unit [2].

All these activities are impossible without medicaments and medicines. To monitor the expiration date and to plan their warehouse activities are the key elements of successful maintenance of health organizations. Warehousing is the main element of timely medical services and their quality. Also, do not forget about the laboratories, which are often a part of medical institution. In difficult situations, the processing speed of medical sample data is the basis for a timely diagnosis.

The research of WHO had shown that fatal diseases are a bottleneck in this area, as they require an immediate and instant response from medical staff. The speed and quality of services provided often depend on the preparation of the necessary materials to assist the patient.

Effective management of medical institutions is often linked with common economic success of a country [3]. A large number of medical institutions are trying to keep up with other economic sectors and trying to actively introduce new technologies to provide their services. With respect to the warehouse management system, not every institution is able to fully automate this process. This is due to several factors:

- Lack of financial resources.
- Lack of staff training.
- Technical basis for introducing innovations is very weak.

Medical institutions are trying to keep up to date, introducing information management systems, but those most often include managing a database of clients as well as registering customer requests to this institution [3].

1.3 Specificities of Warehousing in Medical Facilities

The main tasks of organizing a warehouse activity in medical institutions are:

- Acceptance and storing of materials and products.
- Preparation of materials for use.
- The issuance of materials in the prescribed manner.
- Organization of inventory accounting and regulation of stock standards [4, 5].

To ensure quality of service, it is necessary to provide rational planning of warehouse activities that will ensure:

- Direct transport routes and eliminate unnecessary manipulations.
- Continuity of processes.
- Increase the productivity of processes.
- Exclude contact with products, thereby ensuring sterility.
- Reduce the number of employees and the complexity of their work.
- Will ensure the safety manipulation with materials.

World practice shows top nine innovations in warehousing, such as the following [6]:

1. *Robotics and Automation*

   As robots become smarter, that is, picking, packing and sorting, streamlining assembly operations, loading/unloading, receiving and putting away orders to processing, and more, robotic automation helps to provide higher quality and less downtime and fewer errors.

Globally, robotic automation has experienced strong growth over the past decade, with the International Federation of Robotics reporting that around 254,000 units were sold in 2015—a 41% increase from that of 2013. But it is necessary to note that such technology is very expensive and not every company will be able to invest in it.

2. *Sensors and Automatic Identification*

   Automatic identification technologies help companies to use data for streamline and bring visibility to warehousing processes. By quickly and automatically identifying and tracking customer, it is able to identity location of physical items in a supply chain. Auto ID provides a major opportunity to improve tracking process, control, and inventory management.
3. *Radio Frequency Identification (RFID)*

   Among the most promising automatic technologies is RFID, which is already increasingly being adopted in modern warehouses [6].
4. *Electronic Data Interchange (EDI)*

   EDI tech allows for the seamless flow of information between two different computer systems. EDI is expected to witness significant growth, with a global compound annual growth rate of 27.3% by 2025, according to Grand View Research.
5. *The Internet of Things (IoT) and Telematics*

   The market for the connected devices network known as IoT in warehouse management is predicted to reach 19.06 billion US dollars by 2025, says Grand View Research, from 2.27 billion US dollars in 2015. Innovators are increasingly turning to onboard covert tracking technology—telematics—to improve visibility once shipments are in motion.
6. *Wearables*

   Smart glass hardware and software are being integrated into warehousing and supply chains across many levels.
7. *Moving to the Cloud*

   The cloud—a centralized location on the Internet that stores data, making it accessible anytime, anywhere, from any device—offers businesses greater scalability, security, efficiency, and flexibility. Migrating to the platform will make warehousing easier, safer, and cheaper as a result.
8. *Autonomous Vehicles*

   Driverless vehicles are predicted to become widespread within the next 10 years. Applications include autonomous transport assisted highway trucking and convoying in line-haul transportation and last-mile delivery. Also, it can be autonomous robotic vehicles, and the great example is autonomous warehousing where goods transporting is made by such devices.
9. *Drones*

   Drones also have a 50% projected adoption rate for warehousing and supply chains in the next 10 years, according to MHI and Deloitte. They have a wide variety of applications, from scanning hard-to-reach barcodes to safely navigating warehouses and reducing the need for forklifts and other costly or potentially dangerous machinery.

The first user of a such technology was e-commerce giant Amazon where a patent was created that describes a logistics technology called an "airborne fulfillment center"—a drone capable of flying at altitudes of 45,000 feet or more that would house items the company sells through its online marketplace [7].

Identification systems can be a tool for achieving such results, due to its price and ease of use. There are three main types for matching objects:

- Barcodes (1D codes), especially EAN codes (European article number).
- QR codes.
- RFID tags.

Medicament transporting between hospital departments requires access to information about the type of medicament, its pharmaceutical group, the amount of delivery, and its shelf life, which is transmitted as basic information.

Barcodes are able to contain only information about the country code, manufacturer code, product code, check digit, product symbol, and license type. The most commonly used are:

- EAN-8 (abbreviated)—encoded by 8 digits.
- EAN-13 (full)—encoded by 13 digits.
- EAN-128—encodes any number of letters and numbers merged into adjustable groups [8].

There are four main types of 2D (QR) codes:

- Digital: from 10 to 3 digits, up to 7089 digits.
- Alphanumeric: supports 10 numbers, letters from A to Z, and a few special characters, 11 bits, up to 4296 characters.
- Bit: data in any suitable encoding (ISO 8859-1 standard) up to 2953 bytes.
- Kanji: 13 bits per hieroglyph, up to 1817 characters [8].

QR codes are able to contain more data than barcodes, but they have one important drawback—they are vulnerable to damage during transporting and manipulation with designated products. The principle of 1D and 2D code scanning requires several criteria: scanner should be able to read the information from code; that is why it should be placed near the barcode at a certain angle and distance. Also, such a technology does not eliminate the occurrence of error caused by a human factor, because scanning and manipulation process requires humans [8].

RFID (radio frequency identification) is an automatic identification by radio frequency waves reflected from an RFID tag; that is why it excludes fails provoked by humans. The main types of RFID are:

- Passive high-frequency RFID (UHF) allocates 8 kB of memory.
- Active RFID takes up 128 kB [3, 9, 10].

To choose an appropriate technology, the advantages and disadvantages of each technology should be considered (see Table 1).

Table 1 Comparison of identification technologies [3]

Parameters	RFID	1D and 2D codes
Placement	Scanning past the visual field	Scanning only in the visual field
Amount of memory	10–10,000 bytes	up to 100 bytes
Ability to rewrite information and reuse	Yes	No
Scanning distance	Up to 10 m	Up to 2 m
Scan multiple objects at once	Up to 200 tags/s	Only 1 code
Protection against the influence of the environment: mechanical and chemical injuries, temperature, and humidity	Enhanced tag protection	Depends on the material
Service life	More than 10 years	Depends on label's material
Security and anti-fraud protection	Not falsified	Falsified
Readability in case of damage	Unreadable	Difficult
Identification of moving objects	Yes	Difficult
Sensitivity to electric fields	Yes	No

According to Table 1, the most suitable technology is RFID, but it is more expensive than competitive ones. It is clear that both technologies can be used for different accounted objects, but it will be difficult to compare both technologies and implement them into information system of the medical facility.

Based on mentioned criteria (from Table 1), RFID technology was selected as appropriate due to protection against counterfeiting, ease of use, automatically scanning without contact with an object, and scanning of a large number of medicines located near the RFID scanner [8, 11]. Regarding the specific features of some medicines, labels should be efficient at all conditions of the environment, such as waterproof features or reading ability (information should be readable regardless of the position of the marked object), and RFID tags are suitable for this.

Also, it is necessary to note that health-care institutions work with a large value of information which is required at each stage of patient treatment. All these data cannot be lost, since the correctness of the diagnosis and the quality of the services provided depend on them. These are also criteria to use RFID technology for warehousing system.

By the end of 2019, all EU countries should introduce a single standard for drug labeling, and then by 2020 biological samples, medical devices, and instruments should be labeled. Therefore, it is necessary to determine the correct identification technology [5]. The European Commission determines criteria for accounting system, especially for data written for product designation.

The committee claimed that such designation elements can be barcode, RFID technology, and other types, which are mainly used in logistics; they should be used to assign additional data of products and shipment units [12].

Depending on the type of identifier, the amount of data is varied, as shown in Table 1. Currently, according to the legislation, medical institutions are legally obliged to label all materials and medicines with EAN 13 code [5].

Detailed rules and basic notions addressed by the European Commission and the Commission Delegated Regulation (EU) 2016/161, and published at the beginning of February 2016 in addition to the European Parliament and Council Directive 2001/83/EC, are all set out for the use of secure elements of medicine wrappers for human use.

1.3.1 Elements Used on the Wrapper of Drugs for Human Use

The regulations or rules of labeling of medicines are supplemented by the following information from February 2019:

- GTIN product number.
- Shelf life.
- Batch number.
- Serial number.

All this information should be written on the product label and should be readable in a form of GS1 code format. Mentored specified product data are stored in the European Central Database and the National Registry.

Accordingly, when medicaments are issued, the GS1 data matrix code should be read to confirm the action and verify it with the data of the National Registry. Also, the main criterion is checking and confirming the authenticity of medicine, before it will be given to the patient [3, 5].

The Slovak Republic, as one of many, began preparing for this transition by implementing and launching the e-health portal. It works on the principle of using a passport with a chip (eID) by the patient. This chip stores patient data, which will be used for further identification by general practitioner, nurse, or pharmacy. The chip also provides access to the patient's health electronic book (EZK) [5]. This solution does not meet the requirements of the EU project; that is why here is the bottleneck and it is appropriate to design the accounting system which will fulfill them.

2 Design an Accounting RFID System

2.1 Selection of the Components of RFID Accounting System

To ensure accounting of the objects, special system should be designed. Writing information into tag's chip, information scanning, and saving and receiving the data between tag and information system of a medical facility and other suboperations should be made possible.

Function of data accumulation and data processing is able to fulfill the computer. Receiver or communicator in such situation can be performed by the antenna and RFID reader. The product information provider will be the RFID tag.

Proposed RFID accounting system includes:

- Computer.
- Antennas.
- RFID reader.
- Passive RFID tags.

The detailed representation of the system is taken from previous research of this problem (see Table 2) [18].

It should be noted that components of the system can be replaced with alternative ones. Designed RFID system was used for testing the best ways of reading to improve the efficiency of the system. Testing includes few ways of transporting and three types of packages to show the appropriate conditions for effective reading of the tagged objects while they are moving through RFID gate.

Table 2 RFID system [13–18]

Photo	Component	Model of the component
	Reader	Alien ALR-8800
	Antenna (2 pieces)	Antenna RHCP, AN480-CR66100WR, 865-956 Mhz, 6 dB
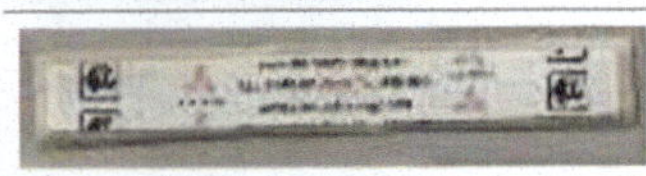	RFID tag (10 pieces)	Alien ALL-9540-02 World Tag 860-960
	Computer	Gaben GBTouch 24″
	Software	Alien RFID Gateway v2.26.03

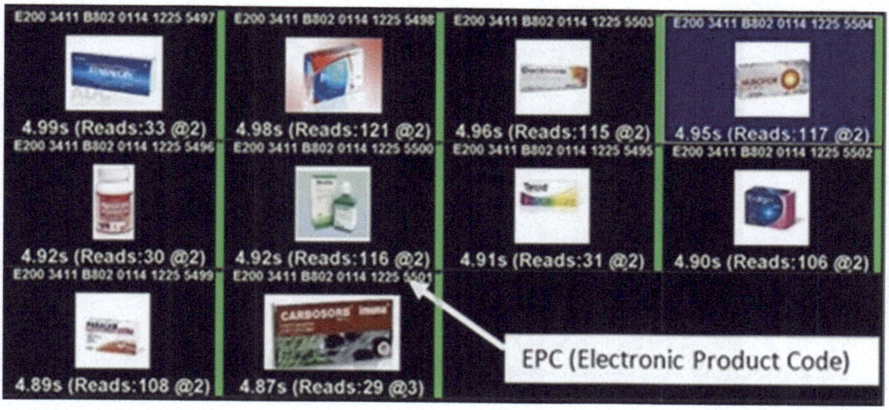

Fig. 1 Medicaments with RFID tags [18]

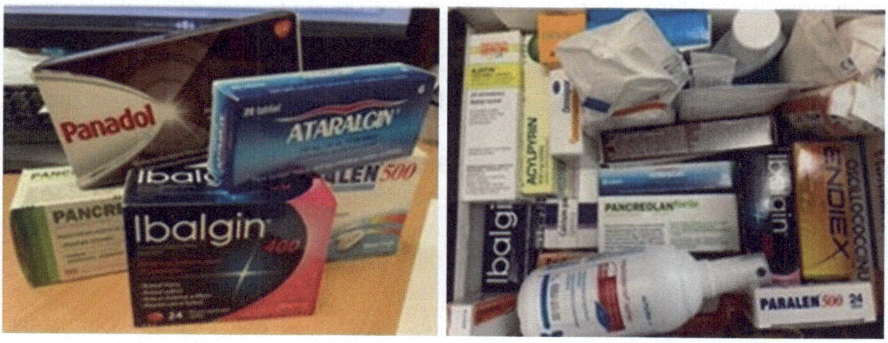

Fig. 2 Medicine in original packages [18]

The medicaments were designated by tags for medicament identification. Each tag was digitally attached by photograph of the medicament, to identify what kind of medicine was detected by the system. Figure 1 shows what medicaments look like in the reading program. Photographs of the medicine are shown in Fig. 2.

Both figures show the drugs in the Alien software and their real appearance. The first figure represents the principle of comparing picture of the medicament and EPC code of the product with the help of Alien software.

Due to requirements of the government, each medicine should have its unique code in every accounting system of the medical facilities in the nearest future; that is why it was decided to include it in the proposed article.

2.2 *Selection of Correct Method of Material Handling*

In the previous article, simulation of conditions of hospital was made. Such conditions were reflected in sizes of RFID gate, and the commonly used transportation materials were selected. The experiment included three basic ways of packaging

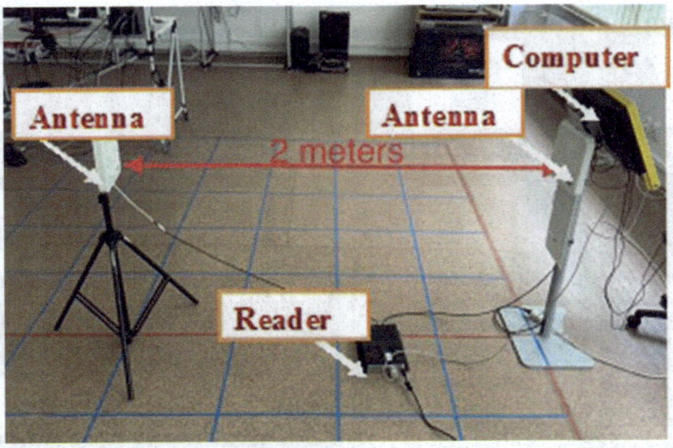

Fig. 3 RFID gate [18]

medicaments and three basic ways of their transporting. The RFID gates had a width of 2 m (Fig. 3).

Such width was selected due to requirement of wider type of the door in warehouses and hospital wards [18]. The RFID gate included two antennas on the tripods, computer, and reader.

The next step was testing of transportation ways for medicaments in different packages. For measurement of quality of reading, Alien RFID Gateway v2.26.03 program and its built-in Readomer were used. This program is able to determine how fast the RFID tag was detected by the system. Transportation packages are shown in Fig. 4.

According to preliminary studies, it was determined that medicines are most often transported open in a trolley, in carton boxes, as well as in plastic containers/boxes (see Fig. 4).

Then, medicaments were transported through RFID gate in each package. The medicament information is readable in the plastic box, which approves proposed RFID system, as can be seen in Fig. 5. The green scale shows quality of reading. In this program, scale can be green, yellow, and red colored. The quality of reading can be:

- Green-colored scale—excellent.
- Yellow-colored scale—sufficient.
- Red-colored scale—no quality reading.
- No scale—tag was not detected.

For another two ways of testing, metal trolley and carton box were used (Fig. 6).

The carton box was designated by 2 and metal trolley as—1, where the yellow scales on the first part of the figure can be noted. These scales indicate sufficient identification of medicine, but it is not qualitative. The major factor which influences quality of reading is iron construction of the trolley [18]. It is obvious that there are more factors that could impact the quality of reading, such as:

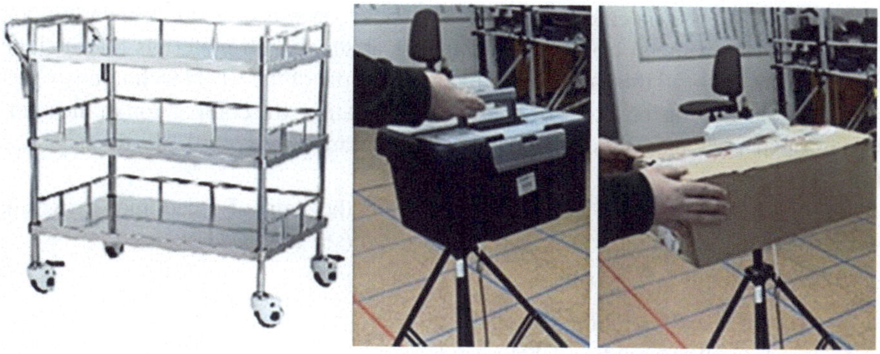

Fig. 4 Packages for medicament transportation [18]

Fig. 5 Detection of RFID tags in a plastic box [18]

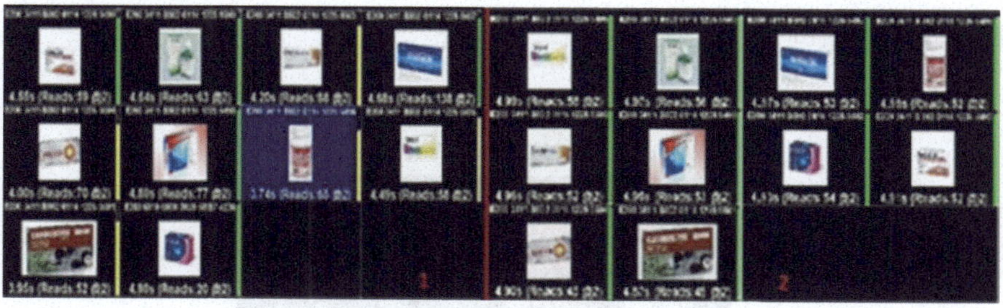

Fig. 6 Metal trolley and carton box [18]

- Speed of moving objects.
- Antenna placement.
- Height of transported package in relation to antenna placement.
- Metal objects on the employee's body.
- Incorrect way of handling and transporting.
- Other external factors.

For example, the speed of reading of the Pankreon was 65 reads/s in carton box and 52 reads/s in metal trolley. Such changes arise with all materials that include metal impurities—all this is triggered by the principle of the RFID technology operation. In selecting the best way of medicine preparation, the comparison of the reading in such packages was made: plastic box, carton box, and plastic bag (see Fig. 7) [18].

The speed of detection was calculated with the help of Readomer program, where unit of measurement is reading per second (see Fig. 8) [18].

According to data from Fig. 8:

- Plastic box allows reading 860 reads/s.
- Carton box allows reading 1094 reads/s.
- Plastic bag allows reading 1319 reads/s [18].

With regard to the data above, it is possible to conclude that plastic bag is the most effective package. The medical facilities should be able to use a nonmetal way

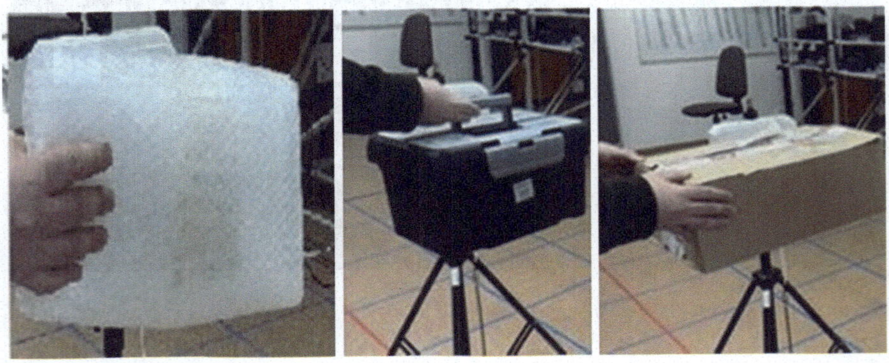

Fig. 7 Packages [18]

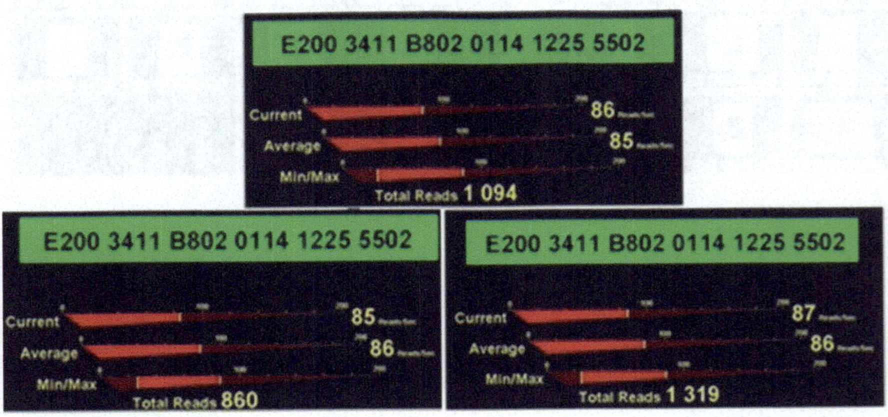

Fig. 8 Speed test by the type of package [18]

of transportation (any package without metal). In previous research, it was considered that such a condition is necessary for quality reading of all medicaments in the transportation set [18]. If the system will not identify any medicine from a whole number, it will lead to a situation where medication will not get to the accounting system. Plastic packages open new ability—using the recycling plastic in this economic field, which will help to save the environment and will enable quality reading of the medicaments [19].

It is worth emphasizing that the RFID system in the management of any warehouse and any health institution will bring ability to:

- See the goods "online," at a distance and in motion.
- Control many objects.
- Monitor the completion and execution of orders.
- Protect goods from falsification and theft.
- Reduce the number of manual operations and the likelihood of human error.
- Automatically perform accounting and control and inventory operations.
- Inventory goods quickly and daily.
- Integrate into databases and create statistical and analytical reports.
- Reduce production and accounting costs.

2.3 *Selecting the Environmentally Friendly Medicine Transporting Package for Medical Facilities*

Selecting the suitable material for RFID accounting system is very important, because the package should make reading of tags possible. The above measurements and testing of material readiness had shown that the most appropriate types of package are plastic packaging and box. Plastic materials fulfill the requirements of effective functionality of proposed system, such as:

- Attachment method (the material and shape of the tagged object are two of the most important aspects to consider when determining the appropriate attachment method).
- Read range (an RFID tag's read range, antenna gain, reader transmit power, and tag orientation).
- Custom printing and encoding (custom printing and encoding do not necessarily narrow down the RFID tag choices; however, both are important to consider before purchasing) [20].

Plastic materials contain no, or a small amount of, metal impurities, but it has no significant effects on the quality of reading, as metal materials.

Due to today's problem with a plastic waste, it is possible to apply modern approaches to the use of this material [21].

It is a well-known fact that the health-care industry uses disposable individual packaging for most medications and medicines. Such packages take the biggest part of plastic waste in this industry. To better understand such problem, it is necessary to have an in-depth discussion of its statistic and research of specialists in this area.

Researches of Practice Greenhealth, a nonprofit organization that works to make hospitals more sustainable, had shown that 25% of the waste generated by a hospital is plastic. The specialists found that some procedures can produce up to 20 pounds of waste, most of them being plastic.

Single-use plastic can be an attractive option for hospitals—cheap, durable, and easily tossed out, and each new fresh plastic container or covering offers a newly sterile environment. This fact becomes an excuse for clinicians, and they cover themselves by using such packages. It is impossible to reject the fact that plastic has revolutionized the medical industry over the past century, but it is necessary to look in the future, where problem with recycling of this material arises [22]. On the other hand, as an example of the importance of the problem, the statistic of US specialists should be considered. They calculated the amount of plastic garbage, and about 14,000 tons of waste is generated daily in the US health-care facilities, with 20–25% of this waste coming from plastic packaging and products [23].

The decomposition period of plastic is more than 400 years. It means that before plastic, which today lies in the garbage cans, completely decomposes, the whole Earth will simply "sink" into plastic waste. There is notion that "microplastics" are pieces of plastic wastes that are found almost everywhere today. Particularly worrisome is the presence of microplastics in water bodies. The presence of microplastics in the seas, oceans, and rivers is growing dramatically every day, and this has a detrimental effect not only on the flora and fauna of reservoirs but also on a person who uses such water and receives a regular dose of microplastic.

A major problem of the industry is the "risk factor" from the point of view of US processors, as companies are concerned that medical plastics are biohazardous. But Steichen noted that 85% of hospital waste is actually harmless, that is, free from contact with the patient and pollution. Waste is heated during the recycling process, so care is based on the safety and exposure of workers, as well as on who will buy the recycled material [23].

There are a lot of views on reducing of plastic waste in health-care industry, but the real ones are:

- Sorting the packages.
- Using alternative materials or recycled plastic.

In a review of 332 hospitals held by Practice Greenhealth, ordinary disposable plastic items were explored in operating rooms that were successfully replaced with reusable items. Instruments such as surgical basins and sterilization wraps can be reused to reduce waste by several tons per year. Practice Greenhealth says that depending on how much of these used materials are reduced, hospitals can save thousands of dollars a year.

"Of course, there are various types of plastics that can be recovered and that cannot be recovered today for a number of reasons," said Kim Holmes, vice president of sustainable development, Association of Plastics Manufacturers.

But it should be considered, as Kim Holmes says: "There are items that are used in caring for patients who do not come into contact with patients, so they are not biohazardous and they can be recycled." Such idea concerns things like packaging and storage containers [22]. Due to this idea, there arises the opportunity to suggest that RFID accounting system will be environmentally friendly, because the idea of such system supports the use of RFID tags only by picking up on package.

Completely refusing the use of plastic materials is very difficult, as noted an engineer at the University of Massachusetts Lowell: "Plastics for biomedical applications have many desirable properties, including low cost, ease of processing, and [ability] to be sterilized easily" [22].

The use of RFID technology in the packaging sector is an important logistical improvement in terms of the benefits offered by this technology over barcodes. However, the question arises: Can the presence of these devices in plastic packaging and, consequently, in plastic waste be a cause of some problems in the processing of tagged materials?

In the research made by César Aliaga, Beatriz Ferreira, Mercedes Hortal, and a group of scientists, mentioned recirculation restrictions were experimentally determined in an experimental recirculation study, but this study concerned the built-in labels in the packaging material.

The quality of the recycled material of each test was studied by injection and by test probes. The results of experimental disposal tests did not show a decrease in the quality of recycled plastic due to the presence of RFID tags [24]. Nevertheless, some operational problems were observed during the recycling process, but the proposed article suggests the use of RFID stickers, which are proposed to be peeled off or reused. In this regard, this warning is proposed to be taken into account but not taken as a basis for rejecting the proposed technology.

Supporting the use of eco-friendly materials should be implemented not only by health-care facilities but also by their suppliers.

Brand owners are increasingly paying attention to the reuse and recycling of finished products. "In the health-care industry, flexible plastics are a unique challenge for manufacturers" said Tristan Steichen, senior consultant of Antea, and Nicholas Pack, DuPont packaging engineer, at HealthPack 2019 in Portland, Oregon [23].

The example of this idea can be presented by Henkel's approach to this problem (see Fig. 9).

Continuing green packaging initiatives, Henkel is taking another step toward a waste-free economy. The company has set itself additional goals (see Fig. 9) that will be integrated into its sustainable development strategy. By 2025, all packaging used by Henkel will be recyclable, reusable, or biodegradable. At this point, Henkel intends to increase the share of recycled plastic in the packaging of its consumer products in Europe to 35% [25].

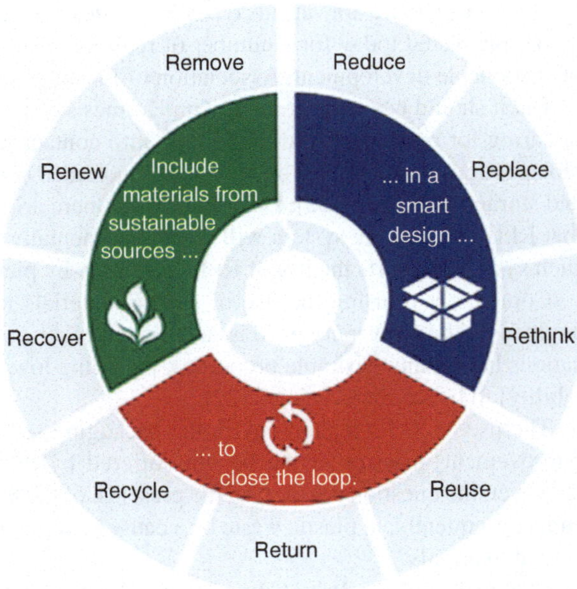

Fig. 9 Henkel's strategy [25]

According to research, it is possible to approve that RFID technology would not hinder plastic recycling process. This idea is supported by research of Kamil Janeczek and his colleagues. In their article, the following idea was supported:

> "In RFID systems tags are utilized in various forms. This paper addresses investigation of a recycling process of RFID tags. Different attachment methods of these tags were tested and then, a recycling procedure was prepared based on the obtained results in which adhesives are recommended to be used as an attaching material between a tag and an identified object. The mentioned procedure was evaluated with using paper-face RFID tags. It was confirmed applicability of the proposed solution. Moreover, the conducted research revealed that an adhesion force of the adhesive layers in the prepared RFID tags decreased in the most extent during the first recycling cycle. It was found that in order to preserve a high adhesion force of these layers it is suggested to place the tags on a clean surface" [26].

It is obvious that scientists try to develop idea of recycling and to expand the use of smart technologies, and RFID technology is able to cover this requirement.

2.4 *New Opportunities of RFID System in Medical Institutions*

The research has shown that RFID warehousing system is very attractive nowadays. When cheap tags are used in such a system, it requires the use of the plastic or non-metal materials to support the best workability of the system. This fact opens the opportunity to implement proposed system to laboratories for monitoring the movement of biological materials and samples.

This group of products was selected as strategic due to its weighting for functionality of medical facility. The speed and quality of transportation of such objects directly affects the quality of medical services for the patient, and often his life may depend on this.

Market research had shown that the biggest part of containers and packaging for the collection of biological materials is plastic (Fig. 10).

Biological materials are not transported one at a time but by a whole group on special tripod stands (Fig. 11).

There are many universal tripods made of high-strength polypropylene or polycarbonate, of various designs and functionality, for installing from 18 to 100 tubes of different diameters.

Storage of biomaterial at low temperatures (cryopreservation) is the most common and effective method for long-term preservation of the initial characteristics of biomaterial for the purpose of conducting deferred studies. It is recommended to use low negative temperatures from −20 °C to −196 °C.

The company uses special racks made of polycarbonate (up to −196 °C) or polypropylene (up to −90 °C) to store test tubes with biomaterial samples at low temperatures [28].

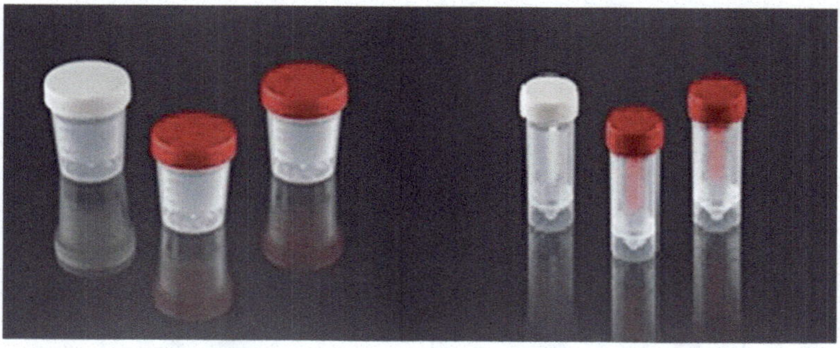

Fig. 10 Example of the most popular containers [27]

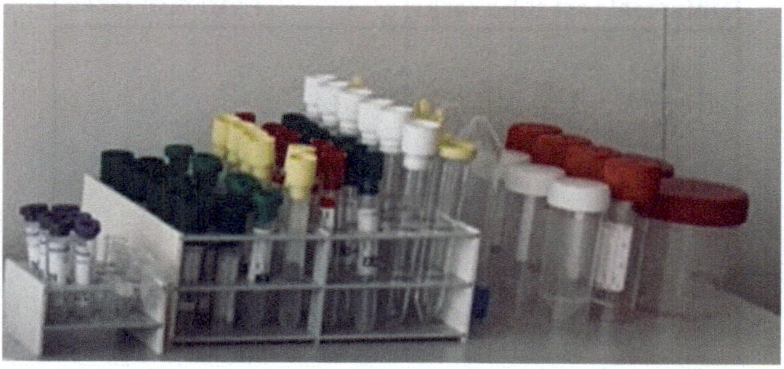

Fig. 11 Example stands for containers

Transport cases/boxes provide convenient and safe transportation of biomaterial in full compliance with the requirements of the international packing instruction P 650/ADR [29] for biological materials. Cases for remote transportation and for transportation on the territory of a medical institution are optimally compatible with block tripods.

Containers, envelopes, and mailing boxes are designed for the safe transportation of samples of biomedical material with maximum protection from environmental factors. Containers, envelopes, and boxes must fully comply with the requirements of packaging instructions P 650/ADR for biological materials. Envelopes and boxes are recommended as external packaging and containers as secondary packaging for primary test tubes and containers with biomaterial. These transporting packages are shown in Fig. 12.

These packages are paper or plastic, and they should not preclude RFID tag reading by the system. In case when transportation includes metal components, it is feasible to use one more tag for labeling the package.

Laboratory samples till now are labeled manually in most of medical facilities (see Fig. 13).

Such way of labeling is not very efficient due to the fact that stickers are used either as paper for handwritten information or barcodes. Both methods are easily exposed to external factors that can damage the label, such as:

- Temperature.
- Moisture.
- Liquids.

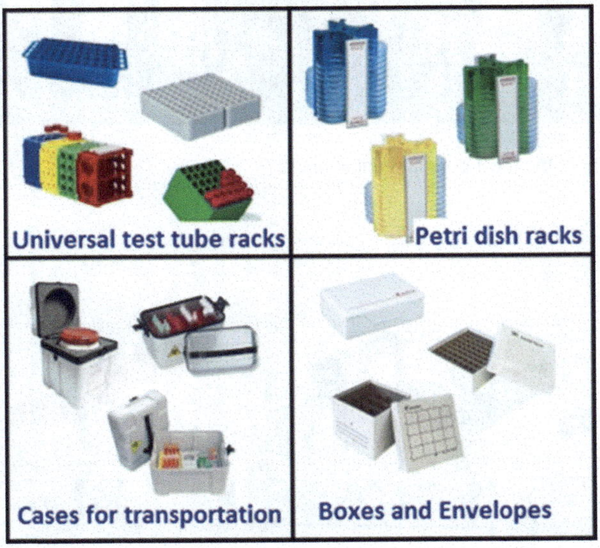

Fig. 12 Examples of packages

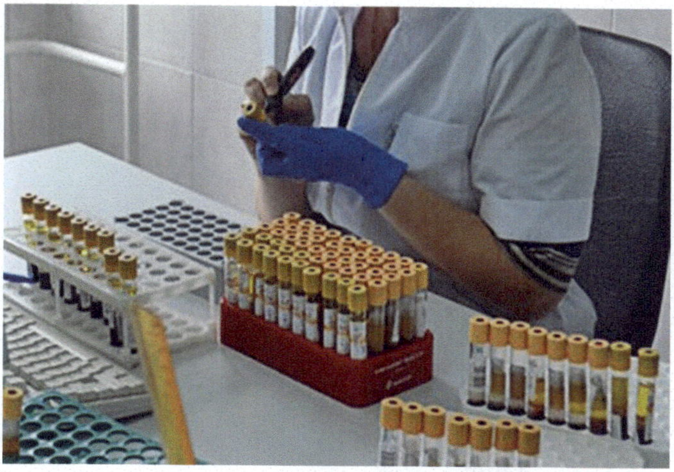

Fig. 13 Manual labeling

It is important to note that damage to such marks will not give complete information about the sample, as well as the movement of such objects requires a large amount of supporting documentation.

This fact once again confirms the need for such tags for biomaterials that would not be exposed to external influences and also help to eliminate the loss of information due to its digitalization (Fig. 14).

There are two opportunities to track biomaterials, and it is linked with a question: where we want to observe the movement of materials: in a medical institution or outside?

The cheapest tags can be used for indoor tracking, like those used in previous studies (Table 1).

The tags ALIEN SQUIG RFID INLAY (ALN-9610, HIGGS-3) were used [30]. Such kind of tags has similar parameters like the tag from Table 1. Donor blood is stored on average at temperatures of less than minus 25 °C up to 12 months and at temperatures up to minus 30 °C for up to 24 months [31]. Therefore, the technical parameters of RFID tags must comply with such storage conditions.

Such tagged samples can be detected by RFID accounting system in some ways:

- When an employee passes through the RFID gate.
- When the product or sample is put on the shelf.
- When the sample is put on the laboratory table.

This information serves for such reader placement options:

- On laboratory desk.
- Into warehouse adjustable shelves.
- Into the entrance (RFID gate).
- Hand RFID reader.

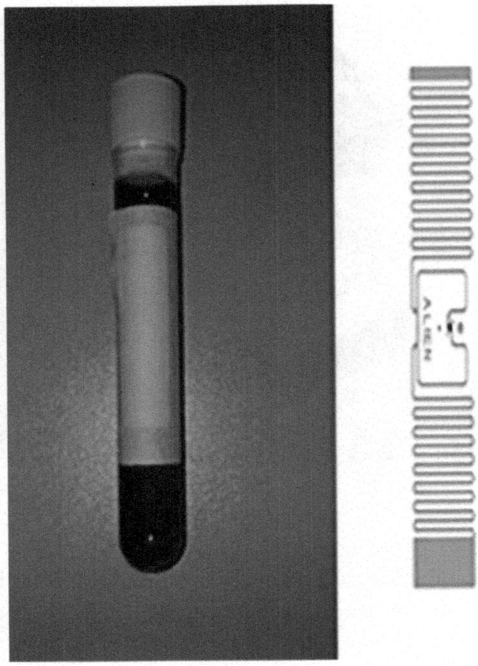

Fig. 14 Tagged blood sample and used RFID tag

The graphical illustration of component placement is illustrated in Figs. 3 and 15.

The adventure of proposed RFID system concluded in the flexibility of variation of component placement.

For outdoor tracking of the materials, it is feasible to use RFID tags in combination with GPS chips; such tags are expensive. That is why these tags will be useful only on a transporting package.

3 Conclusion

Proposed research has shown the relevance of the problem. RFID accounting system was represented as attractive not only as solution for warehousing in the industry but also for health-care industry. Such system can be integrated to existent accounting system and open new opportunities to plan and to manage material and medicament flow and also to immediately react on expiration date of the medicine [18].

It is obvious that the system has various installation ways: can be built on room entrance, shelves, adjustable shelves, and so on. RFID system in medical facilities will help scan a large number of goods and thereby speed up the accounting and eliminate a large amount of documentation and errors provoked by employees [18].

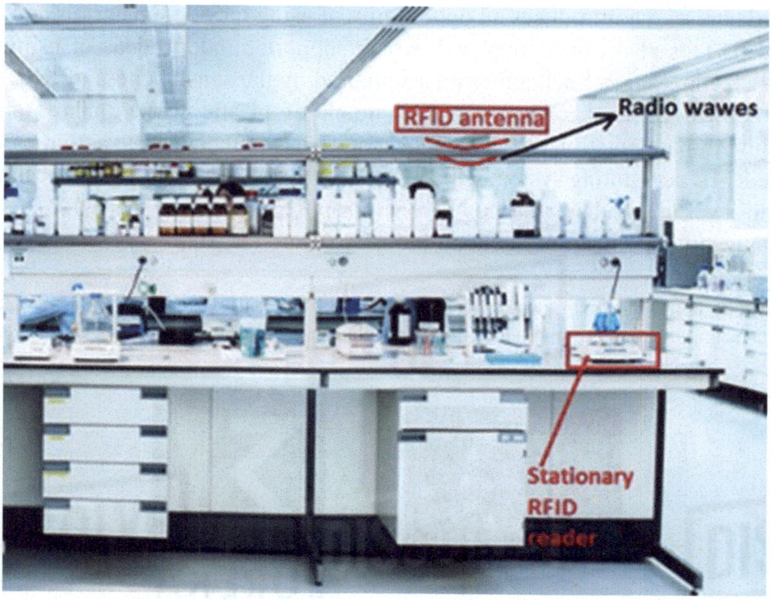

Fig. 15 Method of possible placement for RFID system components

The main types of labeling accounted objects were compared, and the appropriate one was selected—tagging by RFID stickers. Selected tags meet the requirements of sufficiently saving the information about object, and they are protected from influence of external factors. Introduced research demonstrates the perspectives of using RFID system in different medical facilities. The price of proposed system was reviewed in previous research [18]:

- Reader Alien ALR-8800—58.50$ [14].
- Antenna RHCP, AN480-CR66100WR, 865–956 Mhz, 6 dB (2 pieces)—$500 [16].
- Tags Alien ALL-9540-02 World Tag 860–960 (20,000 tags)—$2540 [15].
- Computer—$0.
- Software Alien RFID Gateway v2.26.03—for free [32].

Such type of labeling satisfies the requirements of the criteria established by the European Commission, due to their ability to write satisfied information about medicine.

Rewriting ability is one more advantage of using such technology. Another competitive edge is a potential to build in the existent management system; also, it can be expanded by the use of real-time location system for external tracking of the materials.

Implementation of such system into laboratory accounting system will allow to link biological materials with patient card system, which will reduce the use of paper documentation and make data processing faster and avoid loss of information.

Market research introduces an opportunity of using recycled materials and notes that RFID tags also can be recycled, which supports modern ideas of using eco-friendly technologies. Such a fact was supported by the sensitiveness of RFID system to metal elements. Also, arising amount of plastic waste makes relevant providing implementation of the recycled packages to all economic activities.

The RFID accounting system will open new ways of managing all medical facilities and level up all their services [4, 33]. Proposed system will provide management in health-care institution due to Industry 4.0 trends [6].

Acknowledgments The paper was created within the EUREKA (E!) project 11158 U-health—Auto-ID technology and the Internet of things to enhance the quality of health services.

References

1. WOS: WHO Calls for Urgent Action to Reduce Patient Harm in Health Care. News, WOS (2019). https://www.who.int/en/news-room/
2. Schuller, K.A., Cronin, C.E., Nicks, S.E., Jing, X., Kingori, C., Morrone, M.: Development and application of a rubric to compare strategies for improving access to health care in rural communities in the United States. Eval. Prog. Plann. **74**, 61–68 (2019). https://doi.org/10.1016/j.evalprogplan.2019.02.013
3. Balog, M., Husar, J., Iakovets, A.: RFID system for management of supplies. In: Healthcare, Automation and Management in Theory and Practice, 13th Anniversary Conference of Universities, High Schools and Practice, Stará Lesná, SR, February 2019. ISBN 978-80-553-3250-5
4. Ferreira, D.C., Marques, R.C.: Do quality and access to hospital services impact on their technical efficiency. Omega Int. J. Manag. Sci. **86**, 218–236 (2019). https://doi.org/10.1016/j.omega.2018.07.010
5. Husar, J., Iakovets, A.: Proposal for monitoring the movement of biological samples in hospitals. In: Automation and Management in Theory and Practice, 13th Anniversary Conference of Universities, High Schools and Practice, Stará Lesná, SR, February 2019. ISBN 978-80-553-3250-5
6. Husar, J., Knapcikova, L.: Principles and Characteristics of "Industry 4.0", Management of Manufacturing Systems 2018. TUKE, FVT, Dubrovnik, Croatia (2018)
7. ContentLive: Top 10 Tech Innovations in Warehousing. Natwest. https://natwest.contentlive.co.uk/
8. William, T.D.: Parcel and Small Package Delivery Industry,. ISBN 13:9781461021544. CreateSpace, North Charleston, SC (2011)
9. Erick, C.J., Christopher, A.C.: RFID in Logistics: A Practical Introduction., ISBN 978-0-84-93-8526-1. CRC Press, Boca Raton, FL (2007)

10. Lopez, Y.A., Franssen, J., Narciandi, G.A., Pagnozzi, J.: RFID technology for management and tracking e-health applications. Sensors (Basel). **18**(8), 2633 (2018). https://doi.org/10.3390/s18082663
11. Tiwari, V., Quinlan, J.J., Levine, W.C.: Scaling perioperative services across health care systems. Int. Anesthesiol. Clin. **57**(1), 1–17 (2019). https://doi.org/10.1097/AIA.0000000000000218
12. Hrehova, S., Fechova, E.: Design of system for evaluating and sorting products with tools of statistical process control. In: MMS Conference 2017, pp. 1–8. EAI, Gent (2018). ISBN 978-1-63190-158-4
13. AtlasRFIDstore.com, homepage. https://www.atlasrfidstore.com/alien-f800-rfid-development-kit/. Accessed 03 May 2019
14. Alien ALR-8800 price. https://www.ebay.com/itm/Alien-ALR-8800-Enterprise-RFID-Reader-/332120063306. Accessed 03 May 2019
15. Alien RFID tag. https://www.atlasrfidstore.com/alien-squiggle-rfid-white-wet-inlay-aln-9740-higgs-4/. Accessed 03 May 2019
16. RFID antenna. https://www.atlasrfidstore.com/zebra-an480-rhcp-rfid-antenna-865-956-mhz/. Accessed 02 May 2019
17. LNCS homepage. http://www.springer.com/lncs. Accessed 21 Nov 2016
18. Balog, M., Iakovets, A., Husar, J.,: RFID Monitoring and accounting system in health-care facilities, 4th EAI International Conference on Management of Manufacturing Systems: MMS 2019, Springer Nature, 2020 ISBN 3030342727, ISSN 2522-8595, 19–31 pages.
19. Knapčíková, L., Rimár, M., Fedák, M.: New trends in waste management in the selected Region of the Slovak Republic, 2018. In: MMS Conference 2017, pp. 1–9. EAI, Gent (2018). ISBN 978-1-63190-158-4
20. Smiley, S.: The Best RFID Tag For Your Application: 7 Key Factors to Consider, RFID Insider, October 2014. blog.atlasrfidstore.com
21. Balog, M., Matiskova, D., Semanco, P.: System of Workpiece Automation with RFID Identifier Published Patent Application SK 50063-2016 A/Darina Matisková, Michal Balog, Pavol Semančo Banská Bystrica: IPO SR (2017). 7 sec
22. Gibbens, S.: Can medical care exist without plastic. ScienceThe, Story of Plastic. National Geographic, October 2019. www.nationalgeographic.com/
23. Study: Flexible Healthcare Packaging Materials Can Be Recycled, 7 Mar 2019. Healthcare packaging. https://www.healthcarepackaging.com/issues/sustainability/article/13295965/study-flexible-healthcare-packaging-materials-can-be-recycled
24. Aliaga, C., et al.: Influence of RFID tags on recyclability of plastic packaging. Waste Manag. **31**(6), 1133–1138. https://doi.org/10.1016/j.wasman.2010.12.015
25. Henkel Presents New Ambitious Goals for Sustainable Packaging. unipak.ru. https://article.unipack.ru/72112/. Accessed 16 Oct 2018
26. Janeczek, K., Araźna, A., Stęplewski, W.: Circular economy in RFID technology: analysis of recycling methods of RFID tags. J. Adh. Sci. Technol. **33**, 1–12 (4, 2019)
27. Odberové kontajnery a nádoby. https://www.vidra.sk/odber-biologickych-vzoriek/odberove-kontajnery-a-nadoby/n47,135,0
28. Firm galen cjsc: Storage and Transportation of Samples of Biological Material, Galen (2019). http://www.galen.ru/
29. https://adrbook.com/en/2017/Sutun/8/P650
30. RFIDPOINT. http://rfpoint.com/
31. Blood Service: Rules for the Transportation and Storage of Blood Components. http://www.kkck.ru/
32. Gaben homepage. http://www.gaben.cz/cz/stacionarni-terminaly/gbtouch. Accessed 02 May 2019
33. Husar, J., Iakovets, A.: Fourth Industrial Revolution Technologies, Management of Manufacturing Systems 2018. TUKE, FVT, Dubrovnik, Croatia (2018)

Evaluation of Business with Earth Resources from an Economic Indicator Viewpoint

Katarína Čulková, Andrea Rosová, Michal Cehlár, and Samer Khouri

1 Introduction

Raw materials provide the basis of any region with the potential value that can be extracted from that area, creating profit for the owner, state and mining company, as well as the specific region [1]. Profit for the state can be achieved through the reasonable and rational use of deposits, as defined in the legislation of use of raw materials. Profit for the mining organization is created through maintaining sustainability. Profit for the region is derived from the creation of wealth for the region's inhabitants through creation of new employment, increases in regional or other taxes related directly to the region, and availability of raw material sources that are necessary for the region's development. In Slovakia reserved deposits of raw materials are owned by the state, but the state does not have the specific obligation to evaluate economically all recorded deposits of reserved raw materials [2]. This means that problems with the use of raw materials exist for both the business sector and the private sector [3]. Our goal was therefore to evaluate the development of the most important social and economic indicators that influence the use of raw materials mostly from the stance of sustainable development of the Slovakian industry.

The objective of the paper was to evaluate and compare the business environment in the area of natural resources in Slovakia according to specific indicators with the aim of identifying obstacles that limit business development.

K. Čulková (✉) · A. Rosová · M. Cehlár · S. Khouri
Technical University of Košice, FBERG, Košice, Slovakia
e-mail: katarina.culkova@tuke.sk; andrea.rosova@tuke.sk; michal.cehlar@tuke.sk; samer.khouri@tuke.sk

L. Knapcikova et al. (eds.), *New Approaches in Management of Smart Manufacturing Systems*, EAI/Springer Innovations in Communication and Computing, https://doi.org/10.1007/978-3-030-40176-4_2

2 Present State of Problem Solving

Business in the single sectors is very complex, and plays an important role in the creation of new possibilities for economic development and social stability. The business environment is very closely linked with business, including various indexes that directly or indirectly influence the effectiveness of any business, including the mining industry, because businesses involved with natural resources represent the basis of the industrial sector in Slovakia, as they do in other countries; mining of raw material is a base element, providing important materials for any industrial sector for the production of various products.

Nowadays products are created on the basis of demand-management of production, and this transformation has to be managed in a way that assures the expected results are achieved. Adoption of the process of management philosophy is also possible for the mining industry [4]. The mining process is a process that involves great financial demands. Understanding and application of process management is necessary in order to achieve market success of the mining organization.

Underestimation of natural resources and environmental damage distorts the indicators of economic development and progress, which leads to inefficient social and economic pathways. To assess the relationship between the quantity and quality of economic growth, the trends of social and economic development of countries and regions, sustainability indicators are needed for assessing the effectiveness of this development [5]. Worldwide, the mining industry is in constant development. The reason for this is increasing demand for raw materials. Despite many positive aspects, there is widespread concern from the general public and politicians about the negative effects of mining activities. The greatest relationship is between GDP, average wage, employment and wages in mining [6].

There is a lack of research in the mining and quarrying industry into efficiency in the Central and Eastern European region. This can be illustrated by search results in Scopus, where the literature from 2015 to 2018 on earth's resources are only relate to earth and planetary sciences (459 publications), physics and astronomy (327 publications), environmental science (228 publications), engineering (178 publications), and agricultural and biological sciences (85 publications). With regard to business in relation to earth's resources there is gap in the literature [7]. But business and the economic aspects are key issues to be considered in order to understand future changes in energy systems. The quantity of the energy supply is set to meet the price of the energy demand of end users. This occurs on the condition that the price of the energy supply equates with the price on the demand side according to market conditions [8].

Earth's resources have an impact on economic growth. Their price and the precise channels through which for instance oil prices affect economic activity are only partially known; more needs to be studied in this area. The overall effect of energy prices on growth is negative. Arshad et al. [9] found evidence that high energy prices decrease real interest rates, investment and stock prices. They also decrease the real value of local currency. A high energy price puts pressure on government expenditures and increases local unemployment. Furthermore, Brown and Yücel [10] investigated linking fluctuations in energy prices to aggregate economic activity. Cunado and Perez de Gracia [11] studied the impact of oil price shocks on both economic activity and consumer price indexes, and suggested that oil prices have a significant effect on both economic activity and price indexes, although the impact is limited in the short term and more significant when oil price shocks are defined in local currencies. Many societies today express their desire to participate in the decision-finding on the development of their economic environment. In this area a sustained and sustainable mining development requires the collaboration of the host communities concerned, which means that it has to be developed within a process commonly termed *social licensing*. Falck [12] argues that social licensing should be made an integral element of economic (business risk) management for mining companies. The last decade and a half has witnessed a dramatic growth in mining activity in many developing countries. The relationships between mining and political economic change have also been studied [13].

At present, energy conservation and pollution reduction have become two of the most significant factors in economic development and social conditions [14]. An economics-energy-environment (3E) system is needed. A number of countries are struggling to understand, plan and realize the 3E nexus [15]. Coal in particular has always been an important commodity and is one of the oldest mining enterprises. The energy generated, especially electricity and synthetic fuels, impacts all sectors of the economy and society. It is therefore necessary to determine the key players in the coal mining industry, those who are the beneficiaries of the process underpinning the socio-economic attributes of the industry [16].

3 Methodology

The object was to search for businesses involving earth's resources in Slovakia. The mining industry is usually evaluated from a point of view of macro-economic analysis—GDP, measure of unemployment, measure of inflation, number of companies and employment in the mining sector. We extended the research to include businesses involving black and brown coal. As the basis of the business environment analysis we used Slovakian and worldwide organizations in a comparison of the following:

1. *Development of employment*: Economic growth is a fundamental requirement for the development of the country, but it is not possible without increasing employment. Development was assessed on the basis of annual change, using data from the main mining office in Slovakia from annual reports of mining activities.
2. *Gross value added:* Measure of the value of goods and services produced in an area, industry or sector of an economy. It is expressed as output minus

intermediate consumption, resulting in balancing items of the national accounts' 'production account'.

3. *Global competitiveness index* (*GCI*): From a methodological point of view Schwab [17] defines the index as a "system of institutions, policies and factors that determine the level of economy productivity that on the other hand determines level of prosperity that the country could achieve". Rating by the GCI uses a publicly accessible (statistic) dataset (WB, IMF, etc.) and results of the findings are provided by World Economic Forum. These findings are made annually with the support of partner institutions (research institutions and business centres).
4. *Doing Business:* 'Doing Business' is a statistic evaluation of a large sample of economies, evaluated according to a questionnaire from the World Bank. The methodology of the evaluation consists of the creation of a questionnaire by the World Bank that presents the base for comparing individual countries and includes knowledge of the legal form of the business, volume of business, location of business and character of the business activity. The score of individual economies ranges from 1 to 190 [18]. Data are used to analyse economic outcomes and identify which reforms of business regulation have worked, and where and why.
5. *Index of business environment:* The business alliance of Slovakia generates the index for mapping the business environment that searches for positive changes or finds areas where certain intervention measures are necessary. The alliance has its own methodology for the creation of the index, where the base period of the index calculation is 1 July 2001, when the reference value was 100 points, and since then business environment is evaluated quarterly. For evaluation of the index it is necessary to follow the methodology [19].

4 Results

The mining sector represents a base element for a fluent and developed industry, since the sector provides raw materials for production and participates in the state's economy. In each country the mining sector is different, since the availability and type of raw materials differ [20]. The main assumption of competitiveness is an available and qualitative business environment. Presently a number of organizations deal with competitiveness, following and comparing specific business environments not only in Slovakia but in the whole of Europe. Comparison of business environments in Slovakia can be carried out by the main macro-economic indexes—GDP, measure of unemployment or inflation in given period [21]. The use of base raw materials directly and indirectly influences not only environmental areas of sustainable development—mainly by negative impacts through environmental risk, presenting sources of sustainability instability—but also social and economic areas. Use of raw materials influences employment status, and thereby contributes to the increase in household incomes and also positively influences purchasing power and living standards. At the same time it contributes to the creation of income for state budgets and the budget of administrations through payment for mining areas, mined minerals or stocking of gases, as well as local taxes, fees or obligatory payments arising from dependent activities, for example.

Employment in mining organizations has recorded a more or less decreasing trend in development, which is connected logically with the development of raw materials (RM) mining and the use of technical and technological equipment during RM mining in Slovakia. Over the entire analysed period higher employment was recorded for employees working in surface mining, which is again logically connected with the prevailing way of surface RM exploitation in Slovakia. The highest employment was recorded in 2000, with 13,968 employees (6231 in deep mining and 7737 in surface mining); the lowest level was achieved in 2016, when employment decreased to 5476 (of which 1666 were underground and 3180 were in surface mining). In the following period total employment decreased to 45.5% of employment in 2000 with an average number of employees of 8273/year. In underground mining we recorded in the analysed period an average employment of 3169/year with a decrease in 2017 compared with 2000 of 62.3% (i.e. 3884 persons) from original employment. With surface mining the decrease in employment was steady, with a decrease of 48.2% (i.e. 3731 persons) from the original employment in 2000, when an average in surface mining of 5105/year had been employed (Fig. 1).

Creation of employment is influenced by the development of mining companies in the time horizon 2012–2018. This development is illustrated in Table 1.

A number of companies in the mining sector could help to analyse macroeconomic indexes. Companies dealing with mining of raw materials must go through a long life-cycle before mined and processed material is sold. Therefore the initiation of business and single operation of a mining company take certain amount of time, but also requires sufficient capital for all these activities. The total number of companies in the mining sector includes all companies, including mining of ore and non-ferrous, construction or other mineral resources. According to available

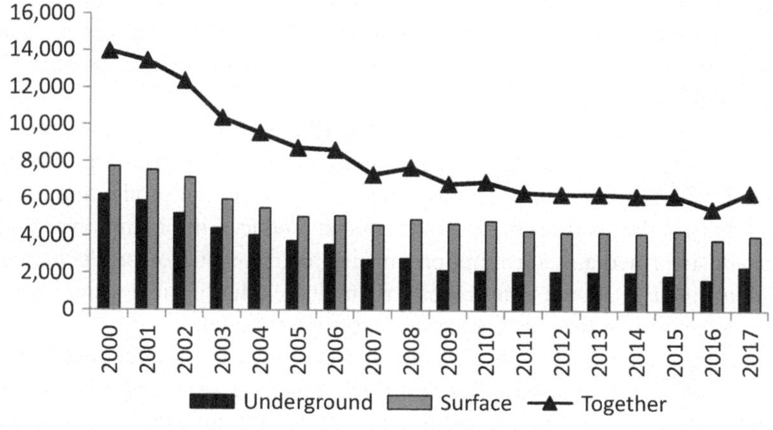

Fig. 1 Development of employment in mining organization of Slovakia. Source: processing according [22]

Table 1 Number of mining companies by year

	2012	2013	2014	2015	2016	2017	2018
SR	108	138	125	157	182	187	223

Source: own processing according [23]

Table 2 Development of employment in the mining sector (%) by year

	2010	2011	2012	2013	2014	2015	2016
SR	13.9	11.6	12.6	11.5	9.5	11.8	10.9

Source: own processing according [23]

figures we can state that the number of companies also depends on the volume of deposits in the state, as well as on which deposit it is possible to mine after gaining a mining permit. In Slovakia, there is rapid registered growth every year; in 2018, 223 mining companies were registered. Development of employment expressed as a percentage is shown in Table 2.

Development of employment in Slovakia was characterized by a variable tendency in 2000–2017 (Fig. 2), with greater employment creation in the mining sector; the rate of the mining sector also recorded an analogical trend in development, which meant a fluctuating tendency (Fig. 3). Average employment recorded in the analysed period in Slovakia was 2,178,705/year, with the highest measure of employment in 2016 (2,390,000 employed) and the lowest measure in 2000 (2,024,848 employed), which when compared with employment development in the exploitation of raw materials means that the average rate of Slovakian mining reached a level of 0.39% in the analysed period.

The highest rate of mining was recorded in 2000, when a 0.69% rate of mining had been achieved, which was in keeping with the lowest level of employment in mining and the lowest rate of employment in Slovakia. On the other hand, the lowest rate in 2016, with a level of 0.22%, resulted logically from the highest level of employment in Slovakia and the lowest employment in exploitation of raw materials. In general, the level of the rate of mining with regard to total employment of Slovakia was characterized by a fluctuating tendency with a decrease of 0.42% in 2017 in comparison with 2000 (Fig. 3).

Exploitation of the raw material base contributes similarly to the creation of value added, which can be characterized as the difference between production of commodities in basic prices and intermediate consumption of production factors in purchase prices. Production consists of products, produced in a common accounting period, and intermediate consumption consists of the value of products and services, consumed in production process as inputs, excluding investment property, which consumption reflects in consumption of fixed capital.

Gross value added as recorded in the analysed period also fluctuated, with an average value of 56,616 million Euro/year, while the highest value was recorded in 2017, when this macroeconomic index achieved 73,602 million Euro. On the other hand, the lowest value was recorded in 2000, when value added was 37,698 million Euro (Table 3).

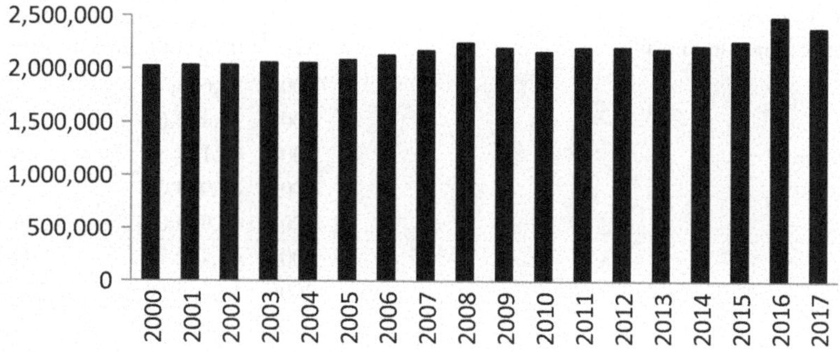

Fig. 2 Development of employment in Slovakia. Source: processing according [24]

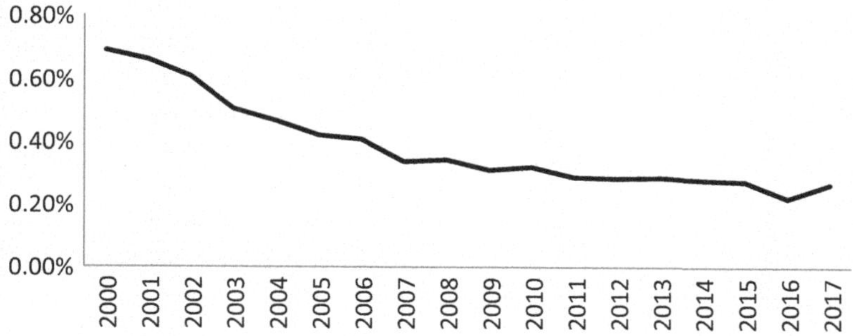

Fig. 3 Effect of development of mining rate on employment of Slovakia. Source: processing according [22, 24]

Similar to the development of employment, in the context of creation of gross value added in the mining sector, we recorded a fluctuating tendency in the analysed period (Fig. 4), while the mining recorded average value added a level of 295.81 million Euro/year, which means a 0.53% average rate. Generally, we can state that the highest value added in mining had been recorded in 2009 with a level of 372.83 million Euro; on the other hand, the lowest value was recorded in 2000, when mining recorded value added at a level of 186.65 million Euro. Comparing 2000 and 2017, we found gross value added, produced by mining, decreased in 2017 compared with 2000 by 144.56 million Euro. The highest rate of mining on value added was recorded in 2003 with a level of 0.64%. On the other hand the lowest rate of mining on value added decreased in 2017 compared with 2000 by 0.05% (Fig. 4).

Development of mining and associated single exploitation of raw material base is also determined by domestic and foreign investments that support increasing economic effectiveness. It is mainly foreign direct investment (FDI) in Slovakia that positively influences sustainable development, since this increases employment by creating new positions and GDP creation; it contributes to the implementation of innovative technologies or increases in qualification of domestic human capital. FDI can be generally characterized as investment of money or other monetary values or rights with the aim of basing, obtaining or extending permanent economic

Table 3 Development of gross value added in Slovakia

Year	Slovak Republic (SR) [mil. Eur]	Mining [mil. Eur]
2000	37,698.76	186.65
2001	39,486.60	187.48
2002	41,118.53	246.22
2003	42,973.05	275.94
2004	44,973.02	236.28
2005	47,492.91	249.22
2006	52,266.60	204.73
2007	58,021.92	328.49
2008	61,691.94	368.03
2009	58,314.00	372.83
2010	61,368.25	320.89
2011	62,898.59	344.24
2012	64,433.97	336.65
2013	65,167.74	337.71
2014	66,708.76	318.14
2015	69,295.49	352.96
2016	71,585.46	326.86
2017	73,602.75	331.21

Source: processing according [24]

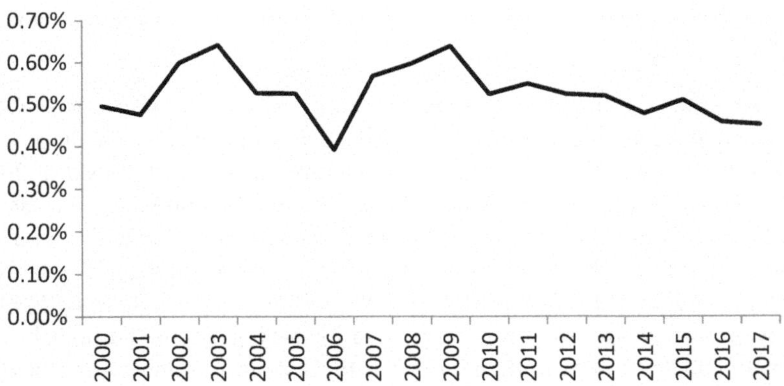

Fig. 4 Effect of development of rate of mining on gross value added in Slovakia. Source: processing according [24]

relations of domestic investors with business abroad, i.e., foreign investors investing in business in Slovakia. Mainly as a result of FDI, the rate of mining in connection with FDI is considered a direct indicator of dynamic development in Slovakia in interaction with investment attractiveness of Slovakia for foreign investors. FDI development in Slovakia is characterized more or less by an increasing trend with a slight decrease in 2013 by 2.62%. According to available data we can state that the average FDI level in the analysed periods was approximately 27,140,447 thousand. Euro/year, while exploitation of raw material base recorded an average FDI level of 232,371 thousand Euro/year, with an average rate of total FDI in the analysed period of 0.88%.

Exploitation of raw material base also plays a role in indexes of social and economic development in relation to GDP development. GDP development has been ranked among the basic indicators of economic effectiveness of Slovakia. This index, as mentioned earlier, also showed a fluctuating trend. According to available data we can state that in the analysed period of 2000–2017, GDP recorded an average of 58,313.5 million Euro/year, with the highest value recorded in 2015 at a level 78,896.44 million Euro. The lowest value was recorded in 2000 at 31,601.29 million Euro. Exploitation of the raw material base in Slovakia played a part in the GDP creation in the analysed period with an average of 58,313.5 million Euro/year, while the highest level of raw material base using GDP in Slovakia was recorded in 2008 (371.3 million Euro) and the lowest value was in 2002 (154.3 million Euro) (Table 4). With evaluation of the rate of the raw material base on GDP creation we can state that this was directly determined by development of a single GDP in interaction with development of the mining sector that in the analysed period recorded an average rate on a single GDP creation in Slovakia of 0.54%/year.

The rate of the raw material base using GDP creation, similar to the previously mentioned analysed indexes, recorded a fluctuating trend, determined by GDP development in Slovakia, as well as by demand and the offer of development of a single raw material base. The highest rate of mining, which means using the raw material base in Slovakia, was recorded in 2000, when the rate of the mining on

Table 4 GDP development in Slovakia and mining

Year	Slovak Republic (SR) [mil. Eur]	Mining [mil. Eur]
2001	34,310.622	158.0
2002	37,279.842	154.3
2003	41,404.346	207.0
2004	46,101.527	160.7
2005	50,415.092	177.7
2006	56,272.653	204.2
2007	63,053.882	263.6
2008	68,491.623	371.3
2009	64,023.061	329.6
2010	67,577.288	358.1
2011	70,627.205	337.9
2012	72,703.513	330.3
2013	74,169.873	334.6
2014	76,087.789	319.8
2015	78,896.443	354.8

Source: processing according [24]

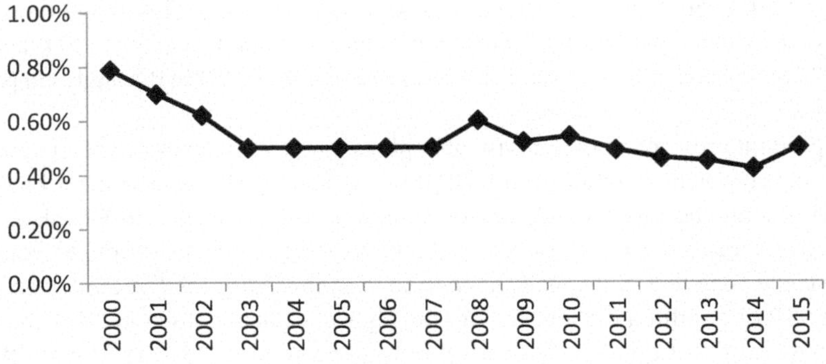

Fig. 5 Effect of rate of mining on GDP in Slovakia. Source: processing according [22]

GDP creation in Slovakia achieved 0.79%, and the lowest rate was in 2014, when the rate achieved only 0.42% (Fig. 5).

5 Discussion

Generally, businesses involving earth's resources face a number of obstacles. Mining companies need from a financial viewpoint additional finances for subsequent establishment of mechanisms and independent operation of the company. Therefore, the most important obstacle is usually a shortage of finance, although financial demands for surface mining are different from those of deep mining of coal, and also of other raw materials mined. Another obstacle could be gaining permits from agencies representing living environments that differ according to mined raw material and locality; for example, if mining were to occur in a protected locality, the company would not obtain a permit. The raw material policy must respect the principles of a socially and ecologically orientated market economy during mining and processing of raw materials and also with regard to effective use of natural resources. Mining activity involves complex technology and a high degree of human input, and in this respect it presents an important source of employment.

It is useful to evaluate mining development from the viewpoint of individual countries. The greatest volume of mining is in Poland and the smallest volume is in Slovakia, which means that Slovakia, but also Hungary, must import coal to cover domestic consumption of coal, while Poland and the Czech Republic have a sufficient volume of coal stocks ([25]; www.mbfh.hu). Effectiveness of raw material use is changing in accordance with internal and external factors. In spite of Slovakia being a small country, it has its own rich sources of raw materials. Mining contributes to the GDP in Slovakia. The majority of a mined raw material is consumed in the domestic market, for example building stone, dolomite, gravel and sand. It is mostly petroleum, earth gas, iron ore and black coal that are imported to Slovakia [26]. Among the positive impacts of mining is the increasing average monthly wage in the sector. As wages increase, people have a need to buy more and more products and services for providing for their needs, which can be met by using raw materials. A decrease in employment in mining can be considered as being a negative social and economic impact of mining activity in Slovakia. New companies purchase new technologies, through which they the increase volume of mining, sales of products, and at the same time the number of employees is decreasing. Another reason for employment decreasing in mining can be the disinterest of people in undertaking such a physically demanding job. Mining activity influences mostly the living environment and inhabitants of the country. Mining of raw materials influences the living environment mainly through a change in landscape relief and soil covering. People in mining works areas are affected by high dust and noise volumes, which can cause increases in various illness. In spite of the negative influence of mining, it is necessary to support the sector from the point of the view of the state, since the demand for raw materials is increasing every year. Estimation of a life-cycle of several stocks in Slovakia is over 200 years. The problems of using raw materials in Slovakia are reflected in available legislation of treatment with raw materials. Such treatment is documented through processes that are included in the raw materials' policy. In spite of the policy being drawn up at state level, it includes only general guidelines that are difficult to implement and realize effectively without institutional tools. However, they do provide a context that is both obligatory and inspiring.

6 Conclusions

According to analysis of the business environment in the mining sector, we have found that there is a need to draw up certain measurements or suggestions to improve the business environment. In business, it does not matter what business the company is in and if the business is prospering, but rather what legislative measurements are involved and what is entailed in the support of starting businessmen [27]. It is suggested that measurements that are necessary for improving the business environment be provided by ministries in specific countries. The basis of such suggestions could be measurements, orientated toward decreasing bureaucracy, improving competitiveness, and support of start-ups. In the context of measurements that could be realized within a certain time period, unexpected situations could arise, when such measurements would not be effective [28]. Since the business environment is always changing and influenced by a number of factors, it is necessary to always follow up the situation in a changing market environment. The next important measure of simplification of the administrative burden can be realized in the context of connected clients' centres, where clients could manage required allowances and permissions in one place. Such measurements include important parts, which are connected with other possibilities, for example, development of infrastructure, transport solutions, development of public institutions, measurements for increasing the effectiveness in the area of business processes etc. Development of infrastructure is important mainly from the point of view of transport and communication, which is necessary for connecting countries to make supply and demand of mining products easier. Transport solutions are very necessary not only for simplification of access to certain places, but they also enable businessmen to have easier access to clients. Development of public institutions could enable access to single clients, while providing services to the public could be more transparent. Measurements for increasing the effectiveness in the area of business processes, present the necessity for development and realization of innovations and technological equipment that helps to simplify and transform certain inputs to outputs.

Evaluation of the business environment in a country varies according various influences, all of which affect the single business environment. Selection of the ideal business environment is very complex due to various limitations. In the mining sector development of the business could be influenced by new technologies and innovations, resulting in growth of mining companies, enabling companies to mine higher volumes of raw materials due to the new technology or innovation from previously heavy available deposits. At the same time business in the mining sector could be improved by: permanent development of innovation for increasing of effectiveness in area of mining and processing, as well as better conditions for new businessmen. A proper business environment needs to be competitive. Competitors could involve only those businessman who are interested in their own business. Therefore it is necessary to find and permanently compare the competition, and also to try to improve each business according new ideas.

Acknowledgments This study was supported project VEGA No 1/0515/18, VEGA No 1/0651/2018 and KEGA No 006TUKE-4/2019.

References

1. Ciruelos, J., Duchene, M.: Modele sur micro-ordinateur pour l'analyse de la faisabilite financiere d'un project minier. Industrie Minérale – Les techniquies. (1983)
2. Zuberec, J., Tréger, M., Lexa, J., Baláž, P.: Raw Material of Slovakia. ŠGÚ DŠ, Bratislava (2005). 351p
3. Cehlár, M., Maras, M.: Economic Evaluation of Mining Projects. TU, FBERG, Košice (2001)
4. Markulik, Š., Cehlár, M., Kozel, R.: Process approach in the mining conditions. Acta Montan. Slovaca. **23**(1), 46–52 (2018)
5. Domaracká, L., Torres, M., Fonseca, N., Sokolova, A., Yazevich, M.: Mining region environmental management in transition to sustainable development. In: E3S Web of Conferences, 41, 020183rd (2018)
6. Taušová, M., Čulková, K., Domaracká, L., Drebenstedt, C., Muchová, M.S., Koščo, J., Behúnová, A., Drevková, M., Benčöová, B.: The importance of mining for socio-economic growth of the country. Acta Montan. Slovaca. **22**(4), 359–367 (2017)
7. Straková, J., Pártlová, P., Váchal, J.: Business management in new global economy. Acta Oeconom. Univ. Selye. **6**(1), 155–166 (2017)
8. Nakata, T.: Energy-economic models and the environment. Proc. Energy Combust. Sci. **30**(4), 417–475 (2014). https://doi.org/10.1016/j.pecs.2004.03.001
9. Arshad, A., Zakaria, M., Junyang, X.: Energy prices and economic growth in Pakistan: A macro-economic analysis. Renew. Sust. Energ. Rev. **55**, 25–33 (2016). https://doi.org/10.1016/j.rser.2015.10.051

10. Brown, S.P.A., Yücel, M.K.: Energy prices and aggregate economic activity: An interpretative survey. Q. Rev. Econ. Finance. **42**(2), 193–208 (2002)
11. Cunado, J., Perez de Gracia, F.: Oil prices, economic activity and inflation: evidence for some Asian countries. Q. Rev. Econ. Finance. **45**(1), 65–83 (2005). https://doi.org/10.1016/j.qref.2004.02.003
12. Falck, W.E.: Social licencing in mining—between ethical dilemmas and economic risk management. Mineral Econom. **29**, 97–104 (2016). https://doi.org/10.1007/s13563-016-0089-0
13. Bebbington, A., Hinojosa, L., Bebbington, D.H., Burneo, M.L., Warnaars, X.: Contention and ambiguity: mining and the possibilities of development. Develop. Change. **39**(6), 887–914 (2008). https://doi.org/10.1111/j.1467-7660.2008.00517.x
14. Liu, D., Yang, X., Tian, X., Wu, R., Wang, L.: Study on integrated simulation model of economic, energy and environment safety system under the low-carbon policy in Beijing. Procedia Environ. Sci. **5**, 120–130 (2011)
15. Ur-Rehman, S.A., Cai, Y., Mirjat, N.H., Walasai, G.D., Nafees, M.: Energy-environment-economy nexus in Pakistan: lessons from a PAK-TIMES model. Energy Policy. **1**, 200–211 (2019). https://doi.org/10.1016/j.enpol.2018.10.031
16. Mathu, K., Chinomona, R.: South African coal mining industry: socio-economic attributes. Mediterr. J. Soc. Sci. **4**(14), 347–358 (2013). https://doi.org/10.5901/mjss.2013.v4n14p347
17. Schwab, K.: The Global Competitiveness Report 2017–2018. Hope Steele, Geneva. ISBN 13:978-1-944835-11-8. http://www3.weforum.org/docs/GCR2017/2018/05FullReport/TheGlobalCompetitivenessReport2017%E2%80%932018.pdf (2017). Accessed 18 Feb 2018
18. Doing Business. World bank: Washington. http://www.doingbusiness.org/~/media/WBG/DoingBusiness/ (2018).
19. Coronto, R., Dlhopolček, J., Kopečný, P., Maxin, R., Tonka, V., Tvrdoň, J., Vajdová, E., Vyšný, T.: Analysis, Monitor of Business Environment Quality in SR and Competitiveness of the Economy. Center of Social Dialogue. pp. 8–23. Centrum vzdelávania MPSVR SR, Bratislava. http://www.azzz.sk/wp-content/uploads/2015/05/Anal%C3%BDza-monitor-kvality-podnikate%C4%BEsk%C3%A9ho-prostredia-v-SR-a-konkurencieschopnos%C5%A5-ekonomiky.pdf (2014). Accessed 5 Feb 2018. ISBN 978-80-89757-18-3
20. Lacko, R., Hajduová, Z., Pavolová, H.: Efficiency of mining and quarrying industry of V4 countries: the impact of investments and selected indicators. Acta Montan. Slovaca. **22**(2), 136–145 (2017)
21. Belanová, K.: Comparison of Business Environment Quality in V4 with Special Accent to the Availability of Financial Sources (in Slovak). BIATEC, NBS, Bratislava. 22(1), 12–17 (2014)
22. HBU SR.: Annual report of activity of mining office in SR in 2018. (2018). http://www.hbu.sk/sk/Vyrocna-a-rocna-sprava/Rocne-spravy.alej
23. Eurostat.: http://ec.europa.eu/eurostat/en/data/browse-statistics-by-theme (2018). Accessed 31 Jan 2018
24. Statistic office of Slovakia. (2017). http://www.statistics.sk/pls/elisw/metainfo.explorer
25. Bońda, R., Brzeziński, D., Czapigo-Czapla, M., Malon, A., Mikulski, S.Z., Miśkiewicz, W., Oszczepalski, S., Siekiera, D., Skrzypczyk, L., Sokolowksi, J., Stawierej, J., Szamalek, K., Szczygielski W., Szuflicki, M., Tymiński, M., Walkuska, A.: Bilans zasobów złóś kopalin w Polsce. pp. 36–53. https://www.pgi.gov.pl/dokumenty-przegladarka/publikacje-2/bilans-zasobow/3845-bilans-zasobow-2015/file.html. (2016). Accessed 27 Feb 2018
26. IEA. Key world's energetic statistics. International Energy Agency, Paris, 2018. https://www.iea.org/publications/freepublications/publication/KeyWorld2017.pdf (2017). Accessed 20 Feb 2018
27. Tóth, Z., Mura, L.: Support for small and medium enterprises in the economic crisis in selected EU countries. In: Hradec Economic Days 2014: Economic Development and Management of Regions, PT V, pp. 424–429 (2014)
28. Balance of raw materials stocks in Hungary. Budapest, 2015. (2015). http://www.mbfh.hu/home/html/index.asp?msid=1&sid=0&hkl=72&lng=1. Accessed 27 Feb 2018

The Indebtedness According to the Property Structure with Respect to the Sector Type

Lucia Domaracká, Marcela Taušová, Mária Shejbalová Muchová, and Marián Albert

1 Introduction

The area of indebtedness and its impact on the viability of the company as well as the search for its causes is a long-term analysed issue. It is dealt with by many authors from all areas of the economy and countries of the world. Analysis of debt of the company in various sectors is an essential part of the financial position evaluation, in which company is located, and also it enables the state structure of the capital available to the undertakings. It presents a tool to assess the way the company manages funds and also serves to locate the areas where the company has an inefficient economy. Such analysis also reflects the short-term and long-term financial development of the company, leaving the owner the image by which the company is developing.

Capital independence of the company means the ability of a company to create the optimal structure of assets and capital. Foreign sources are cheaper than their own capital, and therefore firms tend to borrow more funds to finance their assets. This solution, however, is effective only after reaching the optimum, in which the company is still able to effectively meet its obligations. Calculation of debt indicators serves to know limits when the company is still in a position to acquire external financial sources without threatening its business.

The main stimulus for using foreign sources is their relatively lower price compared with their own sources, enabling to decrease the average total price of used capital. The growth of indebtedness can lead to total profitability.

L. Domaracká (✉) · M. Taušová · M. S. Muchová · M. Albert
Faculty of Mining, Ecology, Process Control and Geotechnologies,
Institute of Earth Resources, Technical University of Košice, Kosice, Slovakia
e-mail: lucia.domaracka@tuke.sk; marcela.tausova@tuke.sk;
maria.shejbalova.muchova@tuke.sk

L. Knapcikova et al. (eds.), *New Approaches in Management of Smart Manufacturing Systems*, EAI/Springer Innovations in Communication and Computing, https://doi.org/10.1007/978-3-030-40176-4_3

2 Literature Review and Marketing Review

The specific corporate indebtedness and capital structure is fundamentally a complex process dependent on a large variety of determinants and the chosen financial strategy; therefore, it depends on the particular decisions of individual firms ([2], p. 136). Authors' research was connected with the analysis of the bureaucratic logic of decision-making process, characterized by nonfunctional stability, inflexibility, individualism, power, authority and hierarchy, centralization, vagueness and fragmentariness. The objective of the research was to analyse to what extent the manager's decision-making is based on bureaucratic tools or organizational learning in either complex problem-solving or non-problem-solving decision-making (MANOVA, method of the repeated measure, intersubject factor—(1) non-problematic and (2) problematic situation). The conclusion of analysis is that there are significant differences in the characteristics of solving problematic situation and non-problematic situation decision-making: the bureaucratic attributes of decision-making are more intensive in problematic situations, while learning approach is more actual in non-problematic situations. The results of analysis have shown that managers who apply the learning organization attributes in their decision-making are more successful in problem-solving. Hasan et al. ([8], p. 647) provide empirical evidences regarding the influence of management practices on financial statements and financial strength of listed companies across countries. The culture, attitudes, beliefs, perceptions, etc., are different from country to country. Falivigna and Ippoliti ([4], p. 534) investigate the relation between the time necessary to define a bankruptcy case and the concentration of enterprises with good financial health. They identify a statistically significant relation between the time necessary to enforce creditors' rights and the concentration of healthy enterprises for several manufacturing industrial sectors. A severe debt overhang problem, either public or private, the analysis of the factors that influence companies' leverage reveals essential, in particular for the high-indebted firms. Also, it is important to define accurately enterprise indebtedness, because it is used in the calculation of financial ratios, serving for evaluation of enterprises' financial performance ([11], p. 58). When determining indicators of liabilities of mining and beneficiation enterprises of Ukraine, the following factors shall be considered: company internal indebtedness (i.e. debt of the company to itself), loan guarantees that are provided to other enterprises of the same vertically integrated structure for which the certain company relates to and enterprise liabilities to affiliates. Internal indebtedness appears at enterprises because of insufficient financing of obligatory programmes. Untimely fulfilment of such programmes may lead to future increasing of enterprise's costs and consequently lower financial and economic results. Another peculiarity of mining and beneficiation enterprises operating as a part of vertically integrated structures is the existence of affiliated liabilities. Two groups of affiliated liabilities shall be distinguished: absolutely affiliated (liabilities of the enterprise to other companies which are fully controlled by the same owners) and conditionally affiliated (liabilities of company to other companies which are partially controlled by the owners of the same company). Absolutely affiliated liabilities shall be transferred from current to long-term liabilities; conditionally affiliated liabilities are current by their essence. The authors have developed a methodology for correcting mining and recipient liabilities, which will enable the accuracy of the definition of its financial and economic results and the assessment of the financial situation to be increased. Teplická [22] examines the indebtedness of the company as well as the cost of capital and financial health as one of the internal factors, which is the result of its own decision-making and significantly affects the performance of the company. Within the framework of the questionnaire research, she analysed the current performance measurement in Slovak industrial enterprises and defined the cause and effect relationships between factors influencing performance measured by the PLS method. The findings of the study by Iatridis and Kilirgiotis [10] provide evidence that also the firm size is positively related to indebtedness and fixed asset revaluation. González and Jareño [6] carry out an analysis of four key financial variables: percentage of indebtedness, volume of equity, overall liquidity, and returns on equity, concluding that some expected relationships are confirmed, such as the logical exchange between equity and borrowed capital, as well as other less obvious relationships, like the positive relationship between the returns and the volume of equity. Kudlawicz et al. [13] study the existence of fit between the capital structure and economic performance leading companies to position themselves on the efficient frontier. The results indicate that the efficiency increases the performance of the company and still more companies, close the efficient frontier, have a higher economic performance and, in addition, lower levels of indebtedness. Kapan and Minoiu ([12], p. 36) study the link between banks' financial health and the strength of transmission of financial sector shocks to the real economy. They find that banks with ex ante stronger balance sheets, in particular higher levels of common equity, were better able to maintain credit supply when faced with liquidity shocks during the crisis. Bank recapitalizations mitigated the lending gap between high and low capital banks but only in countries with strong sovereigns. These findings support the view that strong financial intermediary balance sheets are key for the recovery of credit and economic performance after large financial sector shocks. Fernández-Arias et al. ([5], p. 275) develop an early warning system that is based on a classification model which estimates the Fitch ratings using Bankscope bank-specific data, regulatory data and macroeconomic data as input variables. The authors propose a "hybridization technique" that combines the extreme learning machine and the synthetic minority oversampling technique. Due to the imbalanced nature of the problem, the authors apply an oversampling technique on the data aiming to improve the classification results on the minority groups. The methodology proposed outperforms other existing classification techniques used to predict bank solvency. It proved essential in improving average accuracy and especially the performance of the minority groups. Oehler et al. ([17], p. 310) say that individuals have not only theoretical knowledge about financial products and mechanisms that do not reflect real situations for managing financial literacy. The results show that most exhibit good financial literacy as well as a realistic assessment of risk and return and their impact on (financial) well-being. Marjanovic et al. ([16], p. 153) research whether the stressful psychological experience of financial threat relates to acute more than chronic economic hardship and to making sounder financial decisions that would improve personal financial situations. Their findings contribute to the literature in that they show financial threat – positively correlates with acute financial hardships (e.g. reduced income), and this relation is stronger than its relationship with chronic financial conditions (i.e. current income). Financial threat positively correlates with constructive financial behaviours, such as increased willingness to cut expenses. Economic hardship moderates the positive relationship between financial threat and constructive financial behaviour. When economic hardship is high, individuals with high and low levels of financial threat are willing to engage in constructive financial behaviours. When hardship is low, however, individuals with high financial threat are the most willing to engage in constructive financial behaviour. Potrich and Vieira ([20], p. 1050) point to the fact that the financial health of individuals is affected by their financial literacy and its factors. The study investigated 2487 individuals in Brazil. For an analysis, the authors used confirmatory factorial analysis and structural equation modelling and six research hypotheses. Philip, DJ., Sudarsanam, N. and Ravindran, B. ([18], p. 99) gained a deeper understanding of the various controls using the modified hidden Markov model (HMM). Lee et al. ([14], p. 144) demonstrate the practical feasibility of project owners to evaluate contractor liquidity using financial and nonfinancial information to prepare for possible financial problems. This paper from Baglan and Yilmazkuday ([1], p. 1307) shows that there is a positive and significant relationship between financial health and the intensive margin of trade. The magnitude of this positive relationship is shown to depend on several firm characteristics, where the effects of financial health on firm-level exports are larger for firms with higher levels of export, bigger size (measured by assets), higher productivity (measured by value added per worker) and moderate levels of financial health (measured by cash flow over total assets). The results are robust to the consideration of foreign ownership and country characteristics as well as industry and time fixed effects. The risk of customers affects corporate financial policy by limiting the ability of firms to securitize customer receivables. Liu et al. ([15], p. 454) find that firms with riskier receivables, based on the credit risk and diversification of the firms' principal customers, have lower financing capacity and lower leverage in their asset-backed securitizations. Because securitizations are designed to create a very safe claim by separating the risk of the securitized assets from the risk of the originating firms, increases in the risk of the receivables directly inhibit originating firms' ability to securitize assets and indirectly inhibit the originating firms' access to external finance. The study highlights a novel link between the financing of supplier firms and the financial health of their customers and shows how an increase in risk can limit access to external capital. Piterkova et al. ([19], p. 855) are to examine and evaluate the risk of customers that affects corporate financial policy by limiting the ability of firms to securitize customer receivables. They find that firms with riskier receivables, based on the credit risk and diversification of the firms' principal customers, have lower financing capacity and lower leverage in their asset-backed securitizations. Because securitizations are designed to create a very safe claim by separating the risk of the securitized assets from the risk of the originating firms, increases in the risk of the receivables directly inhibit originating firms' ability to securitize assets and indirectly inhibit the originating firms' access to external finance. They highlight a novel link between the financing of supplier firms and the financial health of their customers and show how an increase in risk can limit access to external capital. Huang ([9], p. 451) introduces macroeconomic monitoring methods into the financial analysis of microeconomy to build a complete panoramic display system of enterprise financial prosperity monitoring. These researches provide a new approach for managers to grasp financial situation, control financial risk and forecast financial trends timely. The phenomenon that macroeconomic cyclical fluctuations make various microeconomic organizations fall into financial risk or crisis is an objective economic discipline. Zhang et al. ([23], p. 384) use the methods of the macroeconomic prosperity early warning to build the enterprise financial monitoring and early warning system which is achieved relying on the comprehensive design of six subsystems and verifies that it is scientific and practical via the applied research. These researches provide a new approach for managers to grasp financial situation, control financial risk and forecast financial trends timely. Siska and Siskova ([21], p. 40) investigate the relation between the time necessary to define a bankruptcy case and the concentration of enterprises with good financial health. In particular, recalling the seminal work of Modigliani and Miller ([3]; Am Econ Rev. 48:261–297, 1958), they test the hypothesis that the enforcement of creditors' rights can affect managers' strategies and the related risks. According to results, judicial delay can affect the industrial spatial dynamics and regional development, confirming the key role of judiciary. Indeed, they identify a statistically significant relation between the time necessary to enforce creditors' rights and the concentration of healthy enterprises for several manufacturing industrial sectors. Presently, the economic effectiveness of the company is becoming decisive alternative to obtain competitiveness at the market. Reliable statement about financial situation can be obtained by analysis of the ability to pay the debts. Development in the industrial companies had been searched by indexes of activity, which express and quantify if the economy of the company is effective by the way of its assets. Results of the analysis show the most effective and less effective areas from the view of payment terms and asset turnover. Strengthening of the less effective areas can contribute to the whole economy of the regions and countries ([3], p. 744). The aim of this work was to show insight into corporate indebtedness by examining the basic financial statements of 100 enterprises over a 5-year period at 20 sectors. Debt ratios have been set in relation to the indicators that characterize corporate core assets, such as liquidity, the gold funding rule and ROA. Through the analysis of variability, the relations between the various indicators at the level of the sector were judged and common features of sectors were defined.

3 The Research Methodology Used

The main research method used is the financial rate indexes, which allow to express rate among own and foreign capital. Indexes measure the extent of debts used.

Indexes serve also as indicators for the determination of risk level, but indebtedness does not always mean negative characteristics. A certain level of indebtedness is profitable for the company, which should try to have a financial structure with minimal cost of capital and correspondent structure of its assets. Searched indexes of indebtedness are as follows:

Total indebtedness (debt ratio):

$$\text{Debt ratio} = \text{foreign capital} / \text{total assets} \tag{1}$$

Debt ratio (total indebtedness) expresses percentage part of foreign sources from total assets of the company. In case total indebtedness is lower, there is an assumption for greater financial stability and profitability. Recommended value is 30–60%, depending on the sector in which company makes business and market situation.

Leverage ratio:

$$\text{Leverage ratio} = \text{total assets} / \text{equity} \tag{2}$$

Leverage ratio testifies about how a large amount of assets forms own capital; value 2 explains the same amount of own and external capital. The acceptable limit of the external capital is 70% (two thirds); according to the given relation, it should be value 3.

In relation to the profitability, the rule is as follows: in case profitability of invested capital is higher than a price of interested foreign sources, financial leverage is positive, and vice versa.

Insolvency:

$$\text{Insolvency} = \text{commitment} / \text{receivables} \tag{3}$$

Insolvency is a measure of a company's ability to pay off its incurred debt. The value bigger than 1 refers to primary insolvency; a value lower than 1 is defined as secondary insolvency.

In addition to debt ratios, it is necessary to examine the relationship between the enterprise's assets and the sources of coverage and, consequently, the ability of the enterprise to generate profits. The most important rules, in terms of maintaining solvency, i.e. long-term payability, include the *golden balance rule*.

The golden balance rule of funding requires that long-term assets be funded by capital that is available to the enterprise on a long-term basis and that short-term assets be financed by capital that is available only for a short time.

The relationship that expresses the golden balance rule can be interpreted as follows:

$$\text{Long} - \text{term asset} < \text{own capital} + \text{long} - \text{term foreign capital} \tag{4}$$

Respectively

$$\text{Long} - \text{term asset} - (\text{own capital} + \text{long} - \text{term foreign capital}) < 0 \tag{5}$$

This means that total long-term assets should be less than the amount of financial resources available over the long term.

$$\text{LI} - \text{liquidity} - \text{available liquidity}$$

Available liquidity compares the relationship of financial assets to short-term liabilities. It points to the solvency of the subject being analysed. It is important that the organization under analysis has as many funds as possible so that it is able to pay its due liabilities. The recommended value varies considerably from one sector to the next, ranging from <0.2; 1>, and that was the subject of our investigation:

$$\text{LI} = \text{financial asset} / \text{short} - \text{term liabilities} \tag{6}$$

ROE – return on equity

Return on equity provides information on equity interest. The level of return on equity depends on the ratio of profitability of total capital to that of foreign capital. We calculate it as follows:

$$\text{ROE} = \text{net profit} / \text{own capital} \tag{7}$$

Optimum is the value that means capital reproduction (own) for the period of 4–5 years at 0.25–0.20.

Mentioned indicators had been searched in companies, analysed in the frame of choosing industrial sectors. Every sector included five chosen companies, where internal data served for interpretation of their financial situation. Consequently, according to obtained data from financial reports during 2009–2016, searched indicators had been calculated for every company individually. Calculated values of giving indexes in choosing sectors had been evaluated by graphs, serving for final evaluation of indebtedness in the 100 companies in the frame of chosen sectors.

The calculated indicators were then analysed by the Wilcoxon/Kruskal-Wallis tests, examining the variability of the indicators achieved with respect to the individual sectors. Subsequently, a multidimensional analysis was carried out using JMP Pro software support, where a pair correlation analysis was carried out at sector level. The results of the correlation coefficients were compared in a sectoral manner in order to identify the common features of selected sectors.

4 Results and Discussion

4.1 Analysis of Financial Indicators

Based on the data collected from the publicly available financial statements of the balance sheet and the profit and loss statement, the financial indicators for the business entities were calculated, the results of which varied widely (see Table 1) and the standard deviation size indicates the need to examine the individual indicators separately for each sector.

4.2 Analysis of Variability

The analysis examined the variability of the resulting values of the calculated financial indicators per sector. For all indicators, the statistically significant variability of the achieved values was confirmed by the choice of the sector. As shown in the analysis sample in Fig. 1 when analysing of the liquidity indicator LI, it can be stated that the average liquidity value for all the examined sectors is 0.6 which belongs to the optimal interval <0.2, 1>.

However, most sectors are located at the lower end of the range or below the optimum level in the period 2009–2016; only five sectors are at the upper end of the range and show excessive liquidity, including realities, information and communication, finance and insurance, forestry and fishing and personalistics.

Each of the indicators analysed showed statistically significant variability of the achieved values by the influence of the industry; at the same time, it can be noted that the impact of the sector on individual indicators is different. Branches that are above average in one pointer are below average in another pointer. In analysing the return on equity indicator, the optimum set at 0.2 exceeds only the information and communication sector, which also shows above average liquidity, but on the other hand significantly lower than average results were recorded in realities, which was significantly above the previous indicator (e.g. Fig. 1). The GBR indicator, represented by

Table 1 Univariate simple statistics

Indicator	Mean	Std Dev	Min	Max
ROE	0.0488	0.2666	−4.3724	0.8838
LI	0.6372	2.0397	−0.001	32.799
GBR	−8.0575	36.048	−624.13	3.0984
Indebtedness	0.6604	0.3206	0	2.8905
Leverage ratio	8.8636	36.0574	−188.08	540.59
Insolvency	1.7833	4.9441	0	79.9369

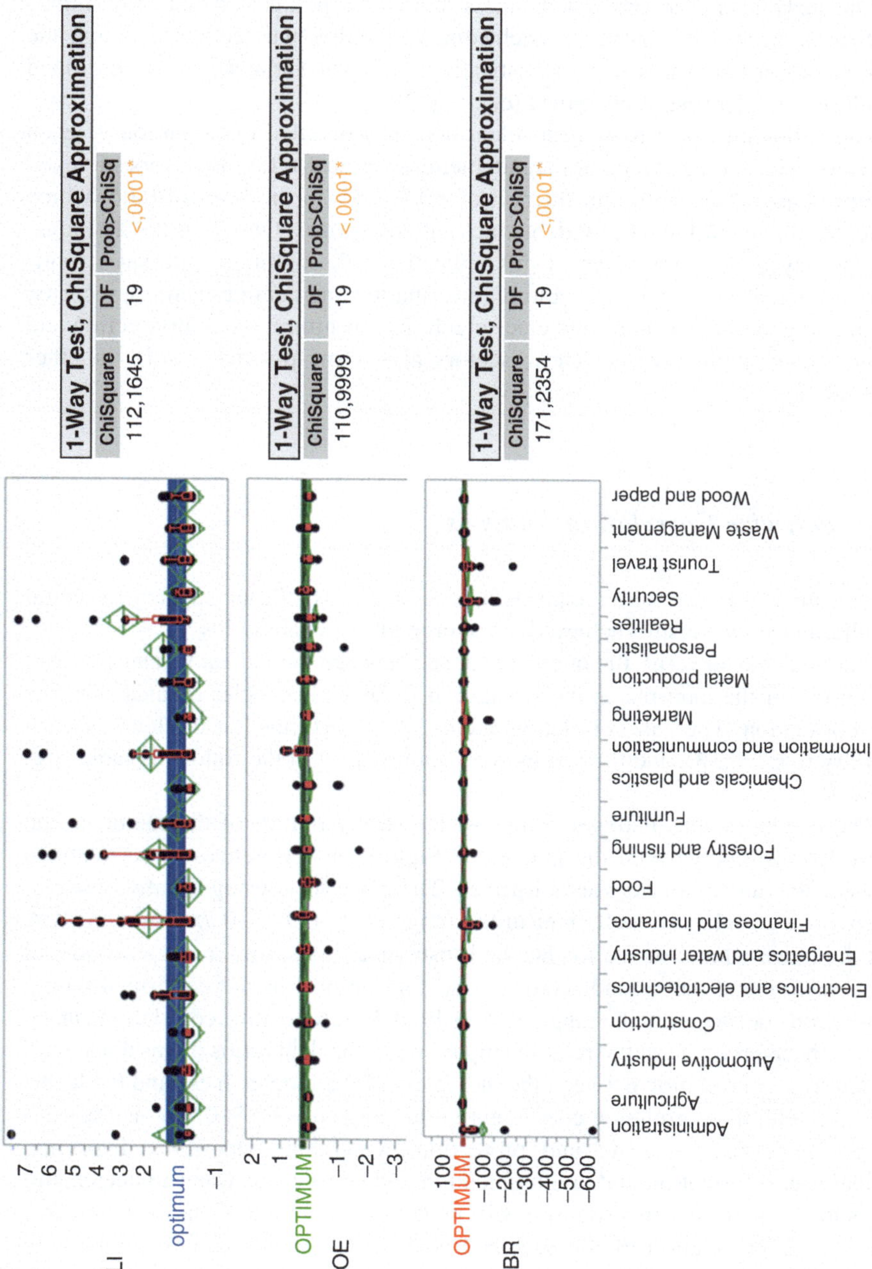

Fig. 1 Analysis of variability—the Wilcoxon/Kruskal-Wallis tests—LI, ROE, GBR

the requirement to finance long-term assets with long-term capital, was met in all sectors, as is evident from Table 1 with values ranging from about <−624; 3>.

The analysis of corporate indebtedness results also points to statistically significant differences in the average indebtedness of individual sectors. Through the Wilcoxon/Kruskal-Wallis tests, statistically significant variability was confirmed for all sectoral debt ratios examined (e.g. Fig. 2).

The following facts have been identified in a detailed examination of each indicator. The average value of the indebtedness indicator for all sectors surveyed is shifted above the optimum interval of <0.3, 0.6> to the level of 0.66, which indicates the overall level of indebtedness of enterprises. Only 8 of the 20 industries surveyed show the value of this indicator at the optimum interval or optimum border. These are the agriculture, information and communication, security, wood and paper, electronics and electrotechnics, furniture, metal production and waste management sectors. These sectors also show positive results in other debt ratios.

4.3 Pairwise Correlation Analysis

The results of the variability analysis have shown a significant impact of sectoral peculiarities on the results achieved in the property and debt ratios.

Our goal was to verify the interdependence between the indicators and compare the impact of the industry on the strength of these dependencies by analysing the pair correlation. The pair correlation analysis was performed at the level of each industry (e.g. Fig. 3), and the results were summarized in the following table (e.g. Table 2).

The results of this analysis also show the strong impact of the sector on the interrelationships between the indicators. Significant differences in correlations between the indicators were demonstrated. There is a very strong negative correlation with a correlation coefficient in the range of <−0.92; −0.7> between ROA and the indebtedness of the finance and insurance, marketing, reality, wood and paper sectors and the second sector whose correlation was not confirmed with a correlation coefficient in the range of <−0.49; 0.49>. A positive correlation can be seen in the assessment of the relationship between the debt ratios themselves, with the strongest correlation between the indicators of the indebtedness and the leverage ratio with the correlation coefficient in the range of <0.85; 0.96> in electronics and electrical, security and protection, wood and paper, metalworking, agriculture, information and communication and energy and water management. The subject of further investigation will be the classification of the common features and segmentation of the sectors based on the analysis of the sensitivity indicators.

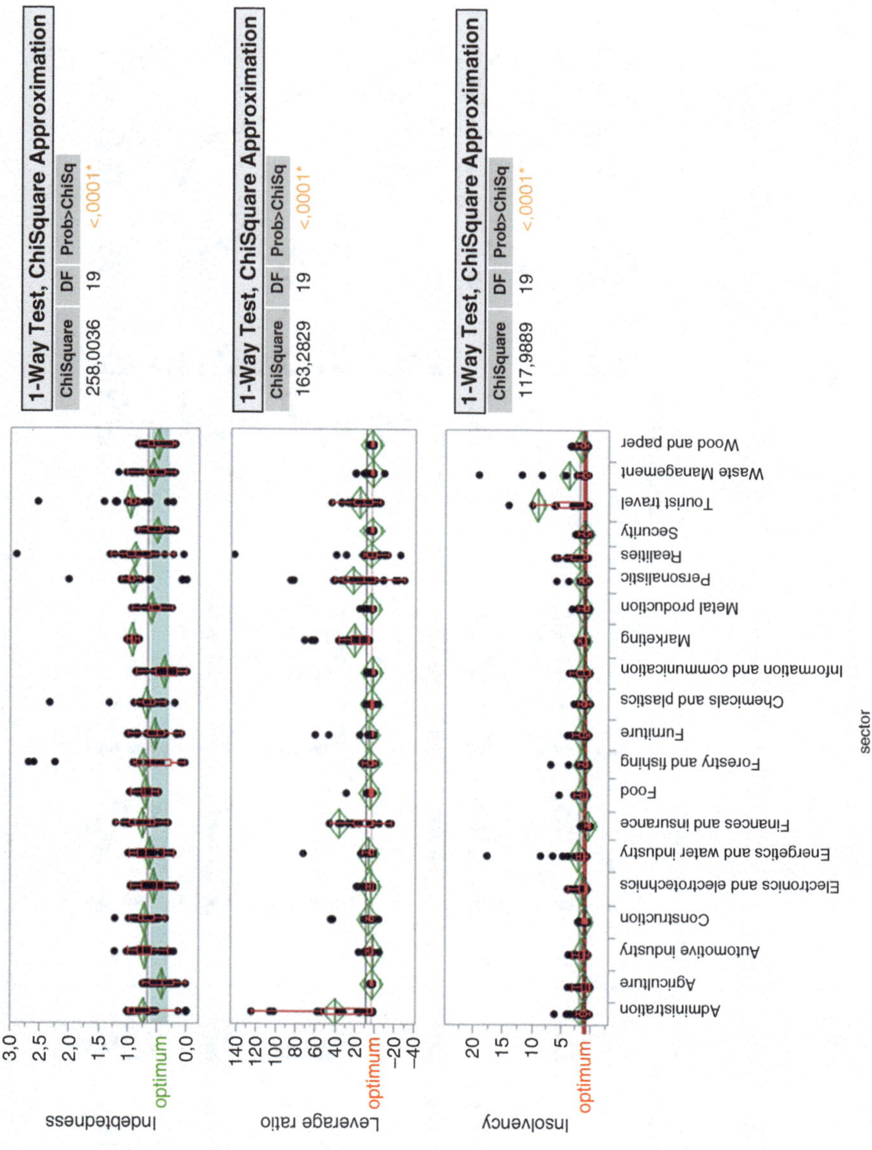

Fig. 2 Wilcoxon/Kruskal-Wallis tests—insolvency, leverage ratio, indebtedness

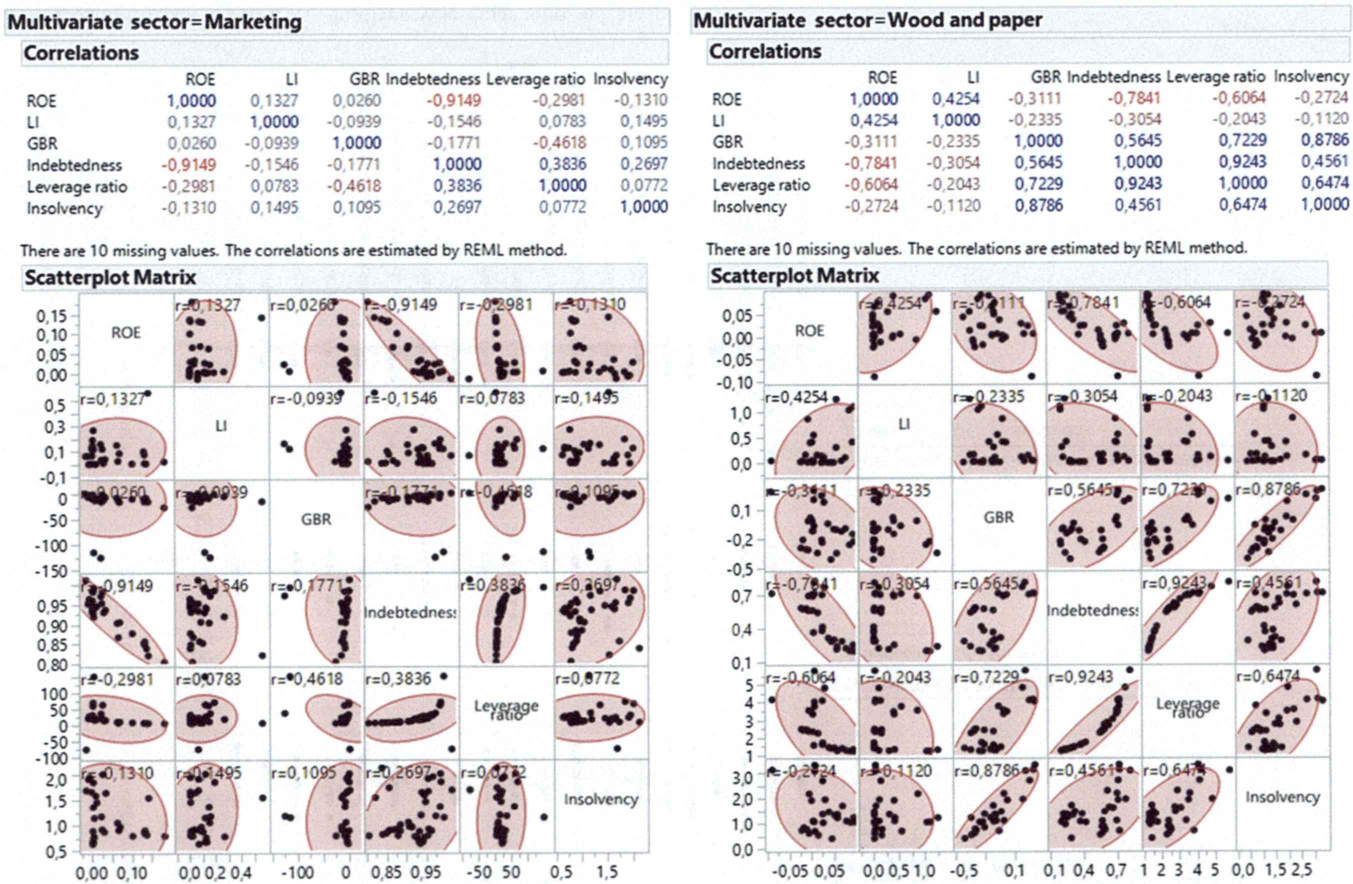

	ROE	LI	GBR	Indebtedness	Leverage ratio	Insolvency
ROE	1,0000	0,1327	0,0260	-0,9149	-0,2981	-0,1310
LI	0,1327	1,0000	-0,0939	-0,1546	0,0783	0,1495
GBR	0,0260	-0,0939	1,0000	-0,1771	-0,4618	0,1095
Indebtedness	-0,9149	-0,1546	-0,1771	1,0000	0,3836	0,2697
Leverage ratio	-0,2981	0,0783	-0,4618	0,3836	1,0000	0,0772
Insolvency	-0,1310	0,1495	0,1095	0,2697	0,0772	1,0000

	ROE	LI	GBR	Indebtedness	Leverage ratio	Insolvency
ROE	1,0000	0,4254	-0,3111	-0,7841	-0,6064	-0,2724
LI	0,4254	1,0000	-0,2335	-0,3054	-0,2043	-0,1120
GBR	-0,3111	-0,2335	1,0000	0,5645	0,7229	0,8786
Indebtedness	-0,7841	-0,3054	0,5645	1,0000	0,9243	0,4561
Leverage ratio	-0,6064	-0,2043	0,7229	0,9243	1,0000	0,6474
Insolvency	-0,2724	-0,1120	0,8786	0,4561	0,6474	1,0000

Fig. 3 Multivariate analysis of sector correlations—marketing, wood and paper

Table 2 Analysis of the pair correlation of selected indicators at sector level

Sector	Correlation														
	ROE: LI	ROE: GBR	ROE: IND	ROE: LR	ROE: INS	LI: GBR	LI: IND	LI: LR	LI: INS	GBR: IND	GBR: LR	GBR: INS	IND: LR	IND: INS	LR: INS
Electronics and electrotechnics	-0.05	0.38	**-0.60**	**-0.57**	-0.17	-0.18	**-0.61**	-0.34	-0.29	-0.39	-0.44	0.46	**0.80**	0.41	0.30
Security	0.45	-0.37	0.28	0.18	-0.07	-0.30	0.13	0.12	0.33	0.21	0.32	0.34	**0.91**	**0.64**	**0.50**
Wood and paper	0.44	-0.32	**-0.75**	**-0.58**	-0.30	-0.22	-0.30	-0.20	-0.09	**0.60**	**0.75**	**0.90**	**0.92**	0.47	**0.67**
Metal production	**-0.53**	0.45	0.21	0.07	0.13	-0.47	-0.36	-0.30	-0.43	0.49	**0.59**	**0.81**	**0.82**	**0.67**	**0.70**
Agriculture	-0.11	-0.02	0.16	0.23	0.04	-0.01	**-0.58**	-0.36	-0.21	0.46	**0.51**	**0.52**	**0.96**	0.44	0.38
Information and communication	**0.75**	**-0.83**	**-0.52**	-0.37	**-0.54**	**-0.72**	**-0.55**	-0.34	**-0.53**	**0.66**	0.44	**0.60**	**0.90**	0.19	0.06
Waste Management	**0.60**	-0.43	-0.37	0.15	**0.56**	-0.44	-0.29	0.16	**0.60**	0.06	-0.08	-0.35	-0.13	0.12	0.06
Administration	**-0.67**	-0.08	0.31	-0.13	0.28	**-0.95**	**-0.71**	-0.20	-0.22	**0.57**	0.33	0.17	0.50	-0.21	-0.12
Automotive industry	-0.05	0.00	-0.20	0.40	-0.07	-0.07	-0.46	-0.20	-0.02	0.23	0.10	0.50	0.40	0.25	0.27
Tourist travel	0.13	0.12	0.14	-0.10	0.26	-0.01	**-0.57**	-0.12	-0.14	0.15	-0.08	0.20	-0.03	**0.56**	-0.37
Construction	0.20	**-0.70**	**-0.55**	0.20	-0.47	-0.11	-0.44	-0.13	-0.30	0.13	0.05	0.17	0.30	**0.77**	0.32
Forestry and fishing	0.09	0.00	**-0.66**	0.24	0.07	0.23	0.00	**0.60**	-0.18	-0.02	0.13	0.11	0.01	-0.22	-0.22
Furniture	0.31	-0.14	-0.19	0.04	-0.37	-0.35	-0.37	-0.15	-0.30	0.14	-0.21	**0.70**	**0.61**	0.03	-0.04
Personalistic	0.04	-0.13	-0.49	-0.03	-0.13	-0.25	**-0.74**	-0.08	**0.60**	0.18	-0.27	-0.12	0.04	0.12	-0.21
Chemicals and plastics	-0.05	**-0.59**	-0.42	0.37	0.11	0.09	0.33	-0.49	0.08	**0.81**	-0.31	**0.64**	-0.18	0.08	0.08
Finances and insurance	-0.10	0.27	**-0.80**	-0.27	0.20	-0.30	0.34	0.47	-0.29	0.06	0.14	0.40	0.35	-0.05	-0.29
Marketing	0.10	0.00	**-0.92**	-0.27	-0.35	-0.12	-0.12	0.11	0.10	-0.15	**-0.51**	0.07	0.38	0.27	0.08
Realities	0.06	**-0.81**	**-0.86**	0.03	0.12	0.11	-0.38	0.00	0.04	0.35	-0.03	0.33	-0.03	0.13	0.16
Food	-0.16	0.25	**-0.53**	-0.46	0.18	-0.02	0.45	0.22	-0.02	0.02	0.03	0.42	**0.64**	0.24	0.12
Energetics and water industry	-0.06	**-0.79**	0.32	0.02	-0.22	0.11	-0.47	-0.29	-0.20	**-0.55**	-0.45	0.32	**0.85**	-0.20	-0.26

5 Conclusion

At present, most V4 companies have a too high value for debt indicators, which may be threatening in the long period. However, indebtedness cannot be perceived globally, irrespective of the specificities of individual sectors, which often display a very different asset structure resulting from different technological claims on the sector itself. There are sectors with a strong impact on the economic cycle, on the other hand, sectors that develop independently of the cyclical development of the economy [7]. Based on this analysis, the impact of individual sectors on debt ratios was assessed, as confirmed by the analysis of variability, and then the intensity of the relationship between debt ratios and asset and profitability indicators was examined. Even in this second part, the sector's marked impact on the intensity of dependence between indicators was evident, opening up further possibilities for review.

References

1. Baglan, D., Yilmazkuday, H.: Financial health and the intensive margin of trade. Emerg. Mark. Financ. Trade. **54**(4), 1304–1319 (2018)
2. Bolfíková, E., Hrehová, D., Frenová, J.: Manager's decision-making in organizations—empirical analysis of bureaucratic vs. learning approach. Zbornik Radova Ekonomskog Fakultet au Rijeci. **28**(1), 135–163 (2010)
3. Čulková, K., Pavolová, H., Cehlár, M., Khouri, S.: Influence of crisis to activity indexes in chosen industrial companies of the country. TEM J. **7**(4), 744–749 (2018)
4. Falivigna, G., Ippoliti, R.: Industrial spatial dynamics, financial health and bankruptcy: evidence from Italian manufacturing industry. Economia e Politica Indusitriale. **45**(4), 533–554 (2018)
5. Fernández-Arias, D., López-Martín, M., Montero-Romero, T., Martínez-Estudillo, F., Fernández-Navarro, F.: Financial soundness prediction using a multi-classification model: evidence from current financial crisis in OECD banks. Comput. Econ. **52**(1), 275–297 (2018)
6. González, C., Jareño, F.: Financial analysis of the main hotel chains of the Spanish tourism sector. Reg. Sect. Econ. Stud. **14**(2), 91–108 (2014)
7. Gavurova, B., Behúnová, A., Tkácová, A., Perželová, I.: The mining industry and its position in the economic cycle of the EU countries. Acta Montan. Slovaca. **22**(3), 278–286 (2017)
8. Hasan, M.S., Omar, N., Hassan, A.B.M.R.: Financial health and management practices: a multi-year cross country analysis of PLCs. J. Financ. Crime. **25**(3), 646–657 (2018)
9. Huang, Y.: Panoramic display system of enterprise's financial prosperity monitoring. Manag. Manufact. Materials Eng. **2**, 451–455 (2012)
10. Iatridis, G.E., Kilirgiotis, G.: Incentives for fixed asset revaluations: the UK evidence. J. Appl. Acc. Res. **13**(1), 5–20 (2012)
11. Ishchenko, M.: Mining and beneficiation companies liabilities figures correction. Econom. Anna. XXI. **11-12**(1), 58–61 (2013)
12. Kapan, T., Minoiu, C.: Balance sheet strength and bank lending: evidence from the global financial crisis. J. Bank. Financ. **92**, 35–50 (2018)
13. Kudlawicz, C., Senff, C.O., Bach, T.M.: The economic performance and capital structure: Brazilian companies to light efficiency frontier. J. Global. Compet. Govern. **9**(3), 40–52 (2015)
14. Lee, J.I., Lee, H.-S., Park, M.: Contractor liquidity evaluation model for successful public housing projects. J. Construct. Eng. Manag. **144**(12), a.n.04018109 (2018)

15. Liu, L.X., Mao, M.Q., Nini, G.: Customer risk and corporate financial policy: evidence from receivables securitization. J. Corp. Finan. **50**, 453–467 (2018)
16. Marjanovic, Z., Fiksenbaum, L., Greenglass, E.: Financial threat correlates with acute economic hardship and behavioral intentions that can improve one's personal finances and health. J. Behav. Exp. Econ. **77**, 151–157 (2018)
17. Oehler, A., Horn, M., Wendt, S., Reisch, L.A., Walker, T.J.: Young adults and their finances: an international comparative study on applied financial literacy. Econ. Notes. **47**(2–3), 305–330 (2018)
18. Philip, D.J., Sudarsanam, N., Ravindran, B.: Improved insights on financial health through partially constrained hidden Markov model clustering on loan repayment data. Data Base Adv. Inform. Sys. **49**(3), 98–113 (2018)
19. Piterkova, A., Toth, M., Serences, P.: The impact of non-financial factors on prosperity of Slovak agriculture sector. In: 9th International Conference on Applied Business Research (ICABR), pp. 851–859 (2015)
20. Potrich, A.C.G., Vieira, K.M.: Demystifying financial literacy: a behavioral perspective analysis. Manag. Res. Rev. **41**(9), 1047–1068 (2018)
21. Siska, L., Siskova, Z.: The relationship between the applied performance measurement and the financial prosperity of the company. Int. J. Knowl. Cult. Change Manag. **10**(1), 39–49 (2010)
22. Teplická, K., Daubner, M., Augustínová, E.: Analysis of causal relationships between selected factors in process of performance management in industrial companies in Slovakia. Ekonomicky casopis. **63**(5), 504–523 (2015)
23. Zhang, Y.T., Huang, Y. & Hong, H.: Monitoring and early-warning system of enterprise financial prosperity. In: Proceedings of the fourth international conference on management science and engineering management, pp. 383–388 (2010).

Decision Support for Municipality Information System Based on Statistical Methods

B. Hamerníková, V. Fábera, V. Beneš, M. Jetmar, M. Jeřábek, J. Kubát, J. Pančík, and M. Plaček

1 Introduction

The analysis of public administration and public sector efficiency is an important area of public economics and public finance theory. The issue of the effectiveness of local authorities is also a question of the theory of regional development and management. The analysis of the effectiveness of municipalities and their comparison is an important tool of public policies and activities of public administration bodies (ministries, regions, municipalities). The statistical methods DEA (data envelopment analysis) and FDH (free disposable hull) are used to analyze the effectiveness of municipalities and their comparison. These methods are used to evaluate the effectiveness of units based on the size of inputs and outputs in the business and public sectors. Units are compared with each other.

We designed and obtained a grant for a project that uses the statistical methods mentioned above [1, 2]. The project will thus improve the quality of decision-making and management processes in public administration and increase the transparency of its activities at the level of municipalities, regions, and ministries. In next chapters, we briefly describe the basis of our project, its objectives, and first results. The main topic described in this contribution is the web application and associated database description.

B. Hamerníková · V. Beneš · M. Jetmar · J. Pančík (✉) · M. Plaček
AMBIS a.s., Prague, Czech Republic

V. Fábera · M. Jeřábek · J. Kubát
Faculty of Transportation Sciences, Czech Technical University in Prague, Prague, Czech Republic

L. Knapcikova et al. (eds.), *New Approaches in Management of Smart Manufacturing Systems*, EAI/Springer Innovations in Communication and Computing, https://doi.org/10.1007/978-3-030-40176-4_4

2 Project of Information System for Decision Support of Municipalities

2.1 Performance Management in Public Sector

According to last bibliographical analyses, e.g., Osborne [3] and Corrado [4], performance management is still one of the most popular topics of research focusing on public management and public administration. Performance management is action based on performance measures and reporting, which results in improved behavior, motivation, and processes and which promotes innovation. The key aspects of performance management are deciding what to measure, how to measure it, interpreting the data, and communicating the results [1]. In the last decade, the potential of big data for performance management has also been discussed as well as the related problems of data misinterpretation and attribution problems [5]. Following the performance management literature, a general performance management model for local authorities has been developed. The performance management tools play an important role in this model (top of Fig. 1). Developing such "user-friendly" software tool dedicated for common users from municipalities was the goal of our project.

2.2 Performance Management in Czech Municipalities

Local government revenues in the Czech Republic amounted to 11.2% of GDP. The EU average is 10.9% of GDP and 9.9% of GDP in the Eurozone [6]. Local government revenues in the Czech Republic consist mainly of shared taxes. The highest share is the value-added tax. According to the Eurostat data of 2016, local governments spend 10.2% of GDP on public spending. The EU-wide average was 10.8% of GDP in 2016; the share of the Eurozone countries was 9.7% [6]. These expenditures are realized in order to carry out the mission of the local government, which is defined in the Municipalities Act No. 128/2000 Coll. as follows: "The municipality takes care of the comprehensive development of its territory and the needs of its citizens; in the performance of its tasks it also protects the public interest."

2.3 Performance Analysis with DEA

For our performance analysis, the data envelopment analysis was chosen. We are currently experiencing a rapid increase in the use of this performance analysis tool in municipalities [7–9]. Data envelopment analysis (DEA) determines the most efficient municipalities in the sample of our research. These form the "best practice frontier" in a multidimensional space defined by inputs and outputs. The relative

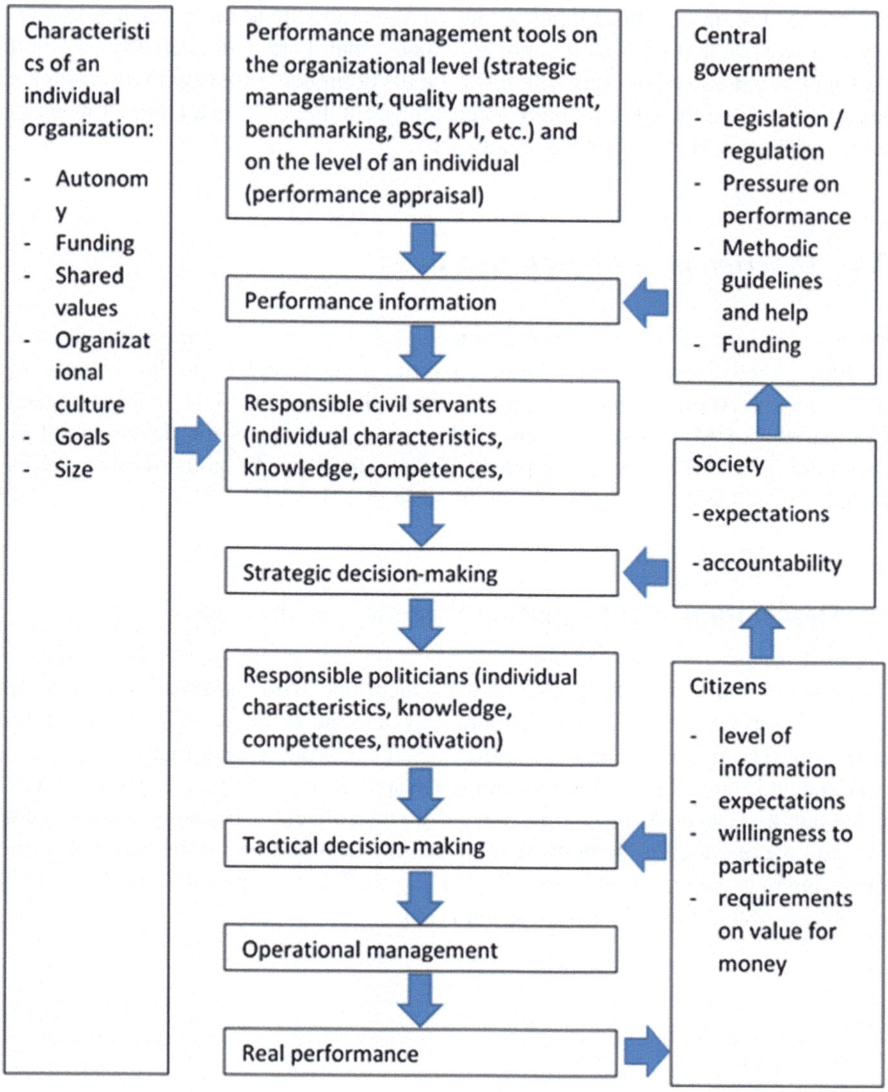

Fig. 1 General performance management model for local authorities

efficiency of municipalities lying under this best practice frontier is computed by their deviations from the frontier [10]. This method is suitable for evaluating the efficiency, performance, or productivity of homogeneous production units—i.e., units that produce identical or equivalent effects, which we will refer to as outputs of this unit. Outputs are, by their nature, maximizing; their higher value results in higher performance of the tracking unit. To produce effects, the production unit utilizes inputs which are minimized by their nature; the lower value of these inputs leads to higher performance of the monitored unit [3]. The DEA method estimates the production units, whose input/output combinations lie at the efficiency boundary and are efficient units as it is not anticipated that there could actually be a unit that achieves the same outputs with lower inputs or higher outputs with lower inputs [11]. When using the DEA method, a constant yield model from a range or a model with variable yields from a range can be used.

2.4 *Description of Project Objectives*

We have designed a practical application of performance management to help analyze the effectiveness of selected municipalities in the Czech Republic. The title of the project is "Application of Nonparametric Methods (DEA, FDH) to Analysis and Comparison of Municipal Efficiency." The project output is the development of methodology and IT software application for ease of use by potential users. This project was awarded by TACR agency to solve in 2018–2021 [2].

3 Description of Information System Requirements

The planned output of the project can be identified with the creation of a fully fledged information system and methodology, and the software application will be part of it. The definition of information system is broader than many people can imagine and states that "…the information system is a collection of people, technical means and methods (programs) that ensure the collection, transmission, processing, storage of data, for the purpose of presenting information for the needs of users operating in management systems…" [12]. Figure 2 defines the nature of the project outputs as part of the information system (IS).

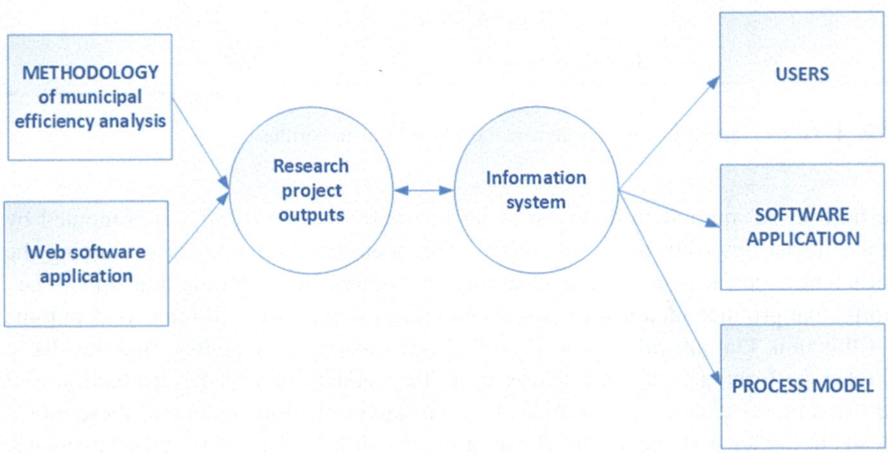

Fig. 2 Nature of the project outputs as part of the information system

3.1 *Definition of Software Application in the Project*

Software application for information system in general is created by [12]:

- Users with different roles.
- Presentation layer for users.
- Data/information processing layer.
- Database management system (database).
- Data.
- Documentation.
- User access to data.

The planned software application will serve for needs of public administration. Within the meaning of Act No. 365/2000 collection of laws [13, 14] which defines the concept of public administration information system (IS), the developed software application will form the basis for the public administration agenda information system. The requirements for the public administration agenda information system are not so strictly as requirements for public information systems from state regulator side.

3.2 *Defining the Procedures for Creating a Software Application*

It has been determined that the software application is part of the IS (which it creates together with the methodology). Standard IS development methods will be used in the development of the software application. These methods define the individual stages of software application development. The following stages are known: specification of requirements, analysis, design, implementation, testing, deployment, using, changes, and their implementation. The development of the software application by the research team, which would be solved without the implementation of the whole IS as a project output, represents a risk of alignment of project objectives. All mentioned stages of software application development can be repeated, and versions of the information system can be created in iteration from the simpler variant to the final version of the IS, namely, functional IS model, IS prototype, and final IS version.

3.3 *Definition of the Task of the Assignment and Specification of the Functional Requirements of the Software Application*

The software application is based on defined project outputs where software application is an integral part of the information system. The specification of software application requirements is based on the division of requirements between functional requirements and (nonfunctional) system requirements. Functional requirements describe individual user roles that interact with a software application (user, government worker, database administrator, etc.) and also describe the individual function roles that the software application will perform based on roles. Functional requirements consist from:

- The IS process model is based on the methodology that is part of the project output. The process model includes a description of individual processes, process owners, and process inputs and outputs.
- A list of roles and a list of their assigned use cases (and usage scenario for each use case) can be created based on a process model.
- Individual functions of the software application can be assigned to individual use cases.

Software application system (nonfunctional) requirements describe the system from the perspective of software architecture, applied technologies, security, standards used, and so on. Individual functional and system (nonfunctional) requirements will be gradually filled in the system and software versions of the functional model, e.g., in functional prototype and final version. All the abovementioned requirement documents will be part of the IS documentation and software application. Figure 3 shows the hierarchy of requirements as was described in the text above.

3.4 WebDEAr Software Application Architecture

In the case of a software application for a project, one will be kind of a so-called web application. The user of the output of the project (the output is an information system which consists of methodology and software application) will be communicated with the software application via a web browser. The working name of the web application is WebDEAr. The WebDEAr web app is like any web app and consists of (Fig. 4 WebDEAr software application architecture):

1. Presentation layer, a part of software application called the front-end WebDEAr, which is the application's user interface and allows user interaction in one of the defined roles.
2. Information layer of software application, the so-called back-end WebDEAr, which represents the program implementation within the project developed algorithms DEA and FDH and represents also other activities necessary for running the software application.
3. Database (DataDEAr) that represents a database management system (DBMS).
4. The user data itself.

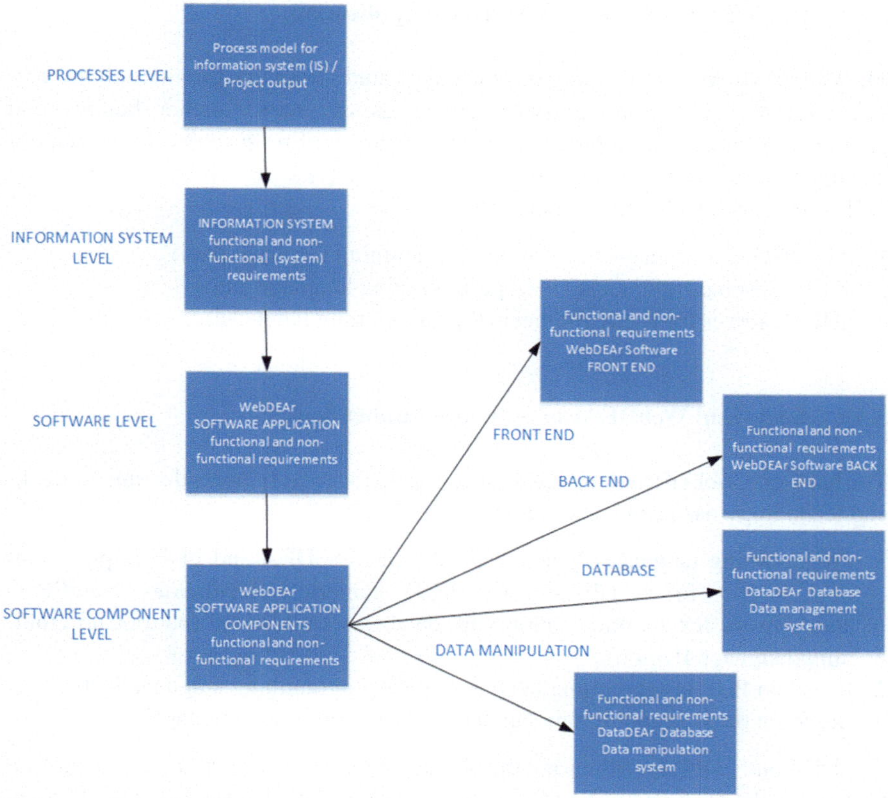

Fig. 3 Hierarchy of requirements

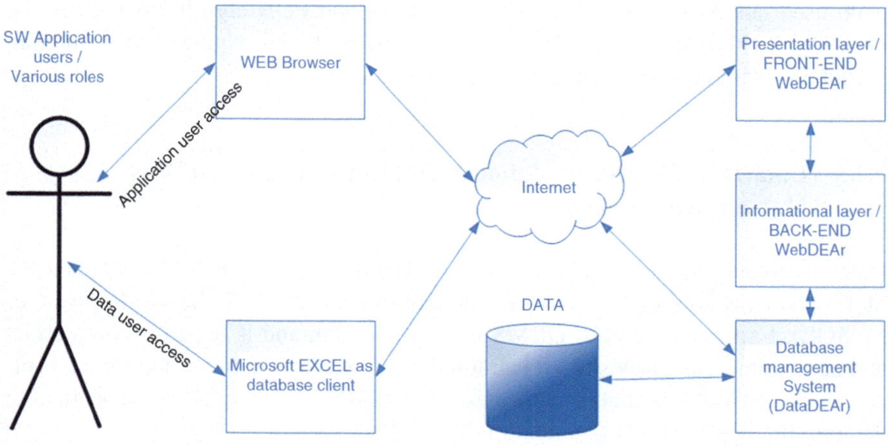

Fig. 4 WebDEAr software application architecture

3.4.1 Front-End WebDEAr of Software Application

The WebDEAr web application runs on a dedicated server that provides all the necessary services. The front-end user interface has user inputs (such as buttons) and graphical visualization outputs. The user interface will be specified in the request document in the front-end section.

The front-end technology can be based on:

1. DJANGO web framework (Python programming language path).
2. PHP for web development (PHP programming language path).
3. SHINY server for R environment (R programming language).

3.4.2 Back-End WebDEAr of Software Application

The back-end WebDEAr is in details specified in the requirements document. Back-end technology used can be based on:

1. Programming language Python with libraries for DEA and FDH implementation, libraries for visualization (SciPHY), numerical calculations (NumPHY) and libraries for communication with database (DataDEAr) (path of programming language Python).
2. Based on PHP programming language (PHP programming language path).
3. Based on the R programming language (R programming language).

The DEA and FDH algorithms are the outputs of the project economist team and are ported into the back-end of WebDEAr application by being delivered to the IT team in the form of R-written functions along with a validation set of data. The IT team also verifies the correctness of porting of in R language written algorithm into the by team used language (Python or PHP). The DEA and FDH algorithms used in the back-end WebDEAr are specified in detail in the back-end requirement document, including their verification method (see schematic in Fig. 5).

3.4.3 DataDEAr Database Solution Architecture: Relational Data Management System

Database management system management (DBMS) is based on relational database SQL server (Microsoft MS SQL Server Express, Oracle SQL Server Express, or MySQL). Express version of DBMS is a free version and is used for commercial and noncommercial purposes—it is limited to a few GB of data needed for the project. Back-end WebDEAr communicates with DBMS. The DataDEAr is specified in detail in the requirement document in the database section.

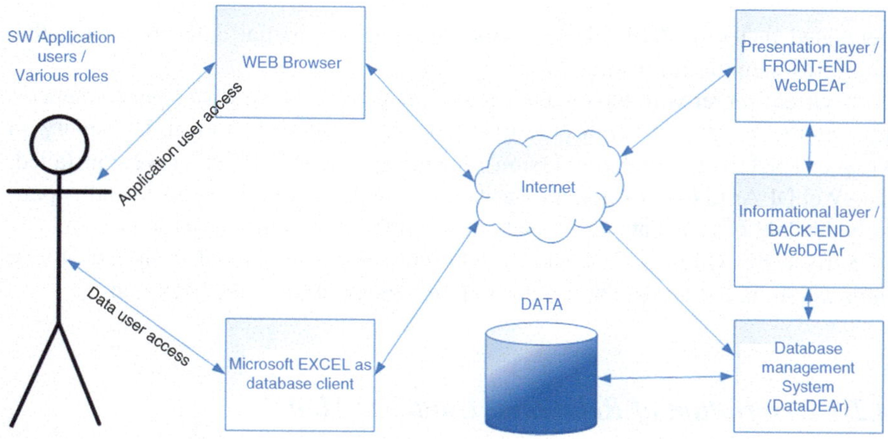

Fig. 5 Process of transfer R language code from economist team to the IT team

4 Description of Developed Functional Prototype of Information System

4.1 Description of Implementation of Functional Prototype Architecture

The requirements for a functional prototype of a software application are:

- Access to the application without authentication.
- Displaying information about the municipality or type of municipality.
- Displaying the results of the municipality and categories of municipalities including inputs and outputs, respectively.
- Calculation according to the chosen method with entered data (inputs/outputs) for a particular municipality.
- Logging user requests and collecting information about their work with the application.

It is a classic three-tier architecture. The data layer is represented by the Oracle relational database management system. The application layer that performs data verification and processing is made up of several modules. The lowest level of PL/SQL procedures and database query functions create basic modules of the database system. The calculation using DEA and FDH methods is solved by "plug-in software modules," which are available as extensions directly to the Oracle database system. Because it is necessary to perform other statistical calculations, the research team decided to use the R language, which is directly designed for statistical analysis of data and their graphic representation; the module is also available for Oracle. The presentation layer is provided by the Apache web server with dynamically generated pages in PHP; PHP's scripts also include partial data processing that interferes with the application layer.

A virtual server with the Oracle Linux 7.5 operating system was put into operation at the Faculty of Transportation Sciences, Czech Technical University in Prague. The Oracle Database Version 18c and Apache HTTP Server were installed. The WebDEAr.fd.cvut.cz domain has been set up to access the website and is publicly available. Basic data has already been imported into the database system.

Screenshots (Figs. 6, 7, 8, and 9) represent various outputs of front-end of web application including the DEA algorithm implementation in the back-end.

4.2 *Description of Relational Database Model*

Based on the analysis of the requirements for a software application for the analysis of municipal efficiency (the output of the project TL01000463-V002), a relational scheme of the database was created, which forms the core of data structures for storing the necessary data. The scheme currently contains 23 tables (Fig. 10).

The central table is the MUNICIPALITY table, containing basic information about municipalities, such as the code according to the CZSO, code NUTS5, IČO, name, GPS position (also in WGS84 format), altitude, catastral area, type of municipality (TYPE_MUNICIPALITY), reference to district (district municipality),

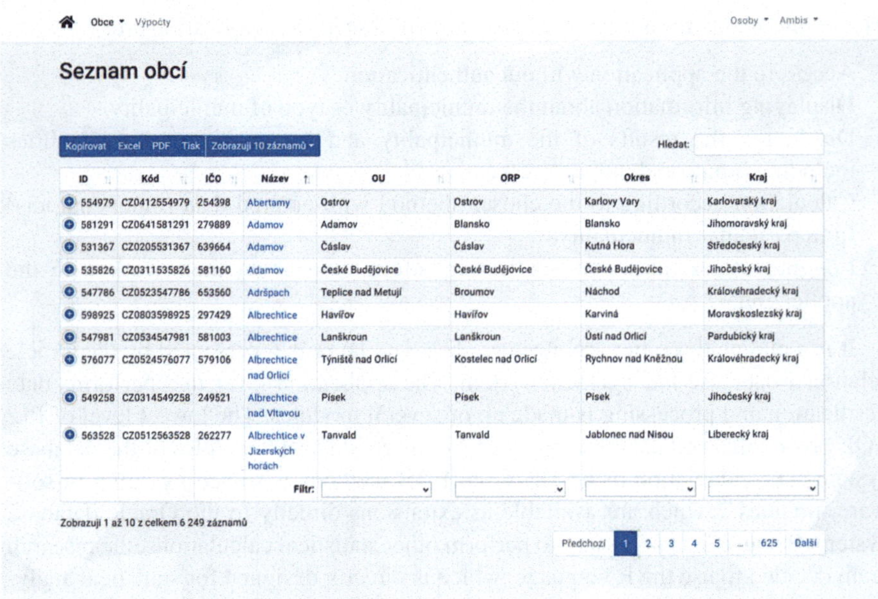

ID	Kód	IČO	Název	OU	ORP	Okres	Kraj
554979	CZ0412554979	254398	Abertamy	Ostrov	Ostrov	Karlovy Vary	Karlovarský kraj
581291	CZ0641581291	279889	Adamov	Adamov	Blansko	Blansko	Jihomoravský kraj
531367	CZ0205531367	639664	Adamov	Čáslav	Čáslav	Kutná Hora	Středočeský kraj
535826	CZ0311535826	581160	Adamov	České Budějovice	České Budějovice	České Budějovice	Jihočeský kraj
547786	CZ0523547786	653560	Adršpach	Teplice nad Metují	Broumov	Náchod	Královéhradecký kraj
598925	CZ0803598925	297429	Albrechtice	Havířov	Havířov	Karviná	Moravskoslezský kraj
547981	CZ0534547981	581003	Albrechtice	Lanškroun	Lanškroun	Ústí nad Orlicí	Pardubický kraj
576077	CZ0524576077	579106	Albrechtice nad Orlicí	Týniště nad Orlicí	Kostelec nad Orlicí	Rychnov nad Kněžnou	Královéhradecký kraj
549258	CZ0314549258	249521	Albrechtice nad Vltavou	Písek	Písek	Písek	Jihočeský kraj
563528	CZ0512563528	262277	Albrechtice v Jizerských horách	Tanvald	Tanvald	Jablonec nad Nisou	Liberecký kraj

Fig. 6 List of municipalities

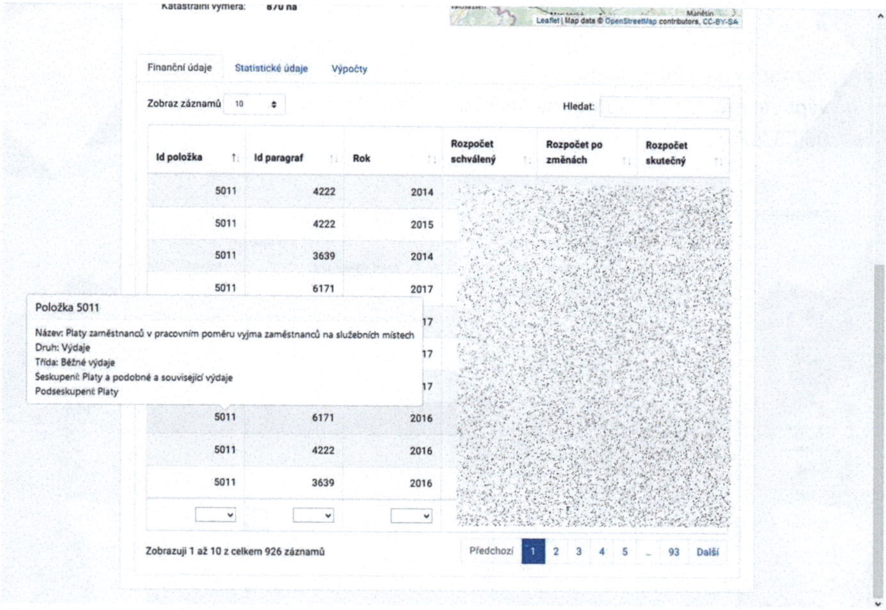

Fig. 7 Municipality financial details (anonymized data)

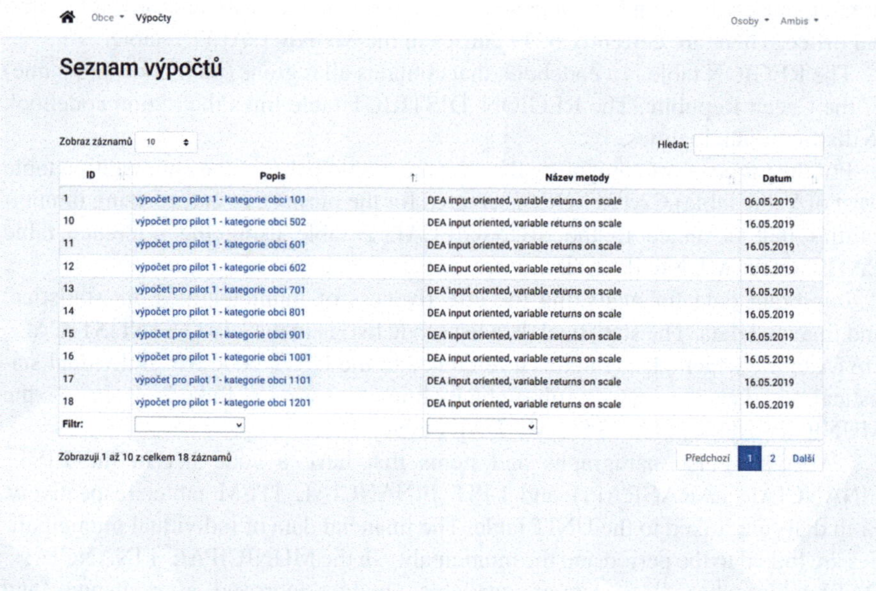

Fig. 8 List of statistical calculation with DEA algorithm

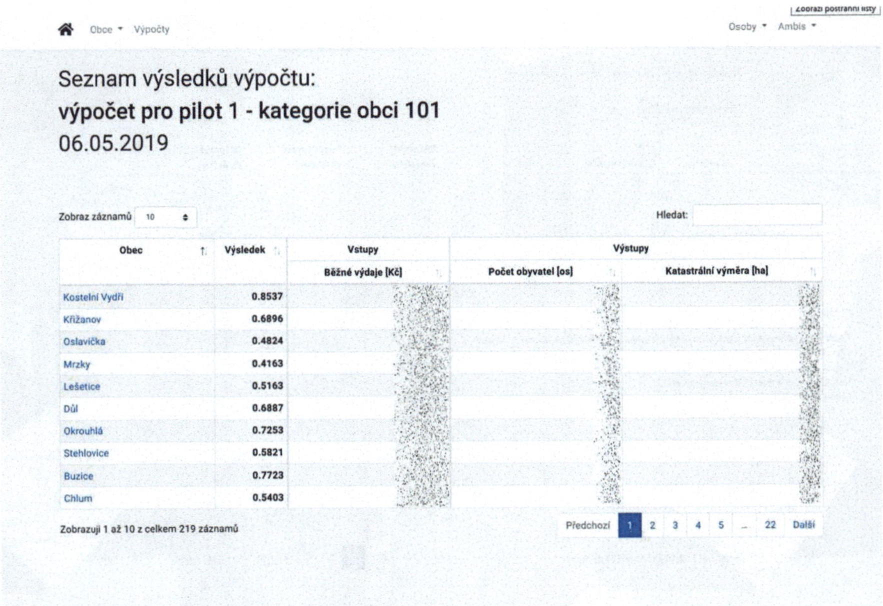

Fig. 9 Statistical calculation with DEA algorithm—detail (anonymized data)

municipality with extended competence, and municipality with authorized municipal office. There are currently 6249 entries in the MUNICIPALITY table.

The REGION table is a codebook that contains all regions (their code and name) of the Czech Republic. The REGION_DISTRICT table links the region codebook to district municipalities.

For the purposes of comparing the relevant municipalities, the enumeration table (List of Value table) CATEGORY is created for the purpose of categorizing municipalities that is linked to the MUNICIPLAITY table using the reference table CATEGORY_MUNICIPALITY.

Important data for analyzing the effectiveness of municipalities are statistical and financial data. The statistical data list (code list) is in the LIST_STATISTICAL_DATA table, which also contains a reference to the UNIT code list. Individual statistical data on sub-municipalities are linked in the MUNICIPALITY_STATISTICAL_DATA table.

We distinguish paragraphs and items that have a code list in the LIST_FINANCIAL_PARAGRAPH and LIST_FINANCIAL_ITEM table, respectively. Both dials are linked to the UNIT table. The financial data of individual municipalities are linked to the period and the municipality in the MUNICIPAL_FINANCIAL_DATA table, where three data are available—budget approved, after changes, and actual. Currently, there are a total of 5,263,845 entries in the MUNICIPAL_FINANCIAL_DATA table, 524 entries in the LIST_FINANCIAL_PARAGRAPH table, and 527 entries in the LIST_FINANCIAL_ITEM table. Financial data is from 2012 to 2017.

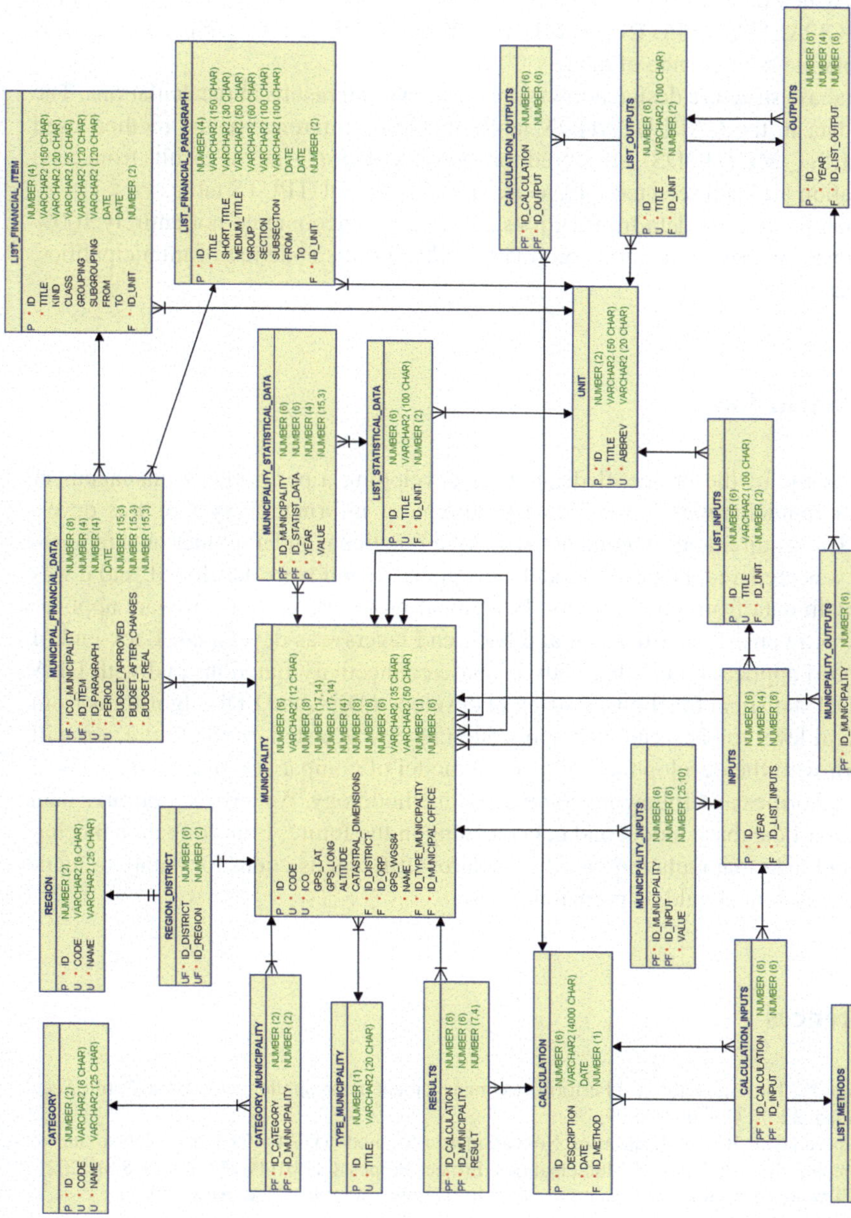

Fig. 10 DataDEAr database model

Various inputs and outputs of statistical and financial data are used to analyze the effectiveness of municipalities using the DEA and FDH methods. The INPUTS and OUTPUTS tables referring to LIST_INPUTS and LIST_OUTPUTS are used to record which data was used as inputs and outputs. Specific data are captured in the MUNICIPALITY_INPUTS and MUNICIPALITY_OUTPUTS tables, with a view to a possible later change of data.

Database structure design allows for retention of all results in recalculations. The basic data in the CALCULATION table, including reference to the method used (table LIST_METHODS), are stored for each calculation. And the results from each calculation for each selected village are stored in the OUTPUTS table.

When processing data from various sources, we encountered not entirely 100% functional integrity constraints on financial data on the code list of municipalities, sections, and items.

5 Conclusion

In the frame of the project dedicated for developing a performance management tool for municipalities, a functional prototype of information system was developed. The requirements document to describe functional and nonfunctional requirements was prepared. Database model and database itself were developed, and it was filled with data from public sector. Functional prototype of web software application which consists of front-end and back-end layers was developed. Back-end of developed application is filled with database connectivity functions and with DEA and FDH statistical functions. Porting of developed DEA and FDH algorithms from statistical level to back-end web application level was tested. In the future, we will concern with the development of process model of comparison of effectiveness of local authorities and following comparison methodology. We have to optimize data structure of database model and data cleaning in the future. Importing data mechanism and data manipulation based on Microsoft Excel as client for database management system should be prepared.

References

1. Fryer, K., Jiju, A., Ogden, S.: Performance management in the public sector. Int. J. Public Sect. Manag. **22**(6), 478–498 (2009)
2. Hamerníkova, B.: Application of Nonparametric Methods (DEA, FDH) to Analyse and to Compare the Efficiency of Municipalities. Project Proposal TL01000463, 02/2018–01/2021. s.l. TAČR, Technological Agentury of Czech Republic, program ETA, 2018. (2018)
3. Osborne, S.P.: Public management research over the decades: what are we writing about? Public Manag. Rev. **19**(2), 109–113 (2017)

4. Corrado, C., Massiomo, A., Sarto, F.: A twenty-five years bibliometric analysis in business and public administration domains. 201. Foundations and trends in performance management. Scientometrics. **108**(2), 595–611 (2016)
5. Lavertu, S.: We all need help: "Big Data" and the mismeasure of public administration. Public Adm. Rev. **76**(6), 864–872 (2016)
6. Eurostat. http://ec.europa.eu/eurostat/web/government-finance-statistics/overview (2016). Accessed 15 July 2017
7. Emrouznejad, E., Yang, G.: A survey and analysis of the first 40 years of scholarly literature in DEA: 1978-2016. Socio Econ. Plan. Sci. **61**, 4–8 (2018)
8. Perpiňa-Narbon, I., De Witte, K.: Local governments' efficiency: a systematic literature review—part 1. Int. Transact. Operat. Res. **00**, 1–38 (2017)
9. Perpiňa-Narbon, I., De Witte, K.: Local governments' efficiency: a systematic literature review—part 2. Int. Transact. Operat. Res. 1, 1–30 (2017)
10. Šťastná, L., Gregor, M.: Local government efficiency: evidence from the Czech. IES working paper. 14 (2011)
11. Boruvkova, J., Kuncova, M.: Porovnání očních oddělení kraje Vysočina pomocí DEA modelů. **5**, 75–84 (2012)
12. Gála, L., Šedivá, Z., Pour, J.: Počítačové aplikace v podnikové a mezipodnikové praxi—3, aktualizované vydání. Praha, Grada (2016)
13. Lunter, J.: Co je a co není informační systém verejné správy. http://www.mvcr.cz/soubor/co-je-a-co-neni-isvs.aspx (2018). Accessed 19 May 2019
14. Zákon 365/2000, Zb. Zákon o informačních systémech verejnej správy. http://www.mvcr.cz/clanek/legislativa-zakon-c-365-2000-sb-o-informacnich-systemech-verejne-spravy.aspx (2000). Accessed 19 May 2019

Description of Possible Approaches to Create and Analyse Data Model

Stella Hrehová

1 Introduction

Data analysis is the process of evaluating data (data, information) using various logical methods and considerations. We can consider data as the basis of knowledge and good decisions. Effective data management helps minimise potential errors and reduce damage caused by bad data. The analysis of this data makes sense only if it is relevant, complete, accurate, meaningful and applicable. Otherwise, they can lead to bad decisions and potential losses. The division for well-managed data is [1]:

- In the right place—moves efficiently between multiple systems.
- At the right time—immediate reactions, streaming sensor data, overnight batch processes.
- In the right format—validated, standardised or enriched, data is made usable.
- For all users – usage governed, business semantics applied.

We can find a many various approaches of data managing in the literature. These approaches have one common attribute, and that is the term data life cycle. The data life cycle is a term coined to represent the entire process of data management. The goal of a data management life cycle is to ensure that data are collected with enough rigor to support the intended use, to support basic data management, to enable reuse and repurposing of the data and to allow for the eventual long-term preservation. The various steps associated with the data life cycle are on some level self-explanatory. However, the basic process involves the creation, validity control, storage and analysis of the required data. Throughout the process, additional metadata and documentation are added [2].

S. Hrehová (✉)
Faculty of Manufacturing Technologies, Department of Industrial Engineering and Informatics, Technical University of Košice, Presov, Slovak Republic
e-mail: stella.hrehova@tuke.sk

L. Knapcikova et al. (eds.), *New Approaches in Management of Smart Manufacturing Systems*, EAI/Springer Innovations in Communication and Computing, https://doi.org/10.1007/978-3-030-40176-4_5

Proper data analysis is a systematic process of successive steps, namely:

- Ask the question we want to answer.
- Acquisition and completion of available data that does not always come from the same source.
- Editing data and ensuring its quality and accuracy.
- Convert data to a form that is applicable to subsequent statistical methods.
- Analyses when data is transformed into tables and graphs. This is a descriptive phase of data analysis that can provide you with first understandable results.
- Advanced statistical algorithms, such as classification trees and regression. More advanced analyses sometimes turn into data mining, which is of great importance for predictive purposes.

2 Process Description

The possibilities for collecting and storing the monitored data are currently at a very high level. The reason can be that information technologies allow greatly increased accessibility of this data through web-based user interfaces [3, 4]. Data stored in databases and real-time sensor data from production processes can be analysed and used as a source of information and knowledge for future purposes.

The impact of the development of information and measurement technologies has spread to many areas. One such area is the area of building heating. The outdoor temperature control method was used to control the heating. The supply water temperature is the desired variable for control and is controlled by the outside air temperature according to the present heating curve. Subsequently, the heating curve describes the dependence of the supply water temperature on the outside air temperature. This achieved a balance between the energy supplied and the heat loss of the building. In some cases, temperatures were also measured in reference rooms for inspection purposes. Such regulation of the heating process is called room temperature compensation and may provide additional compensation for other heat gains or losses in the heated space [5]. The role of weather compensation in the heating system is to adjust the heat supply to help maintain the desired room temperature while reducing energy costs despite weather conditions [6]. PID controllers were used to control the feed water temperature. The controller increases or decreases the flow set point calculated from the present heating curve and the actual outdoor temperature according to the control deviation of the set point and the actual temperature in the reference room. The data analysed in this paper were obtained through the SCADA system (supervisory control and data acquisition). A SCADA system is a type of industrial control system that allows remote monitoring and control and works with coded signals via communication channels (Fig. 1).

Data sets consist of several parameters which are measured at individual time intervals. The most typical parameters are outdoor and indoor temperature, heat

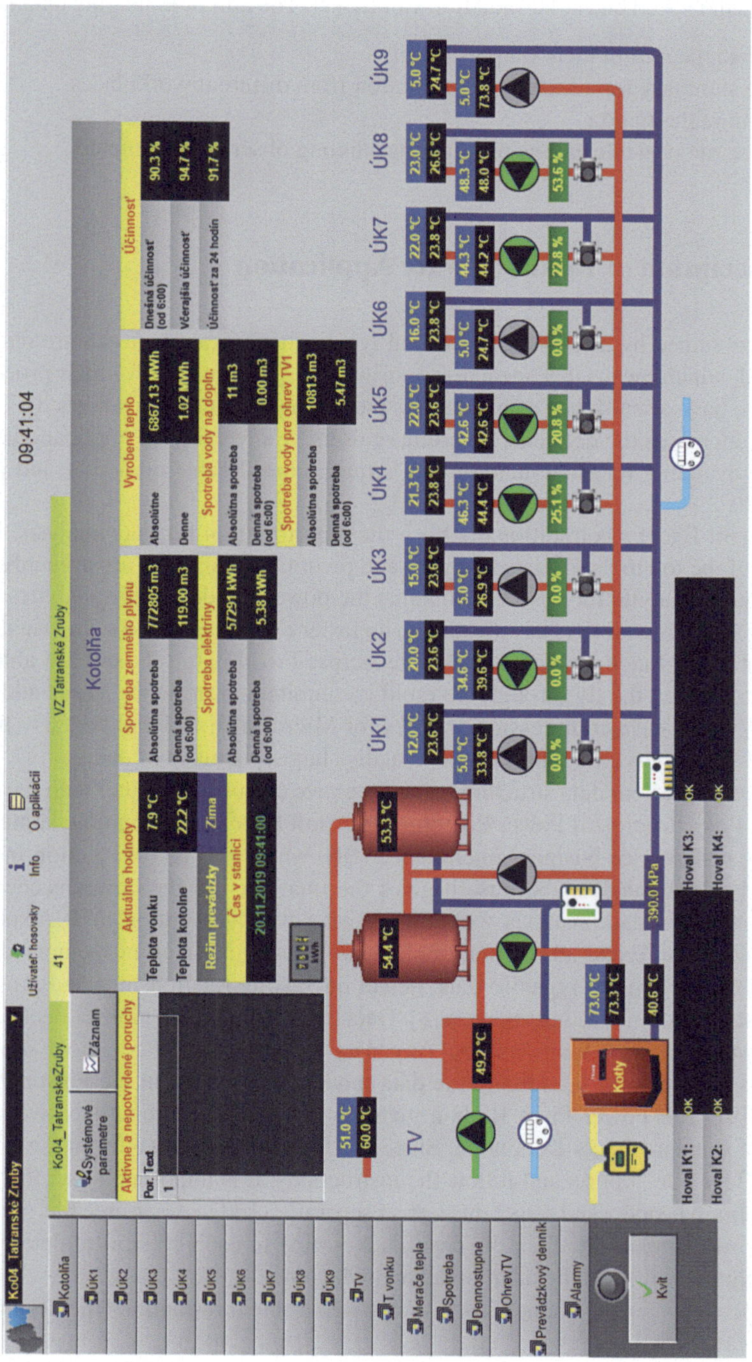

Fig. 1 SCADA interface for data export

consumption, gas/electricity consumption, heating medium temperature and flow, possibly wind speed and solar radiation values [7]. The aim of managing data is to:

- Visualise the requirement outputs.
- Create summary report of gas consumption from different workplaces.
- Determine the trend.
- Analyse the possible causes of difference among observed parameters.

3 Description of Used Software Application

The data obtained by measuring individual types of technical processes are mostly structured, which means that specialised software is not necessary for their processing. As we have a set of data available, it is increasingly required to analyse it. This, however, increases the demands on ordinary users. Many software applications are currently available for such analyses but require considerable computer and program skills.

Microsoft Excel has traditionally been the reporting tool for data analysis. It is also one of the top tools for data analysis, and pivot tables in Excel are arguably the most popular analytic tool. Now it contains the powerful add-ins for analytics and reporting features as well. They are able to do faster experimentation with visualisations, statistical functions and calculations across broad data sets and are able to derive answers on the fly through the rapid recombination of fields. These add-ins have very impressive compression abilities for Microsoft Excel and text/.csv files, which allows you to view, analyse and visualise huge quantities of data.

With respect to our data structure and its size, we can use MS Excel with its add-in Power Pivot. Microsoft Power Pivot for Microsoft Excel is a technology aimed at providing self-service business intelligence (BI), which is a real revolution inside the world of data analysis because it gives the final user all the power needed to perform complex data analysis without requiring the intervention of BI experts. Power Pivot is an Excel add-in that implements a fast, powerful, in-memory database that can be used to organise data, detect interesting relationships and provide the fastest way to browse information [8]. Data model becomes a table collection of tables connected with relationships. Any data which is imported into Excel or already is in Excel, once added to the data model, is available in the Power Pivot window [9]. The Power Pivot Ribbon gives us the additional functions over and above the standard Excel Data table. Since the file is saved in csv format, we will use the "Text File" option and follow the instructions. It is important to set the first line as column headers and select the correct separator used in the source file. Within this view, we can simply click the check boxes next to each column to determine which column will be in the data model and which not.

Distance between the elements is given as well as the information whether or not reading was carried out.

4 Data Preprocessing

Before any analysis of a data set, it is very important to assess their accuracy. This step is commonly referred to in the literature as data preprocessing. Preprocessing is a step or phase in the identification, selection or problem handling of the data [10]. Data preparation encompasses all activities to construct and clean the data set. It is the process of systematically reviewing data and removing suspected erroneous data based on rules of thumb and criteria [11]. It is mainly an examination of the following [12]:

- Eliminating duplicate rows.
- Missing or invalid values.
- Formatting properly.

4.1 Duplicate Values

The elimination process is very important to produce a cleaned data. Duplicate record detection is the process of identifying different or multiple records that refer to one unique real-world entity or object if their similarity exceeds a certain cutoff value [13]. Duplicate values can significantly affect some basic statistical indicators and also complicate the creation of data models when there is a requirement to analyse data from multiple sources. After elimination, only one copy of the exact duplicate records is preserved, and the other duplicate records are deleted. Currently, there are many software applications that are able to detect duplicate records. They contain tools that detect duplicates in a data set relatively easily and quickly and that can be applied to databases with millions of records. In this case, the duplicate values can occur mainly when the data is written in very short time intervals.

4.2 Missing Values

Because almost all standard statistical methods presume complete information for all the variables included in the analysis, missing values become one of the problems. Missing data present various problems. First, the absence of data reduces statistical power. Second, the lost data can cause bias in the estimation of parameters. Third, it can reduce the representativeness of the samples. Fourth, it may complicate the analysis of the study. Each of these distortions may threaten the validity of the trials and can lead to invalid conclusions [14]. From the point of view of source data assessment, it was not permissible to remove records with missing values as this would lead to incorrect assessment. Finding an optimal access to missing values can be challenging as it requires careful examination of the data to identify the type and pattern of missingness (Fig. 2).

Fig. 2 Patterns of missing values [15]

Missing monotone				Missing arbitrarily			
v1	v2	v3	v4	v1	v2	v3	v4
X	x	x	x	x	x	.	x
X	x	x	x	.	x	x	.
X	x	x	.	x	.	x	.
X	x	.	.	x	x	.	.
X	.	.	.	.	x	x	x

Conventional methods that address missing values in time series data will add or ignore missing values [16]. These methods are simple but effective only with a small percentage of missing data. If there is a higher percentage of missing data, the result of the analysis can be affected, or we will get a biased result [17]. This information therefore implies that those who carry out empirical research will have to decide how to handle missing data [18].

A widespread approach to the classification of missing data is given in the literature [19]. The term mechanism in this context refers to dependencies within a variable or variables without missing data and variable or variables with missing data that specify the three missing data classes.

Let the set of measured values be in the form of a cross-sectional data matrix **Y** ($n \times K$), where n is the number of cases and K is the number of variables. $\mathbf{Y}_{\text{obs}}$ contains the observed scores and $\mathbf{Y}_{\text{mis}}$ contains the hypothetical scores that are missing.

It is defined the matrix **R** (Eq. 1) for each variable, which takes on the value "1" if the variable has data for that case and "0" if the value is missing for that case. This leads naturally to descriptions of the missing data, that is, patterns of missingness [20]:

$$R = \begin{cases} 1 & \mathbf{Y}_{ij} \text{ is know} \\ 0 & \mathbf{Y}_{ij} \text{ is unknow} \end{cases} \tag{1}$$

where i is the order of the case and j is the order of the variable. The matrix **R** is of the same dimension as matrix **Y**. Furthermore, θ is defined as a parameter affecting the **Y** distribution and an unknown parameter ψ affecting the conditional distribution **R** given by **Y**. These relationships can be formally written as a common conditional probability in the form (Eq. 2):

$$p(\mathbf{Y},\mathbf{R} \mid \theta,\psi) = p(\mathbf{Y} \mid \theta) f(\mathbf{R} \mid \mathbf{Y},\psi), \tag{2}$$

After replacing **Y** with its components $\mathbf{Y}_{\text{obs}}$ a $\mathbf{Y}_{\text{mis}}$ and integrated according to $\mathbf{Y}_{\text{mis}}$, it is possible to express from Eq. (2) the distribution of known data according to Eq. (3):

$$p(\mathbf{Y},\mathbf{R},|\theta,\psi)=\int p(\mathbf{Y}_{obs,}\mathrm{Y}_{mis}\,|\,\theta)f(\mathbf{R}|\mathbf{Y}_{obs}|,Y_{mis}\,|\,\psi)d\mathbf{Y}_{mis}, \quad (3)$$

Based on Eq. (3), it is then possible to define three missing mechanisms. Part p ($\mathbf{R}|$ $\mathbf{Y}_{obs}$, $\mathbf{Y}_{mis}$, ψ) can be expressed in the following ways [21]:

4.2.1 Missing Completely at Random (MCAR)

The formal definition of MCAR requires that the probability of missing data on a variable Y is unrelated to other measured variables and is unrelated to the values of Y itself. Put differently, the observed data points are a simple random sample of the scores you would have analysed had the data been complete [21]. Examples from technical monitoring of process data could be regular or irregularly sampled multi-stream data, missing data due to sensor failure, etc. [19].

The MCAR mechanism requires that missingness is completely unrelated to the data. Consequently, both $\mathbf{Y}_{obs}$ and $\mathbf{Y}_{mis}$ are unrelated to $\mathbf{R}$, and the distribution of missing data simplifies even further to

$$p(\mathbf{R}\,|\,\psi), \quad (4)$$

4.2.2 Missing at Random (MAR)

Data are missing at random (MAR) when the probability of missing data on a variable Y is related to some other measured variable (or variables) in the analysis model but not to the values of Y itself. Said differently, there is no relationship between the propensity for missing data on Y and the values of Y after partialling out other variables. The term missing at random is somewhat misleading because it implies that the data are missing in a haphazard fashion [21].

In this case, the missing absence depends only on the observed $\mathbf{Y}_{obs}$ component and not on the missing $\mathbf{Y}_{mis}$ component:

$$p(\mathbf{R}|\mathbf{Y}_{obs}|,\psi), \quad (5)$$

4.2.3 Missing Not at Random (MNAR)

The probability of missing data on a variable Y is related to the values of Y itself, even after controlling for other variables [22]. In this case, the underlying reason that caused the missing data has an effect on the inference and has to be included in any analysis [21]:

$$p\left(\mathbf{R}|,\mathbf{Y}_{\text{obs}}|,\mathbf{Y}_{\text{mis}}\,|,\psi\right) \tag{6}$$

The methods for handling missing data generally fall under three categories: deletion, direct estimation and imputation techniques. Listwise and pairwise deletion techniques are encompassed in the first category. These techniques discard cases during an analysis if they contain missing data. Because it is not possible to omit data from the analysis, in the next part, we will focus only on approaches based on replacing missing values.

Imputation is the process of replacing missing data with substituted values. Imputation method comes as a solution to give a better result; it solves the missing values problem, and the number of data that is expected to be proceeded remains to be the same. But the drawback of this method is the bias caused by so many values on the data that have a similar value. Imputation or estimation is the procedure that handles missing values problem by replacing each of the missing values with certain values [8].

- Mean Imputation

One of the easiest ways to impute or estimate missing values to get a complete sample is replacing each of the missing values with the mean of the observed data for that variable [15]. Arithmetic mean imputation (also referred to as mean substitution and unconditional mean imputation) takes the seemingly appealing tack of filling in the missing values with the arithmetic mean of the available cases.

Mean imputation is fast and simple for the missing data. However, extensive theoretical research has shown that this method has many disadvantages. The biggest disadvantage is that there are misaligned estimates. Mean imputation should only be used when a handful of values are missing, and it should be avoided in general [22, 23].

- Regression Imputation

Regression imputation (also known as conditional mean imputation) replaces missing values with predicted scores from a regression equation. The first step of the imputation process is to estimate a set of regression equations that predict the incomplete variables from the complete variables. A complete-case analysis usually generates these estimates. The second step is to generate predicted values for the incomplete variables. These predicted scores fill in the missing values and produce a complete data set [21].

5 Design of Data Model

Data has been recorded every hour throughout the year. The raw data consists of 8181 rows and 55 columns. Data set has been exported to a .csv file through the user interface which allows data processing in multiple software applications.

Because we have a set of historical data, we determine mean values for this data. The table structure for this purpose allows using of the MS Excel "pivot table", which determines the required values very effectively.

5.1 *Elimination of Duplicate Values*

Using a simple tool called "conditional formatting", we can find duplicate values. This tool allows only visual inspection followed by elimination of duplicate values. Based on this check, we can see that the duplicate rows are exactly the same and can be deleted without compromising the analysis goals. An effective tool is the possibility to make a new query that ensures elimination and then creates a new data set without duplicate values. The duplicities are shown in Fig. 3.

5.2 *Missing Values*

In the context of continuous monitoring of selected parameters of the heating process, there were states where this data was not recorded (server failure, accidental failure of the sensor record, etc.). This was caused by source data files containing incomplete data for the period 31/8/2018–15/11/2018 due to server failure. The example of missing values is shown in Fig. 4.

Because the data is important for further analysis, we will try to use the described methods to find at least approximate values of the selected parameter. The monitored parameter is a variable, which is identified in the file as SPZPLYNDEN and

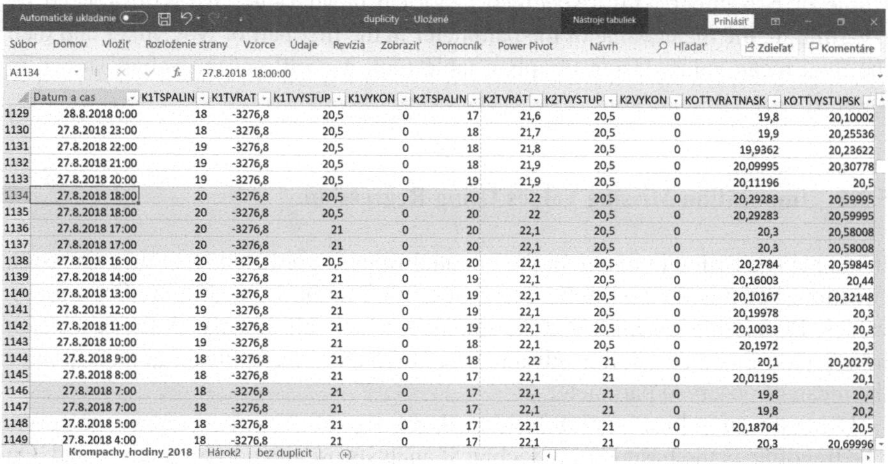

	Datum a cas	K1TSPALIN	K1TVRAT	K1TVYSTUP	K1VYKON	K2TSPALIN	K2TVRAT	K2TVYSTUP	K2VYKON	KOTTVRATNASK	KOTTVYSTUPSK
1129	28.8.2018 0:00	18	-3276,8	20,5	0	17	21,6	20,5	0	19,8	20,10002
1130	27.8.2018 23:00	18	-3276,8	20,5	0	18	21,7	20,5	0	19,9	20,25536
1131	27.8.2018 22:00	19	-3276,8	20,5	0	18	21,8	20,5	0	19,9362	20,23622
1132	27.8.2018 21:00	19	-3276,8	20,5	0	18	21,9	20,5	0	20,09995	20,30778
1133	27.8.2018 20:00	19	-3276,8	20,5	0	19	21,9	20,5	0	20,11196	20,5
1134	27.8.2018 18:00	20	-3276,8	20,5	0	20	22	20,5	0	20,29283	20,59995
1135	27.8.2018 18:00	20	-3276,8	20,5	0	20	22	20,5	0	20,29283	20,59995
1136	27.8.2018 17:00	20	-3276,8	21	0	20	22,1	20,5	0	20,3	20,58008
1137	27.8.2018 17:00	20	-3276,8	21	0	20	22,1	20,5	0	20,3	20,58008
1138	27.8.2018 16:00	20	-3276,8	20,5	0	20	22,1	20,5	0	20,2784	20,59845
1139	27.8.2018 14:00	20	-3276,8	21	0	19	22,1	20,5	0	20,16003	20,44
1140	27.8.2018 13:00	19	-3276,8	21	0	19	22,1	20,5	0	20,10167	20,32148
1141	27.8.2018 12:00	19	-3276,8	21	0	19	22,1	20,5	0	20,19978	20,3
1142	27.8.2018 11:00	19	-3276,8	21	0	19	22,1	20,5	0	20,10033	20,3
1143	27.8.2018 10:00	19	-3276,8	21	0	18	22,1	20,5	0	20,1972	20,3
1144	27.8.2018 9:00	18	-3276,8	21	0	18	22	21	0	20,1	20,20279
1145	27.8.2018 8:00	18	-3276,8	21	0	17	22,1	21	0	20,01195	20,1
1146	27.8.2018 7:00	18	-3276,8	21	0	17	22,1	21	0	19,8	20,2
1147	27.8.2018 7:00	18	-3276,8	21	0	17	22,1	21	0	19,8	20,2
1148	27.8.2018 5:00	18	-3276,8	21	0	17	22,1	21	0	20,18704	20,5
1149	27.8.2018 4:00	18	-3276,8	21	0	17	22,1	21	0	20,3	20,69996

Fig. 3 Duplicities in data source

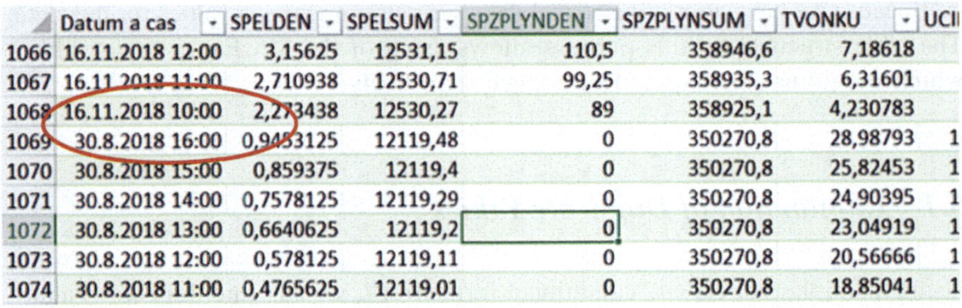

	Datum a cas	SPELDEN	SPELSUM	SPZPLYNDEN	SPZPLYNSUM	TVONKU	UCI
1066	16.11.2018 12:00	3,15625	12531,15	110,5	358946,6	7,18618	
1067	16.11.2018 11:00	2,710938	12530,71	99,25	358935,3	6,31601	
1068	16.11.2018 10:00	2,273438	12530,27	89	358925,1	4,230783	
1069	30.8.2018 16:00	0,9453125	12119,48	0	350270,8	28,98793	1
1070	30.8.2018 15:00	0,859375	12119,4	0	350270,8	25,82453	1
1071	30.8.2018 14:00	0,7578125	12119,29	0	350270,8	24,90395	1
1072	30.8.2018 13:00	0,6640625	12119,2	0	350270,8	23,04919	1
1073	30.8.2018 12:00	0,578125	12119,11	0	350270,8	20,56666	1
1074	30.8.2018 11:00	0,4765625	12119,01	0	350270,8	18,85041	1

Fig. 4 Missing data

contains gas consumption for a certain hour per day. For future purposes of the analysis, we will consider the mean consumption per month.

5.2.1 Imputation Missing Values Using Mean

Based on this method, the missing values are replaced by the mean obtained from the complete set. According to historical data, gas consumption is zero during August, and therefore we can assume it to be the same for the missing item (since this is summer month, we will accept this assumption). Therefore, the following steps will focus mainly on the months September, October and November. The values from September and October 2018 are missing for the whole month; they will be replaced by mean obtained from complete data.

The month of November is specific. We have a data of first half of the month, and so only half of the month is missing. It is therefore not appropriate to replace the whole of November with a specified historical mean value. We will proceed by determining the mean value of the parameter in the first half of November and then the total mean value. Using the pivot table tool, we will get the values shown in Table 1.

5.2.2 Imputation Missing Values Using Regression

In simple regression imputation, the imputed value is predicted from the regression equation. For this method, full observation information is used to predict missing values. Regression assumes that the imputed values fall directly on the regression line. To use this method, it is advantageous to display a graphical interpretation of the mean of observed parameter.

Using the "trendline" tool in MS Excel, we determine the best regression curve. The trendline is the basic tool of technical analysis used to determine the trend. On the basis of the graphical form, the polynomial regression was used. The following equation represents the general equation of polynomial regression:

Table 1 Added values according to mean

2017		2018	
Month	Values of observed parameter	Month	Values of observed parameter
January	294.0476619	January	215.8339235
February	200.2344441	February	226.1651786
March	131.425813	March	176.6522529
April	81.93051676	April	42.09805285
May	22.34543011	May	0.629263302
June	0.004874652	June	0.246175243
July	0	July	0
August	0	August	**0**
September	23.14315642	September	**23.14315642**
October	83.20614035	October	**83.20614035**
November	169.5034868	November	**177.5854**
December	192.8219323	December	208.8835425

$$y_i = b_0 + b_i x_i + b_2 x_i^2 + \ldots + b_n x_i^n \tag{7}$$

Using the method of least squares, we get a system of three equations (Eq. 8), which solves the estimation of coefficients of the regression model:

$$\begin{aligned}\sum_{i=1}^{n} y_i &= nb_0 + b_1 \sum_{i=1}^{n} x_i + b_2 \sum_{i=1}^{n} x_i^2 \\ \sum_{i=1}^{n} y_i x_i &= b_0 \sum_{i=1}^{n} x_i + b_1 \sum_{i=1}^{n} x_i^2 + b_2 \sum_{i=1}^{n} x_i^3 \\ \sum_{i=1}^{n} y_i x_i^2 &= b_0 \sum_{i=1}^{n} x_i^2 + b_1 \sum_{i=1}^{n} x_i^3 + b_2 \sum_{i=1}^{n} x_i^4\end{aligned} \tag{8}$$

On the next figure, the chosen regression line and the regression equation are shown. We will compute the values using this equation (Fig. 5).

The following table shows the final estimates obtained by these methods (Table 2).

5.3 *Data Analysis*

In further data processing, we will use the pivot table options. Pivot table is a data processing technique that summarises the data of a larger table. This summary may include the sums, averages or other statistics of the pivot table groups in a meaningful way. They allow the user to rearrange or to highlight useful information.

To meet the analysis requirements, we will either process data from a single source table or create a model with a defined link among multiple source tables.

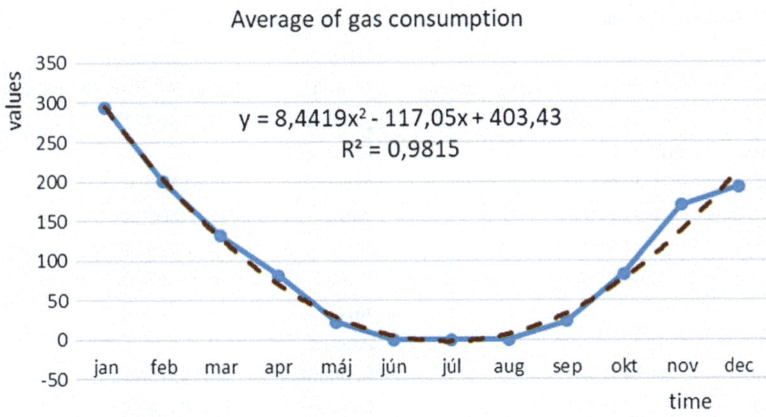

Fig. 5 Value of R^2 and equation

Table 2 Calculated values

Rank	Calculated values
1	294.8219
2	203.0976
3	128.2571
4	70.3004
5	29.2275
6	5.0384
7	−2.2669
8	7.3116
9	**33.7739**
10	**77.12**
11	**137.3499**
12	214.4636

- Assessment of heating quality based on monitored temperatures from one source.

Within the process, the parameters of TVODZ, which represents the required water temperature, and TVODS, which represents the actual water temperature, are observed. Visualisation of these parameters indicates the quality of the process. The smaller the difference between these parameters, the better the process is set up. The graph of the selected month is shown in the following figure (Fig. 6).

In the figure, we can view that the TVODZ parameter is 5 °C. This value indicates that the boiler was shut down at that time. The graph also shows that the TVODS parameter has maintained a certain temperature even during the shutdown and hence the short-term shutdown of the boiler should not have a significant effect on the actual water temperature. To assess the quality of observed process, the standard deviation was also used. Standard deviation is a useful statistical tool for

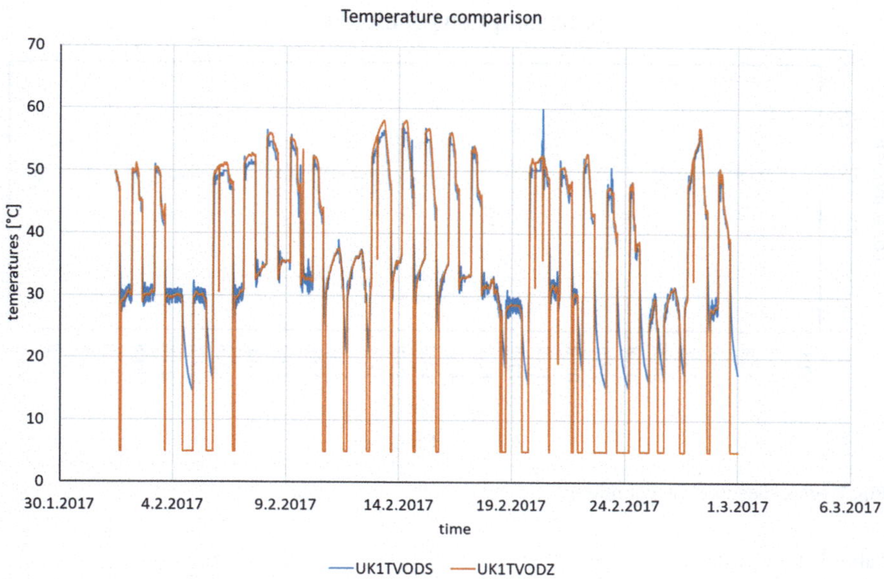

Fig. 6 Visualisation of selected parameters

analysing quality evaluation. By definition, it is a measure that is used to quantify the amount of variation or dispersion of a set of data values. If we have a bunch of numbers that we want to analyse, it is one of the tools that will help us to understand how widely the data points spread from the top to the bottom and how tightly they cluster around the average [24]. Basically, a small standard deviation means that the values in a statistical data set are close to the mean of the data set, on average, and a large standard deviation means that the values in the data set are farther away from the mean, on average [25].

The standard deviation value in the selected month in 2017 was 1.789833, indicating a large variance of values around the arithmetic mean. The standard deviation of the same month in 2018 is 0.978846, which means an improvement in the quality of the process, as the differences between the desired and actual water temperature values have diminished. Values when the boiler was switched off were not included in the calculation.

- Assumption of gas consumption by using regression analysis for data from one source.

By simply using a pivot table and a pivot chart, we create a point graph. The pivot table itself sorts the values by month, and so we get an estimate of consumption for the whole year. In our case, it is an equation:

$$y = 5.8613x^3 + 71.067x^2 - 1864.8x + 8171.5 \tag{9}$$

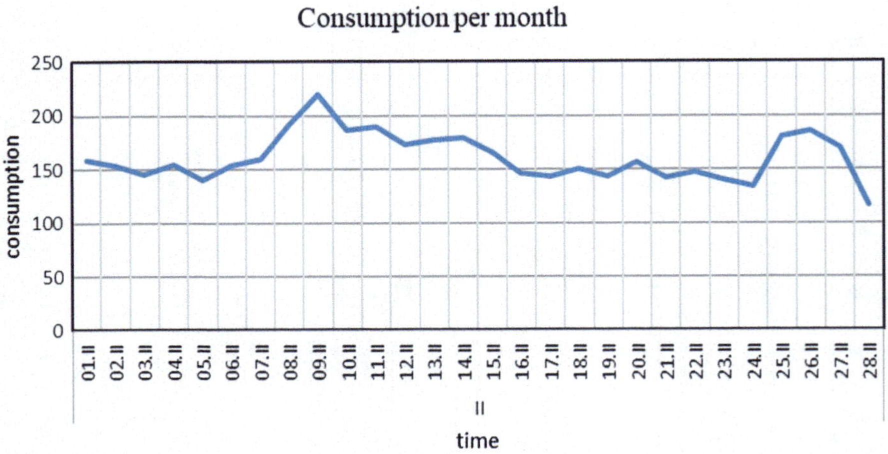

Fig. 7 Gas consumption per month

Table 3 Predicted and real values

Month	Predicted	Real
1	6383.629	**5413.75**
2	4773.06	**5659.125**
3	3374.963	**5608.125**
4	2224.505	**1581.781**
5	1356.855	**678.75**
6	807.1816	**419.125**
7	610.653	**429.1563**
8	802.4376	**392.875**
9	1417.704	**937.875**
10	2491.62	**2864.375**
11	4059.355	**3938.5**
12	6156.076	**5781.75**

The value of *R*-squared in this regression model is $R^2 = 0.9721$. *R*-squared is a statistical measure that represents the proportion of the variance for a dependent variable that is explained by an independent variable or variables in a regression model.

Pivot chart allows visualisation of consumption even after individual days within the selected month. As you can see in Fig. 7, it is not meaningful to assume the development of consumption for 1 month based on trendlines.

We use Eq. (9) and determine the assumption for the following year (Table 3).

- Comparison gas consumption from multiple sources.

The Power Pivot tool is used to create a data model for the required purposes. We will compare gas consumption from five different sources. In order to obtain data from multiple sources, it is necessary to create a relationship among them. As a suitable attribute for all tables, marked PM_TIME is the parameter, which represents

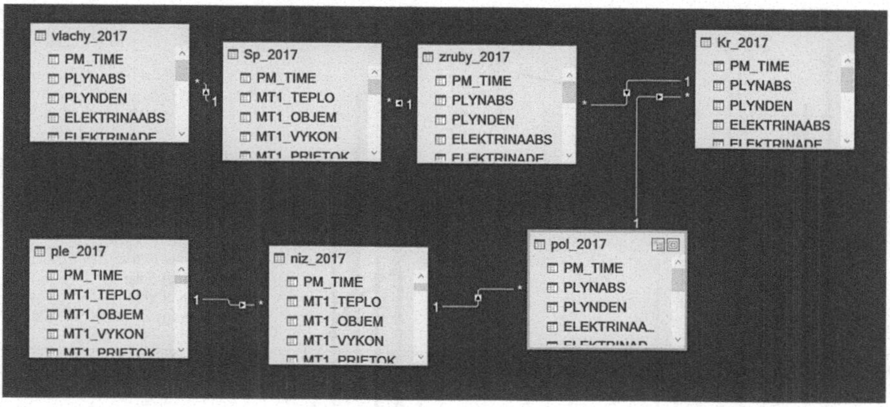

Fig. 8 Relationship among different sources

the date (day, month, year and time) of data entry. So we create a session using this parameter. The next figure represents the relationship among different sources (Fig. 8).

Based on the created relationship, it is possible to use the possibility of creating a pivot table. The Power Pivot tool creates the table that we work with as usual. We can use all its benefits. A minor problem is that it is necessary to select individual fields in the pivot table in a certain order. If we do not follow the order, or select a relational field from the wrong table, the pivot table will request a re-establishment of the relationship. This request indicates that we have selected the field in the wrong order. The following figure shows gas consumption from multiple locations (Fig. 9).

As can be seen from the figure, the largest gas consumption is at the location marked as Zruby. It can be seen that consumption is also in the summer months, which probably indicates the heating of water that is not intended for heating. On this figure, we can also see the values of average of temperatures. It allows us make a better evaluation of gas consumption.

However, the final evaluation should be carried out by an expert or group of experts to make the conclusions relevant.

- Calculation of possible loss.

The knowledge of Taguchi loss function implementation was used for possible economic evaluation. This feature states that any deviation from nominal (target) performance could cause some economic losses. The loss function L therefore has an economic relationship to variability. Any deviation from the ideal set point T results in an economic loss that can be quantified by a quadratic relationship [26]:

$$L = k(X - T)^2 \tag{10}$$

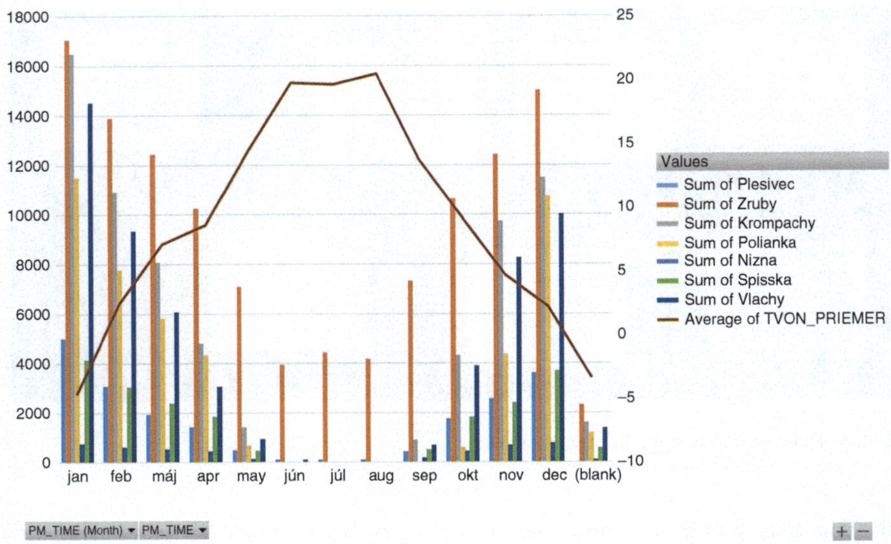

Fig. 9 Gas consumption from multiple locations

where L is loss expressed in financial values (incremental loss).

k is cost coefficient.
T is the required value of the indicator.
X is the true value of the indicator (Fig. 10).

As a rule, the cost coefficient is calculated from the relationship.

$$k = A / \Delta \tag{11}$$

where A is cost of corrective actions that reflect changes by Δ.

Δ is value of change.

As basis for determining possible price losses, the average value of gas consumption for the selected month is taken into account. We will consider this value as our target value. Gas consumption is given in units of m^3, and pricing for 1 m^3 of gas is around 0.44 euro.

For example, it is $k = 44/100^2 = 0.0044$.

Target value is $T = 389.31$.

In Fig. 10, we can see that the biggest value is 539,25:

$$L = k\left(X - T\right)^2 = 0.0044\left(539.25 - 389.31\right)^2 = 98.92$$

From the calculation, we can conclude that if the deviation from the average is ±100 units at the fixed gas price per 1 m^3, the loss per month could be around 100 euros.

As the following table shows, the losses for individual months could be even greater (Table 4).

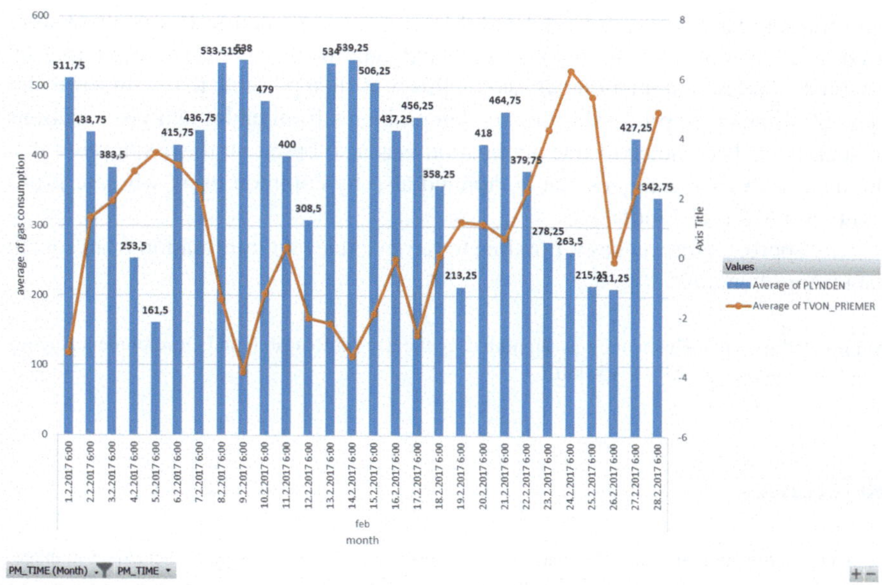

Fig. 10 Data for calculation of loss function

Table 4 Calculated values of observed parameter

Month	Average of PLYNDEN	Max of PLYNDEN
January	583.6895161	1003.625
February	389.3130571	539.25
March	260.6048387	427
April	159.7416667	451
May	45.61290323	289.75
June	0.008333333	0.25
July	0	0
August	0	0
September	30.325	167.75
October	139.0645161	332.5
November	324.6583333	630.25
December	370.3145161	536.75

6 Conclusion

Data is a symbol of the present. However, they are relevant only if they provide the necessary information. Data analysis methods and tools can achieve this goal while respecting the data life cycle. In this paper, the preprocessing part was described in detail, especially the selected procedures for inserting missing data, since records with missing values cannot be deleted. MS Excel and its add-in Power Pivot were used as a software tool. It enables to manage data from several sources based on established relational relations. Based on this, it is then possible to obtain a broader view of the selected process or find hidden problems from tables and visualisations of such data. The choice of this application was also based on the assumption that the data analysis techniques can be applied also by common users who are proficient with MS Excel tools.

Any approach that makes it possible to retrieve relevant information from source data brings benefits.

Acknowledgement This work was supported by the Slovak Research and Development Agency under the contract no. APVV-15-0602.

References

1. Data management. https://www.sas.com/content/dam/SAS/en_us/doc/other1/data-management-why-is-it-important-107421.pdf. Accessed 12 Nov 2019
2. Lenhardt, W.C., Ahalt, S., Blanton, B., Christopherson, L., Idaszak, R.: Data management lifecycle and software lifecycle management in the context of conducting science. J. Open Res. Soft. **2**(1), e15. https://doi.org/10.5334/jors.ax. pp. 1–4. Accessed 13 Nov 2019
3. Piteľ, J., Mižáková, J., Mižák, J.: Control systems of the heat production and supply for concept of industry 4.0. In: Vykurovanie 2017, Podbanské, 6.3.-10.3.2017, pp. 391–394. SSTP, Bratislava (2017)
4. Židek, K., Piteľ, J., Hošovský, A.: Machine learning algorithms implementation into embedded systems with web application user interface. In: Proceedings of the IEEE 21st International Conference on Intelligent Engineering Systems 2017 (INES 2017). October 20–23, 2017, pp. 77–81, Larnaca, Cyprus. IEEE, Budapest (2017)
5. Saloky, T., Piteľ, J.: Adaptive Control of Heating Process with Outdoor Temperature Compensation, XXX. In: ASR ‘2005 Seminar, Instruments and Control, Ostrava, April 29. Available from http://akce.fs.vsb.cz/2005/asr2005/Proceedings/papers/383.pdf (2005). Accessed 15 July 2019
6. Mižáková, J., Piteľ, J.: An analytical dynamic model of heat transfer from the heating body to the heated room. In: 21st International Conference on Circuits, Systems, Communications and Computers (CSCC 2017). Crete Island, Greece, July 14–17, 2017. MATEC Web of Conferences 125, 02047, 5p
7. Čorný, I.: Overview of progressive evaluation methods for monitoring of heat production and distribution. Process. Eng. 619–626 (2017)
8. Russo, A., Ferrari, M.: Microsoft Excel 2013: Building Data Models with Power Pivot, Published by Microsoft Press. http://index-of.co.uk/OFIMATICA/Microsoft%20Press%20Excel%202013,%20Building%20Data%20Models%20with%20PowerPivot.pdf (2013). Accessed 5 Nov 2019
9. Singh, A, Collie, R.: Power Pivot and Power BI. Independent Publishers Group, Chicago, IL. http://www.allitebooks.in/power-pivot-and-power-bi/. Accessed 5 Nov 2019
10. Pratama, I.; Permanasari, A. E.; Ardiyanto, I.; Indrayani, R. A review of missing values handling methods on time-series data. In: International Conference on Information Technology Systems and Innovation, Bandung—Bali, October 24–27, 2016

11. Custodio, C. Y.; Walter T.; Dunn, L.; Mercado, A.; Brown, A. E.; Mathew, P. Data Preparation Process for the Buildings Performance Database. https://www.energy.gov/sites/prod/files/2015/09/f26/bto_Data_Preparation_for_BPD_Aug2015.pdf (2015). Accessed 25 Jul 2019
12. Polong, L.: The Data Science Process. Available from https://www.01.ibm.com/events/wwe/grp/grp304.nsf/vLookupPDFs/Polong%20Lin%20Presentation/$file/Polong%20Lin%20Presentation.pdf. Accessed 25 July 2019
13. Tamilselvi, J.J., Gifta, C.B.: Handling duplicate data in data warehouse for data mining. Int. J. Comput. Appl. **15**(4), 975–8887 (2011). Available from https://pdfs.semanticscholar.org/be0c/a2a8054c2778c2dc2f8c6bbbc1707db34fff.pdf. Accessed 5 Nov 2019
14. Graham, J.W., Cumsille, P.E., Elek-Fisk, E.: Methods for handling missing data. In: Schinka, J.A., Velicer, W.F., Weiner, I.B. (eds.) Handbook of Psychology, Research Methods in Psychology, vol. 2, pp. 87–114. John Wiley and Sons, Hoboken, NJ
15. Berkery, E.M.: Missing data analysis with the Mahalanobis distance, 2016 Available form https://ulir.ul.ie/bitstream/handle/10344/5235/Berkery_2016_missing.pdf?sequence=6
16. Sitaram, D.; Dalwani, A.; Narang, A.; Das, M.; Auradkar, P. A Measure of similarity of time series containing missing data using the Mahalanobis distance. In: Proc. 2015, 2nd International Conference on Advances in Computing and Communication Engineering ICACCE 2015, pp. 622–627 (2015)
17. Aydilek, I.B., Arslan, A.: A hybrid method for imputation of missing values using optimized fuzzy c-means with support vector regression and a genetic algorithm. Inf. Sci. **233**, 25–35 (2013)
18. Soley-Bori, M. Dealing with missing data: key assumptions and methods for applied analysis. Technical report no. 4. Boston University. Available from https://www.bu.edu/sph/files/2014/05/Marina-tech-report.pdf. Accessed 12 Nov 2019
19. Rubin, D.B.: Inference and missing data. Biometrika. **63**(3), 581–592 (1976)
20. Graham, J.W.: Missing data analysis: making it work in the real world. Ann. Rev. Psychol. **60**, 549–576 (2009). by Pennsylvania State University
21. Enders, C.K.: Applied Missing Data Analysis. The Guilford Press, New York (2010). Available from http://hsta559s12.pbworks.com/w/file/fetch/52112520/enders.applied. Accessed 8 July 2019
22. Little, R.J., Rubin, D.B.: Statistical Analysis with Missing Data, Probability and Statistics, 2nd edn. Wiley, Hoboken, NJ (2002)
23. Buuren, S.: Flexible Imputation of Missing Data. CRC Press, Boca Raton, FL (2012). http://pzs.dstu.dp.ua/DataMining/preprocessing/bibl/fimd.pdf. . Accessed 25 July 2019, Accessed 8 July 2019
24. Klenke, M.: What is standard about deviation? (2015). Available from http://www.qatc.org/winter-2015-connection/what-is-standard-about-deviation/
25. Rumsey, D.B.: How to interpret standard deviation in a statistical data set. Available from http://www.dummies.com/education/math/statistics/how-to-interpret-standard-deviation-in-a-statistical-data-set/. Accessed 2 Nov 2019
26. Benková, M., Floreková, Ľ., Bogdanovská, G.: Variability of quality parameters and loss function. Acta Montan. Slovaca. **10**(1), 57–61 (2005)

Design of Concept for Transport and Monitoring of Biological Samples Using RFID Technology

Jozef Husár, Stella Hrehová, and Patrik Kaščák

1 Introduction

The process of transformation is also applied in healthcare. By this, we mean innovation of healthcare facilities, construction of new specialized departments and modernization of the medical documentation system. The Slovak Ministry of Health has launched the project called e-Health, which aims at the electronization of healthcare to improve the quality of healthcare provision. The main element of this project is to share medical documentation among healthcare facilities or individual healthcare professionals. The project is divided into six units, each of which is oriented to one specific part [1].

The article deals with the issue of one of the areas of the e-Health project called e-Lab and focuses on the results and labelling of laboratory samples. Here opens the possibility of using RFID technology as a substitute for the simultaneous labelling of biological samples with barcodes. The RFID system consists of several parts that cooperate with each other to form one functional unit. All its components are important and dimensioned for the industry or application. In practice, we often encounter a number of interfering and unwanted elements that occur during communication between the RFID transponder and the RFID antenna. Therefore, the greatest emphasis and demands are placed on these elements, on which the reliability and operation of the RFID system depends, without gross errors or collisions. RFID antenna and tag have their own specific parameters, such as the frequency at which they work, the reliability and radiation characteristics (range) specified by the manufacturer in the catalogue or documentation of the product. These characteristics are in most cases only in the theoretical plane made according to calculation from the

J. Husár (✉) · S. Hrehová · P. Kaščák
Faculty of Manufacturing Technologies, Department of Industrial Engineering and Informatics, Technical University of Košice, Prešov, Slovak Republic
e-mail: jozef.husar@tuke.sk; stella.hrehova@tuke.sk; patrik.kascak@tuke.sk

L. Knapcikova et al. (eds.), *New Approaches in Management of Smart Manufacturing Systems*, EAI/Springer Innovations in Communication and Computing, https://doi.org/10.1007/978-3-030-40176-4_6

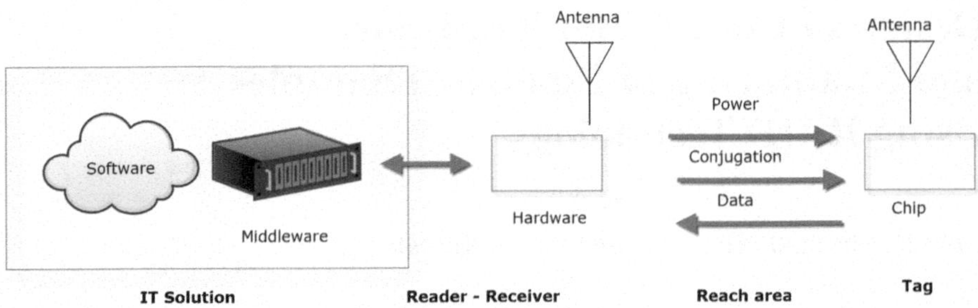

Fig. 1 Structure of RFID system representing the principle of object identification process realization

design parameters of individual components, without subsequent verification and measurement. This article focuses on two basic areas. The first area is the design of a transport system and monitoring of biological samples. The second area is the experimental verification of the proposed model system. If we want to implement radio frequency technology, it is necessary to briefly describe the principle of object identification; in our case, these are biological samples. In simplicity, the RFID principle consists of the following steps [1]:

1. The transmitter of reader generates a signal and generates an electromagnetic field that defines the size and shape of the read range.
2. Assuming that the RFID tag is located within range, its antenna receives the transmitted signal, and the rectified electric current charges the power capacitor. This process creates space for the absence of a power supply.
3. The previous process ensures that the RFID tag sends the programmed information from the built-in memory via a modulated signal to the environment. The content of this signal is a unique EPC (product) number, i.e. information about what object was just scanned (what kind of product from which producer), where and at what time it has been scanned and of course also other data that can be stored in the tag.
4. The reader receiver detects this signal and receives data based on logic levels 1 and 0. The content of this signal is the electronic product code (EPC).
5. The reader then transfers the data to the control computer where it is used in the selected way (Fig. 1).

According to experiences of using RFID technology, it is necessary to define the basic requirements for tags and antennas [2]:

- Frequency—the frequency tag band depends on the location of use. This includes the country where the tag will be used. Alternatively, it depends on the specific application which band will be most appropriate (certain frequencies pass better through some materials).
- Range—this parameter is described below as the maximum reading distance.
- Shape—the shape is important because of the need to attach the tag to an object, as well as the antenna, because it is installed in various places that are not often

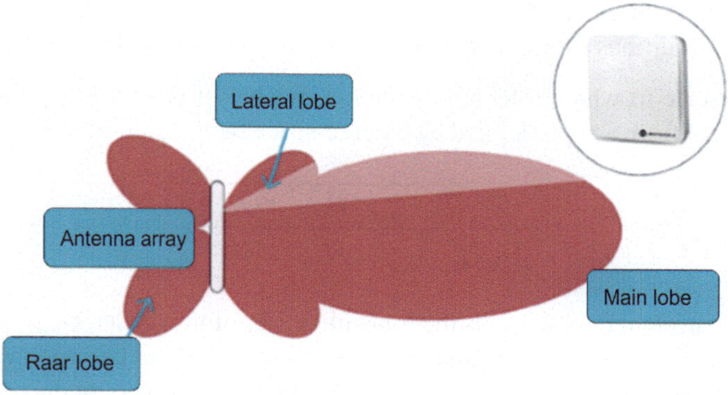

Fig. 2 Ideal antenna radiation characteristic

sufficient. According to these requirements, the antennas on the tags or the antennas themselves are made in different sizes and shapes.

- Reliability—the prerequisite for proper function is the independence from changes in temperature, humidity and, last but not least, some mechanical resistance, which is absolutely necessary in production and logistics.
- Price—this criterion applies mainly to transponders that are used in RFID systems in hundreds, thousands or even millions. Especially from the economic point of view, it should be as cheap as possible, but it also takes into account the material from which the device is made and what are its features.
- Mobility—antennas and tags can be used many times in applications where the device on which the tag is attached or the antenna itself on the mobile device is moving. For safe identification of goods, persons or even vehicles, a certain reading speed is necessary so that the identified element does not disappear from the reading field before reading the necessary information.

The main focus of the whole article was to determine the maximum reading distance, which is the biggest problem. The reading range is biggest in the direction of maximum profit. The readable distance ranges from few millimetres to several tens of metres. In order to reach the reading distance parameter, two conditions must be ensured for reliable communication between the reader and the tag. The first is that the tag has enough power to charge itself. A second condition is the need for sufficient signal strength that the tag has sent to be able to record the transmitted data to the tag without errors or collisions. The common reading distance using frequencies in the 800–900 MHz range is about 1 m for passive tags. The biggest problem of these systems is low efficiency while keeping small dimensions (Fig. 2) [2].

The maximum reading distance for transponders depends on several factors:

- Tag sensitivity
- The relative position, orientation and polarization of the antenna and tag
- The distance between the reader and the tag
- Tag moving speed in reading area

- Use of carrier frequency
- Current conditions (noise, multiple tags side by side, obstacles, etc.)

The distance to which a RFID antenna is capable of detecting a passive back-emitted signal from a tag is defined by the free-space radio communication equation:

$$r = \frac{\lambda}{4\pi}\sqrt{\frac{P_{\text{TX.reader}} * G_{\text{reader}} * G_{\text{tag}} * \tau}{P_{\text{th}}}}$$

where λ is wavelength, $P_{\text{TX,reader}}$ is the transmit power of the reader, G_{reader} is transmit gain of antenna reader, G_{tag} is gain of antenna tag and P_{th} is the minimum power needed to power the tag. This calculation does not take into account the polarization losses that occur in the proposed systems [3].

2 Methodology

The experiment, which was carried out in the laboratory of the Department of Industrial Engineering and Computer Science, was solved on the basis of requirements for RFID system design, where many factors influence reading accuracy. Nowadays, many hospitals pay little attention, when installing a sample monitoring system to important factors such as loading intensity. They are focusing only on whether the system works and identifies the tag. Companies that design custom RFID systems will develop a design. This includes everything needed to operate the system from antenna, reader, software to transponders, but each application to the real environment is unique in its layout and placement of interfering elements. Therefore, it often happens that the system does not work 100% after some time. According to chronology, the elaboration of the design and documentation is followed by the implementation of the proposed system into practice, where this system must fulfil all criteria specified by the client. First of all, the grouping of elements must be functional and reliable. After the next start, it turns out that the system does not work reliably, so that reading the RFID tags is not constant or transponders do not load at all. After this detection, the location of the antenna is changed in most of these cases because its range is not sufficient and there is no RFID transponder loading. In another case, additional antennas are added because of their insufficient coverage of the loading area. Precisely because of problems with implementation in practice, measurements were carried out to verify the ability to read RFID tags in a laboratory environment. The main task was to identify the maximum reading distance of the RFID tag and then to identify the ideal 95% loading limit. The whole experiment was based on the methodology of overlap evaluation of the transponder and antenna. The individual shapes of each antenna or transponder are different, but mainly they are different in terms of their distance to which they can transmit or receive radio frequency waves. These areas can be affected by a number of factors and thus impair the readability. As the unfavourable factors, we can consider the pillar in the reading area, devices that by their operation worsen or disturb the transmission of the waves necessary for reading. There are a large number of such factors in real-world applications; therefore, when implementing an RFID system, it should be kept in mind in which place the reading will be acted [4, 5].

In the laboratory of the Department of Industrial Engineering and Informatics, an area of 0.5 m has been created and graphically marked to create a “chessboard”. The area was optimized for the experiment purposes with respect to using the measuring device parts. In this area size optimization, however, the greatest emphasis was placed on the measured elements, which were RFID passive transponders, based on the theoretical knowledge and information provided by the producers. According to the knowledge of the functioning of the passive RFID system, it was known that the reading distance of these transponders is several metres, which is why the measuring space of 4.5 × 5 m was created. This area was evenly divided into 0.5 m in length and width, giving 90 squares of 0.5 × 0.5 m. So it is possible to use the intersection points of the area distribution, where 110 measuring points were created. Other factors have been identified for the evaluation and adequacy of the conditions:

- Temperature: 21 °C
- Humidity: 53%
- Pressure: 101 kPa

The principle of measurement has consisted in obtaining data realized through system analysis, evaluation of the number of readings described in ISO standards and case studies and supplemented with data from own measurements. The obtained data provide the basis for the design of chronological procedures for performing repeatable measurements [6].

2.1 RFID Measurement and Component Selection

The measurement has been carried out in the laboratory of the Department of Industrial Engineering and Informatics, which is specialized for measurements related to RFID technology. The measurements have been performed due to an industrial computer, model GB TOUCH 24″, and has been manufactured by Gaben. The reader and two alien antennas were plugged in and tested to verify functionality. The decision of installing antennas was selected among three variants of connection. The first version was placing the antennas side by side with a gap of approximately 10 cm. The second version was placing the antennas under one another with the same gap of 10 cm. For these two connections, the antennas should be oriented in the same direction. The last suggestion was to install RFID antennas facing each other with the intention of creating a so-called RFID gateway used in common practice. In a deeper analysis of the issue, the last option was chosen. It will be the most suitable for the experiment. Two RFID gateway configurations were tested in this connection. The different parameter between the antennas was the distance, which was 0.5 and 1 m. The laboratory tripods on which the antennas were attached were set 1 m from the floor [7]. The connection of all hardware elements of the system together with the RFID gateways width is shown in Fig. 3.

Overall, the RFID system is based on Fig. 4. Individual components are discussed in the next text.

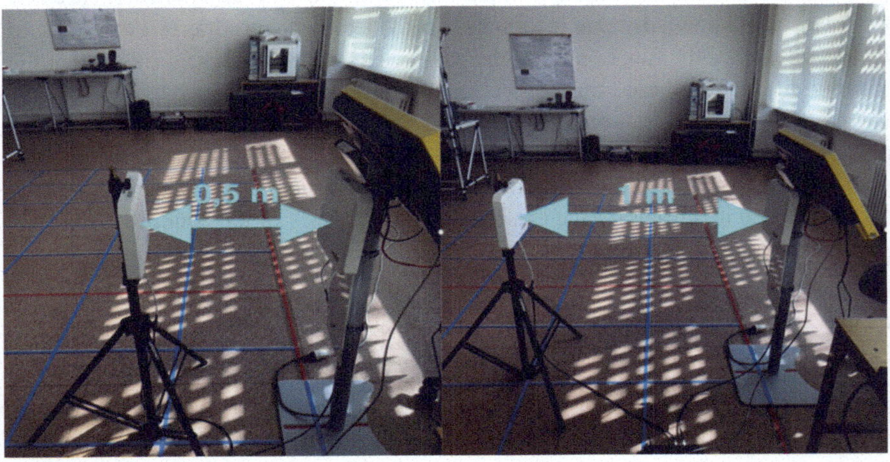

Fig. 3 Distance difference between antennas for measurements

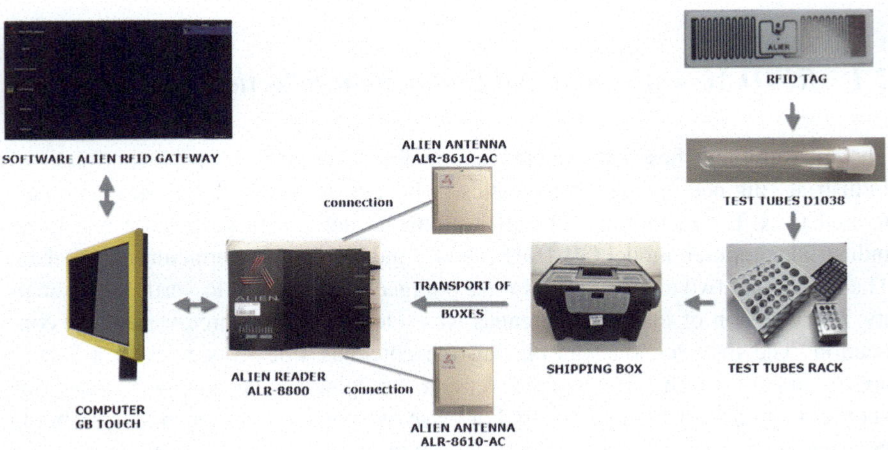

Fig. 4 Diagram of the components and measurement principle

2.1.1 Computer

One of the two computers in the laboratory was used to measure the experiment. The computer has a special design that has been adapted for industrial use and was made by Gaben. This model is marked GB TOUCH 24″ [8]. The industrial computer used in the laboratory has hardware parameters (HDD capacity, RAM size, processor), approximately the same as conventional desktop computers. The computer design consists of a steel stand. It keeps the computer in a suitable working position and height, which is comfortable for the type of measurement carried out in the laboratory. An antenna holder is attached to this structure and is capable of holding two antennas. The construction is inflexible, and the weight of the whole assembly is high for easy handling.

Technical parameters of laboratory computer include:

- Model: GB TOUCH 24″
- Screen type: standard
- Touch area: IR (infrared technology)
- Display: 16: 9 full HD LED
- Monitor size: 24″ 60 cm
- Monitor resolution: 1920 × 1080@60 Hz
- Monitor brightness: 250 cd/m^2
- Monitor cover: safety glass
- Dimensions (H × W × D): 38 × 60 × 8 cm

2.1.2 Static Reading Device

As part of the proposed concept, the Alien ALR-8800 ENTERPRISE RFID READER static reading device can be used, which is part of the laboratory's technical equipment. The device has the ability to control the scanning density and also suppresses the surrounding interference. The device is suitable for application of RFID technology to various areas. Configuring the Alien model is simple and unpretentious.

Another advantage is the universal I/O signals that can control many commonly available external devices. This has a positive impact on the additional costs of technology implementation. The reader is reliable and durable for use in environments with increased humidity, shaking, vibration and high or low temperature. We can store information about 2500 tags in the reader's internal memory [9].

Technical parameters of Alien reader include:

- Model: ALR-8800
- Supported RFID standard: EPC Class 1 Gen 2, EPC Class 1 Gen 1
- Network protocols: DHCP, TCP/IP, SNTP, DNS, SNMPv.3,
- Frequency: 865.6–867.6 MHz
- Number of channels: 10
- I/O: 4 input, 8 output, optically insulated, 0.5 W current capacity

- Dimensions: 280 × 229 × 56 mm
- Operating temperature: −20 to +50 °C
- Compatibility: with antennas ALR-8610-AL a ALR-8610-AC
- Power: AC/DC 45 W max. 120 or 240 VAC, 50–60 Hz

2.1.3 Antenna

Two Alien ALR-8610-AC laboratory antennas were used, which were specially designed to provide optimum performance when used with Alien Technology RFID readers. This means the antenna is fully compatible with the ALR-8800 hardware reader. These antennas are optimized for the frequency range from 865 to 869 MHz. A circularly polarized antenna evenly distributes UHF energy into a radially symmetrical pattern, providing the ability to read RFID tags regardless of orientation.

Technical parameters of the Alien antenna include [10]:

- Model: ALR-8610-AC
- Frequency: 865–965 MHz
- Profit: 5.5 dBiL max
- Ray width: less than 70°
- Cable: 6 m, 50 ohmic coaxial connector
- Dimensions: 250 × 250 × 38 mm
- Operating temperature: −20 to +50 °C
- Compatibility: with reader ALR-8800 and ALR-8780

2.1.4 Software

RFID gateway is a software solution to connect hardware elements from Alien producer that allows for maximum configuration and the best connection to accessories. This software allows the user to easily set up the reading device for their needs and test the configuration to determine the best setting. The programme is divided into five subcategories, each of which is specialized and focused on some kind of RFID system issues or analysis [11].

Subcategories of Alien RFID gateway include [11]:

- Tag Grid—this software section allows us to perform basic functions such as reading tags in range and also allows you to configure various settings such as EPC code filtering and transponder management.
- Command line interface—the command line function provides the user to directly configure the reader using predefined commands. A list of these commands can be found in the Alien Fixed Reader, which is freely available on the producer's website.
- Tag Programmer—the third part of the programme allows the user to read and encode the EPC codes of RFID transponders as needed.

- Readometer—the fourth section provides the user with the possibility to monitor a number of transponder reader parameters such as the total number of reads, the number of reads per second, the signal strength between the reader and the tag and many other functions.
- The fourth section provides the user the possibility to monitor a number of transponder reader parameters such as the total number of reads, the number of reads per second, the signal strength between the reader and the tag and many other functions.
- Tag Doppler—the last subcategory is for duplicating RFID transponder settings and their configurations

2.1.5 RFID Tags

When designing using RFID technology, the effort was to select the best and most affordable hardware solution. The transponder selection was aimed at the same hardware element producer. Of the Alien's broad product range, the most affordable transponder suitable for healthcare, automotive and retail has been selected. One of the best-selling is Alien Squig transponder. The tag size is 44.5 × 10.4 mm. Another important indicator is its reliable functionality on materials such as glass, plastic, wood or paper. Work in different regions such as Europe, America, the Middle East, Asia and Africa is a matter of course.

The second selected RFID tag is the AD-814 model from Avery Dennison. This transponder is primarily made for pharmaceutical and healthcare applications. The tag size is conditioned by an antenna having dimensions of 22 × 22 mm. The operating temperature ranges from −40 to 85 °C and meets Class 1 Gen 2 standards. Its operating frequency is in the range for Europe, and it is 860–960 MHz (the frequency range for Europe is 865.6–867.6 MHz). This transponder was chosen as an alternative in the case of not fully reliable functioning of the first tag, because the focus of the Alien Squig tag is not only directed to the medical segment. This transponder is shown in Fig. 7.

2.1.6 Types of Test Tubes

Today, we can see a variety of disposable medical supplies needed for different types of specimen collection. This description also includes elements such as syringes, needles, gloves and other aids needed to perform various tasks and activities. The subcategories we focus on in our experiment are test tubes and containers designed to their transport and handle them.

A wide variety of test tubes are produced for the collection of biological material. These sampling vessels are manufactured in various volumes from 0.75 ml up to the order of tens of millilitres. The basic biological samples taken for laboratory analysis include blood, urine, stool and various body fluids. This is the reason for producing a great variety of volumes, shapes and sizes of the test tubes. Test tubes

Fig. 5 Test tube D1038

with various additives are also produced for the blood collection that our experiment focused on. These substances are already present in test tubes and assist in the processes associated with a particular consumption. The test tubes also have different colour coding for tube caps. Each colour is determined for a specific type of test, sample, volume and additive. There is also a different colour designation for clinical biochemistry, haematology, blood bank, bacteriology and serology, and generally eight primary colours are used. This division and categorization can be found on the websites of companies engaged in the sale and distribution of such material [12].

For the experiment to be performed, a D 1038 tube of 16 × 75 mm with a volume of 8 ml was selected. The selected tube is not sterile because this property is not necessary for our experiment. A polystyrene tube was chosen for a wide range of applications in common practice. This tube is categorized as a tube for test, and it was exactly required. The tube is translucent with a white cap without labels, markings and other elements (Fig. 5).

We currently know several ways to label test tubes. The basic and the most widespread to date is writing on ordinary paper adhesive labels on clean tubes. In better cases, the test tubes already have a label attached and pre-printed with the information that the laboratories need. This information can include the name, date of sampling and type of sample taken. Some types of test tubes do not have labels attached, but their surface is specially adapted to write information directly on them. Another method is barcode marking. This marking works on the principle of sticking identical barcodes to the test tube and the request form, which is sent together with the material to the laboratory. However, these barcodes have their own specific rules for sticking to the samples. In the case of improper sticking, this sample may not be correctly loaded and identified. Some tubes have a label that indicates the blood level needed for collection [13].

2.1.7 Types of Transport Elements

Transport elements are an integral part of the collection of biological material. These elements are very useful for handling, transporting and working with the collection containers. The important functions of these components are to keep the test tubes in a vertical position and to ensure the necessary stability when handling them and the possibility of transparent and organized distribution of the collection containers as needed. There are a large number of types of transport elements on the market which are tailored to the specific needs of the consumer and are also divided into several categories. The first of them are test tube racks, the range of which is very wide [14]. The variety of racks consists of a combination of the number of holes and their diameter, height, colour and material from which they are made. Another very important thing is the kind of biological material for which the stand is designed. We will use two types of racks in our experiment. Specifically, it will be a rack with the capacity to store 36 tubes with a diameter of 16 mm and four tubes with a diameter of 26 mm. The second fixture will be a rack with the possibility of holding ten tubes with the same diameter of 16 mm.

The second big category for transport elements is transport or shipping boxes. These components are also produced in various combinations of colours, sizes and materials. The most important feature of these boxes is the ability to keep biological material cool. Three types of transport boxes are available on the market. The first type is the classic transport boxes without special features. The second is the so-called thermo boxes, which are commonly used and most widespread. The last subcategory is transport units with an active cooling system that keeps biological material at the necessary temperature level. For the purposes of our measurements, a classic transport box made of plastic material without thermo properties was chosen. Nowadays, producers are also focusing on compatibility between racks and shipping containers. The boxes have internal dimensions which, for example, exactly match the size of two tube racks or other combination with three, four or more devices [15]. An example of such compatibility and the box selected for the experiment can be seen in Fig. 6.

Subsequently, the paper deals with functional design of RFID technology aimed at the identification of blood samples. Other tasks that will be addressed in this section are related to placing RFID transponders on test tubes and configuring the RFID system. The article also includes readability measurements of various RFID tag layouts in racks or shipping boxes.

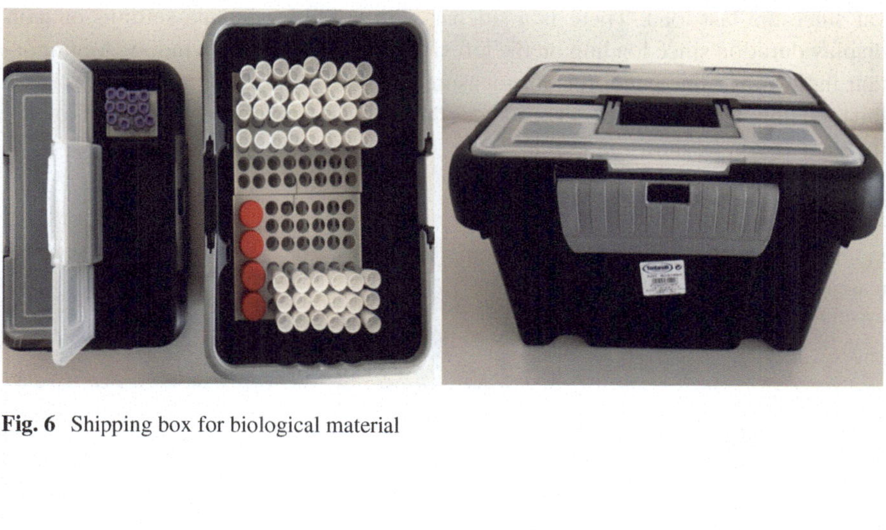

Fig. 6 Shipping box for biological material

2.1.8 Placing RFID Tags on Test Tubes

The Smart Label format has been selected for the RFID transponder application. This type is characterized by the fact that the tag is placed on a self-adhesive paper backing that is practically easy to apply anywhere. In our case, the Alien Squig tag is placed on an area of 60 × 30 mm. The site of attachment was selected at the top of the test tube, namely, 1 cm from the edge. In the transponder orientation, the size of the test tube and the transponder antenna had to be taken into account. Because the tag antenna needs to be free of distortion over its entire width, the only possible solution was to stick the Smart Label vertically oriented. All containers used for measurement were labelled in this way. We can see the test tube along with the tag location in Fig. 7.

2.1.9 Reading Using Alien Gateway

Static measurements were performed using Alien's RFID gateway software, whose capabilities and functions are described in a part of static reader. The choice of measurement application was dependent on the hardware used, which is also from Alien. Its capabilities have been utilized in measurement as this system can very well configure and adapt various parameters. In our case, a library of patient photos has been created that includes information such as first name, last name and date of birth. The information was then assigned to the specific EPC codes of the used transponders. This programme can also export the EPC code that will be processed and necessary for subsequent assignment to certain profiles in the patient database. The example of our configuration can be seen in Fig. 8 showing ten such profiles. Each window represents one of the patients who were created for the experiment, and their identities are fictitious. The bracket at the bottom of the window shows the number of transponder reads per second. At the top, we can see the unique EPC code of each tag. The bars to the right of the photo indicate in colour the time interval since the last load. There is a sidebar with possible antenna settings or icon display duration since loading on the left side of the application window. An important indicator in this panel for our measurements is at the bottom. This represents the total number of transponder read near the RFID antennas. The number of identified tags is ten, as shown in Fig. 8.

Fig. 7 Placement of RFID transponder on experimental test tube

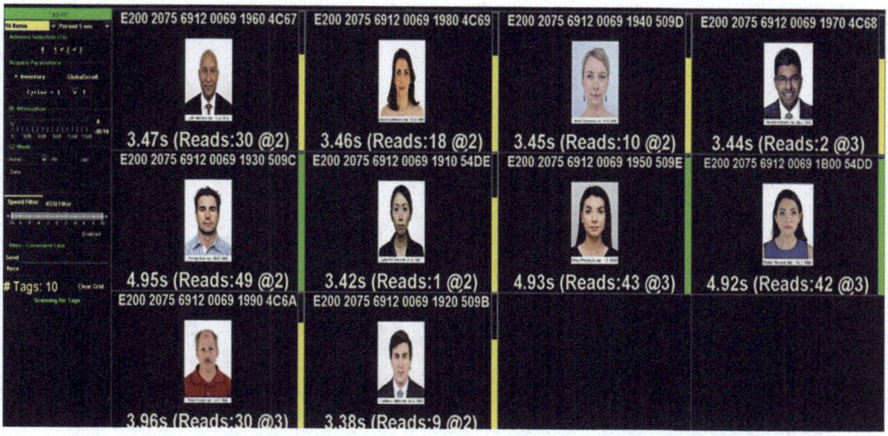

Fig. 8 Example of configuration for RFID gateway

3 Measurement Procedure

For the experiment, it was necessary to verify the reliability of RFID technology on empty and full test tubes. The reason was the concern about the system malfunction due to the presence of liquid. For this reason, measurements were performed not only with empty test tubes but also with test tubes containing the fluid that in our case demonstrated blood. Another object of the research was to determine the magnitude of the effect on the transponder readability of the various tube racks. Therefore, several ways of distributing samples into racks have been proposed.

Before starting, it was needed to carry out the necessary preparation in the form of gluing transponders to test tubes. Detailed information on Smart Label location, orientation and size can be found in the previous chapter. After this application, all ten test tubes were placed into the first of the racks used in the experiment. The first option was to arrange the test tubes side by side in one row. The test tubes thus placed and prepared for the first measurement can be seen in Fig. 9.

The next step was to launch the function of RFID gateway software to detect transponders within a range of antennas. After switching on the programme, unknown transponders were detected in the vicinity of the measurement, which had to be removed from the range. Undesirable transponders have been identified and eliminated from the measurement area. Subsequently, the track containing the tubes was continuously inserted in the vertical direction of the RFID gate space, the width of which in this case was 0.5 m. At the same time, it was necessary to determine the space in which the preparations could move during the measurement. A width of approximately 1.5 m and a height of approximately 0.5 m were chosen. How the tube racks were inserted into the RFID gateway and how much space was reserved for the racks is shown in Fig. 10.

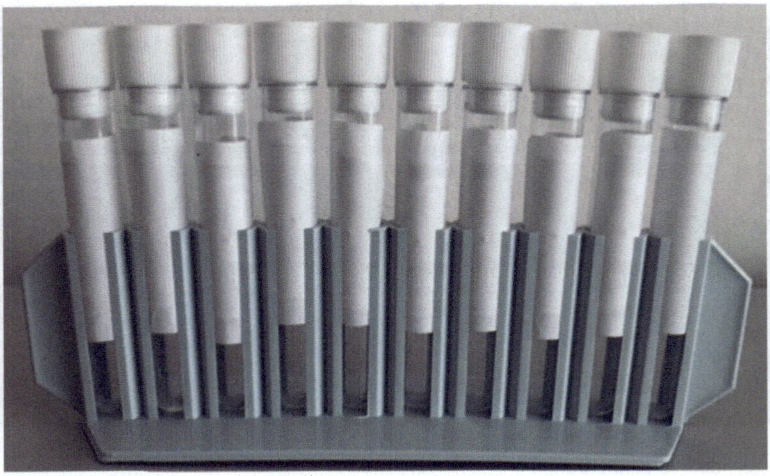

Fig. 9 The first arrangement of the test tubes in the rack

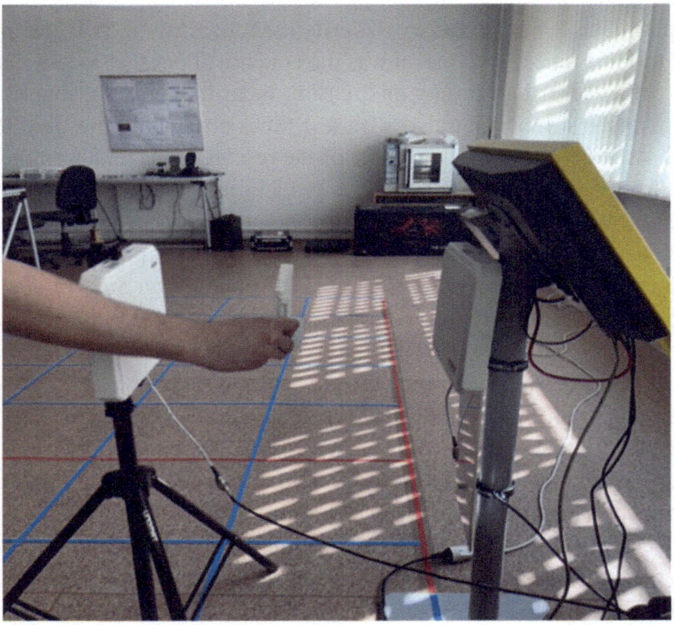

Fig. 10 Inserting the tube rack into the RFID gateway

By this movement of appliance between the antennas, ten patient profiles by the system were identified and displayed. These profiles represent the number of read EPC codes from the rack tubes. In the first measurement, in which the RFID gate was 0.5 m wide and the samples were arranged in a single row, we can say that the technology works reliably and without any problems. The number of transponder read indicates the number ten, which means 100% sample reading capability for such parameter configurations. The results of the first measurement can be seen in Fig. 11.

Another type of test was performed to obtain more detailed data to verify the ability to read transponders. It consisted in launching the function RFID gateway called Readometer. This feature has the ability to measure more accurately because of the introduction of the current, average, maximum and minimum readability of any of the transponder groups. In this particular measurement, we can read in Fig. 12 that the current readability is 74 reads per second (hereinafter referred to as read/s). The average value is 77 read/s. The minimum and maximum values, after subtracting from the scale, are approximately between 46 and 78 read/s.

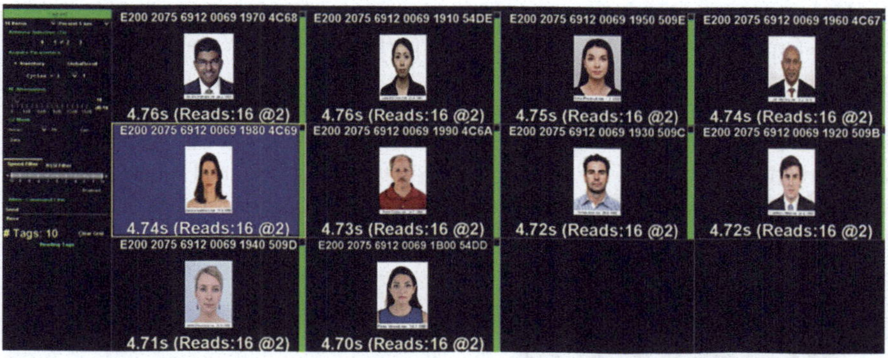

Fig. 11 Screenshot of gateway's first measurement

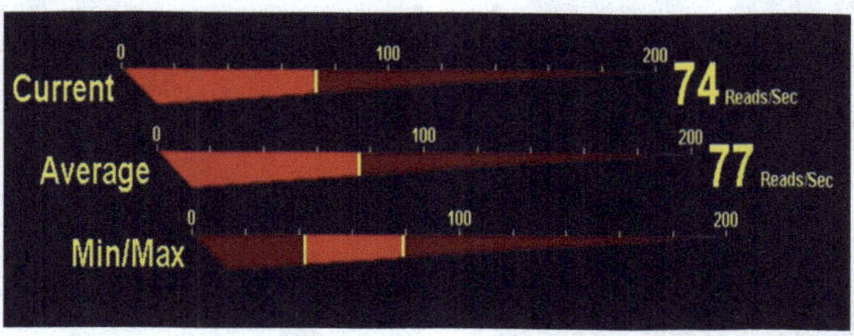

Fig. 12 The result of the Readometer function for the first measurement

4 Results

For the authenticity of the application in practice, an experiment was carried out in which the test tube rack was placed in a standardized transport box intended for biological material. Our goal was to verify loading under normal operating conditions, in which the sample preparations are stored in different boxes. In this case, the tube rack was placed in a box and then placed in the RFID gateway. An example of the placement of the tube rack and the measurement with the biological material box can be seen in Fig. 13.

It can be stated that RFID technology worked without any problems. Again, ten patient profiles could be seen on the computer screen without any complications. However, when the Readometer function was started, there was a certain decrease in transponder readability. This decrease is quite noticeable when looking at the product values outside the box. If the tube rack was placed in the box, the current reading value was 48 read/s, the average readability was 45 read/s and the minimum and maximum values were 22 and 74 read/s. The results can be seen in Fig. 14, wherein both screenshots are taken during measurement.

In these two tests, the characteristics were slightly different. We can observe reliable system functionality. Further tests were performed with the same antenna width of 0.5 m, but the layout of the test tubes was varied. The first two tests were conducted with ten samples arranged in one continuous row. For the second measurement, the tubes were distributed in two rows of five pieces. The last measurement was a version of the random placing of the tubes in a laboratory rack designed for handling and transporting samples. The last two described variants are shown in Fig. 15.

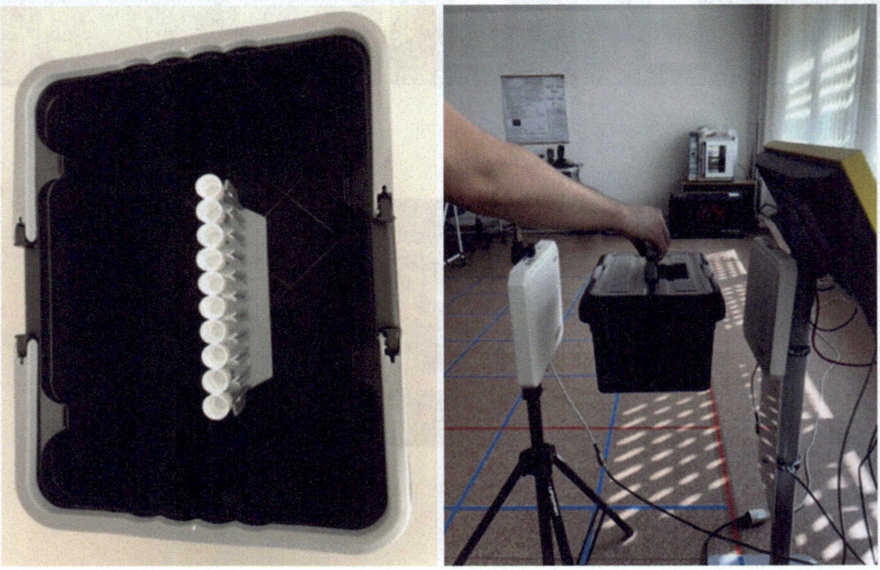

Fig. 13 Storage of the preparation with test tubes in the box and measurement with the box

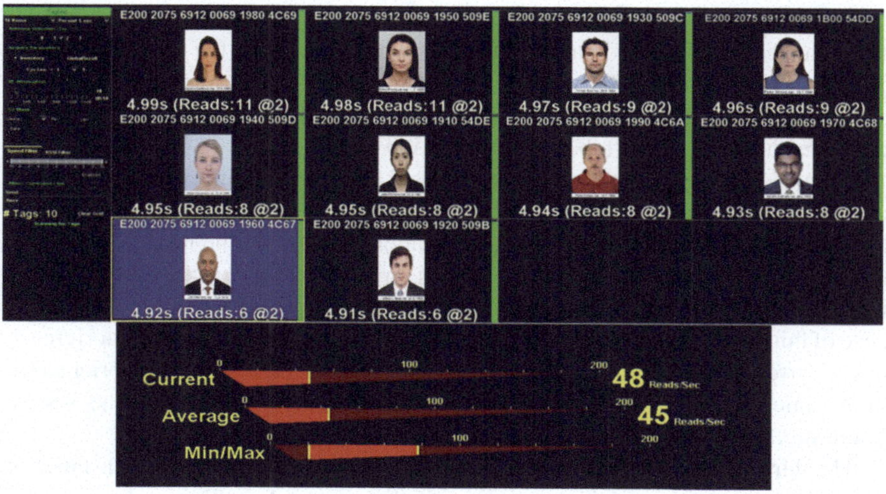

Fig. 14 Measured values when placing the product in the box

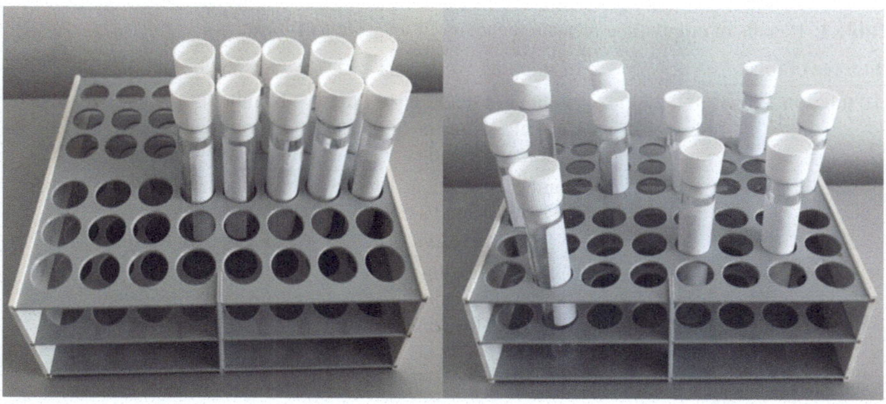

Fig. 15 Arrangement of tubes in rack for further measurements

Subsequent measurements again confirmed the read reliability of RFID gateway settings with 0.5 m antenna width when the tubes were free of liquid. The number of read samples was still equal to 10 in the case of measurements with the preparation stored in or outside the box. Specific results of the experiment can be seen in Table 1, where images from individual measurements are placed. The content of the table adequately shows the number of loaded transponders, which are displayed as profiles of patients requesting a blood assay. The first two tests are described in more detail in the previous paragraphs, where it was also possible to see profile information in more detail in Figs. 11 and 14.

Table 1 shows the results of measurements that were made when the parameter was changed. The changed parameter in this whole assay was the test tube content. The tubes were filled with a solution demonstrating blood measurement that could absorb the propagation of radio frequency waves. The filled tubes in a random distribution and a row of ten tubes can be seen in Fig. 16.

In Table 2, a slight decrease can be observed at the current and average transponder read values. The most important finding is that in random distribution an anomaly was detected, which was one missing patient profile. Of the total of ten tubes, the system only recognized nine tubes. This error could have occurred by decreasing the ability of the transponder to reflect the radio frequency wave due to the presence of fluid or by overlapping individual tag antennas where radio wave interference has occurred. Another reason for not loading a transponder could be the orientation of its antenna, which resulted in the RFID antenna not picking up the waves. Therefore, no information could be sent back.

The third series of measurements was associated with a change in the distance of the RFID gate antennas, which was set to a distance of 1 m. The goal was to see what impact this change would have on the ability to load samples. The tests were carried out in the same way as the two previous groups of measurements. The results

Table 1 Results of empty tube measurements in 0.5-m-wide RFID gate

Storage of tubes	In shipping box	Outside the shipping box
10 tubes in 1 row	Number of reading: 10 tubes Current: 48 read/s Average: 45 read/s	Number of reading: 10 tubes Current: 74 read/s Average: 77 read/s
5 tubes in 2 rows	Number of reading: 10 tubes Current: 48 read/s Average: 48 read/s	Number of reading: 10 tubes Current: 54 read/s Average: 57 read/s
Random distribution of tubes	Number of reading: 10 tubes Current: 53 read/s Average: 51 read/s	Number of reading: 10 tubes Current: 63 read/s Average: 60 read/s

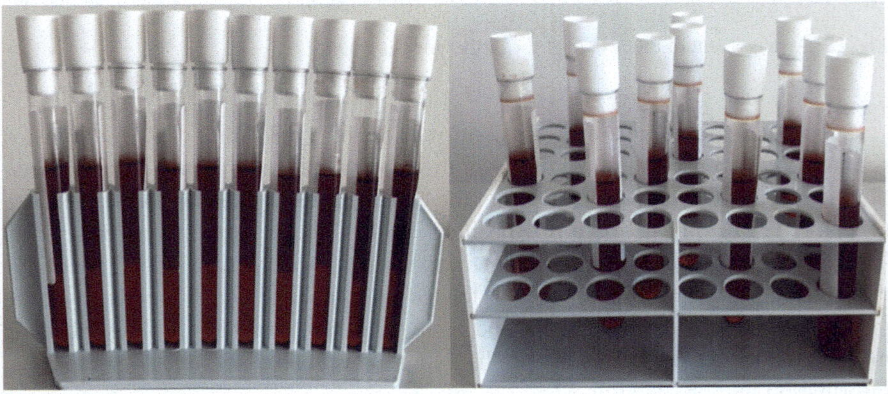

Fig. 16 Liquid-filled tubes placed in racks

Table 2 Results of filled tube measurements in 0.5-m-wide RFID gate

Storage of tubes	In shipping box	Outside the shipping box
10 tubes in 1 row	Number of reading: 10 tubes Current: 41 read/s Average: 43 read/s	Number of reading: 10 tubes Current: 71 read/s Average: 71 read/s
5 tubes in 2 rows	Number of reading: 10 tubes Current: 39 read/s Average: 41 read/s	Number of reading: 10 tubes Current: 54 read/s Average: 55 read/s
Random distribution of tubes	Number of reading: 10 tubes Current: 51 read/s Average: 49 read/s	Number of reading: 10 tubes Current: 59 read/s Average: 60 read/s

Table 3 Results of empty tube measurements in 1-m-wide RFID gate

Storage of tubes	In shipping box	Outside the shipping box
10 tubes in 1 row	Number of reading: 10 tubes Current: 45 read/s Average: 44 read/s	Number of reading: 10 tubes Current: 72 read/s Average: 70 read/s
5 tubes in 2 rows	Number of reading: 10 tubes Current: 50 read/s Average: 47 read/s	Number of reading: 10 tubes Current: 54 read/s Average: 56 read/s
Random distribution of tubes	Number of reading: 9 tubes Current: 51 read/s Average: 51 read/s	Number of reading: 9 tubes Current: 62 read/s Average: 61 read/s

of a third set of measurements are in Table 3, which are again represented by screenshots from individual measurements. Looking at the results, we can conclude that the displacement of the antenna has not yet had a major impact on the ability to read transponders from empty laboratory tubes. When examining the reader results in more detail, you may notice a slight decrease in the current and average readings compared to the 0.5 m width of the RFID gateway. In measurements, it was found that the reading radius of the antenna samples increased compared to the area contained in the RFID gate with an antenna spacing of 0.5 m. This phenomenon is caused by the propagation of radio frequency waves whose width is proportional to the distance from the source. However, it should be noted that the intensity of the waves decreases with this distance. This phenomenon most likely caused a slight decrease in read values. We can say that, despite minor changes, RFID is still very reliable. This is evidenced by the number 178, which represents the number of all loaded samples during the experiment out of a total of 180.

For the last six measurements, the samples were filled with liquid and measured using an RFID gate with 1 m antenna spacing. In this measurement, the greatest error in the reliability of sample loading was observed. The worst measured result during the whole experiment showed placing five tubes in two rows in the transport box. Table 4 shows that only six patient profiles were read in this test. Looking at the right column of application window, we can see loading only eight samples.

Table 4 Results of measurements of filled tubes in 1-m-wide RFID gate

Storage of tubes	In shipping box	Outside the shipping box
10 tubes in 1 row	Number of reading: 7 tubes Current: 32 read/s Average: 30 read/s	Number of reading: 10 tubes Current: 56 read/s Average: 55 read/s
5 tubes in 2 rows	Number of reading: 6 tubes Current: 29 read/s Average: 28 read/s	Number of reading: 10 tubes Current: 44 read/s Average: 45 read/s
Random distribution of tubes	Number of reading: 10 tubes Current: 50 read/s Average: 48 read/s	Number of reading: 10 tubes Current: 52 read/s Average: 54 read/s

Another complication occurred when ten samples were stored in a continuous row where only seven transponders have been read. The resulting defects in the experiment can be caused by the effects mentioned in the paragraphs above. Probably the most likely cause of these anomalies will be the absorption of radio frequency waves by the liquid in the tubes. Another reason may be the wave intensity that has been reduced as the distance between the RFID gateway antennas increases. It means that there is not enough energy supply to the transponder antennas to send the information back. The transport box for biological material also has a certain share in the loading disturbance. The last series of measurements was not entirely reliable at the level as in the previous cases. There have been a number of sample loading issues suggesting that this particular RFID system configuration is not well suited to normal operation, whether in a laboratory or medical environment.

5 Conclusion

After all measurements can be made, an overall assessment of RFID technology can be evaluated for application to the area of blood sample identification. Radio frequency identification is suitable for marking various subjects and objects in a large number of sectors. After the realization of all measurements, it is possible to make an overall assessment of RFID technology for application to the area of blood sample identification. The radio frequency identification is suitable for marking various subjects and objects in a large number of sectors. The application of this identification is also no different in the medical environment, for which this experiment was targeted. The evaluation should also take into account important factors such as the choice of hardware elements, configuration and system setup. The above results show the importance of choosing the right components of the RFID system. Reliable readings can be seen in the measurements but also results that indicate inappropriate hardware choices and settings. After summarizing all the results, we can state that RFID technology is compliant and applicable for the purposes of labelling and identifying blood samples. Overall, this technology is adequate for application.

Acknowledgement The paper was created within the EUREKA (E!) project 11158 U-health – Auto-ID technology and the Internet of things to enhance the quality of health services.

References

1. Husár, J., Knapčíková, L.: Automatic Identification Systems, 1st edn. TUKE, Košice (2017)
2. Rao, K.V.S., Nitkin, P.V., Lam, S.F.: Antenna design for UHF RFID tags: a review and a practical application. IEEE Trans. Ant. Propag. **53**(12), 3870–3876 (2005)
3. Finkenzeller, K.: RFID Handbook: Fundamentals and Applications in Contactless Smart Cards, Radio Frequency Identification and Near-Field Communication, 3rd edn. Wiley, Chichester. 478p (2010)
4. Kascak, P.: Analysis of Characteristics of the RFID Components in the Laboratory. Bachelor work, FVT TUKE. 72p (2016)
5. Sommerova, M.: Fundamentals of RFID Technology: Learning Material (2018). Available from http://rfid.vsb.cz/miranda2/export/sites-root/rfid/cs/okruhy/informace/RFID_pro_Logistickou_akademii.pdf
6. Balog, M., et al.: RFID system for inventory management in healthcare. In: ARTEP 2019 Automatizácia a riadenie v teórii a praxi, pp. 1–9 (2019)
7. Balog, M., et al.: RFID technology in the rail freight management. In: Management of Manufacturing Systems 2018, pp. 5–8 (2018)
8. GB Touch (2019). Available from www.gbtouch.cz/cz/specifikace
9. ALR-8800 ENTERPRISE RFID READER. Available from https://intihalkarlik.files.wordpress.com/2015/09/afk_msc_o9__colored__showimage-do_19303.pdf
10. ALR-8610-AC Circularly Polarized RFID Antenna. Available from https://intihalkarlik.files.wordpress.com/2015/09/afk_msc_o11__colored__alr-8610_highresprinting.pdf
11. Using the Alien Gateway Software (2018). Available from https://support.atlasrfidstore.com/article/41-using-the-alien-gateway-software
12. National Transfusion Service of the Slovak Republic (2019). Available from http://www.ntssr.sk/o-darovani-krvi/preco-darovat-krv
13. Study: Flexible Healthcare Packaging Materials Can Be Recycled Healthcare Packaging (2019). Available from https://www.healthcarepackaging.com/issues/sustainability/article/13295965/study-flexible-healthcare-packaging-materials-can-be-recycled
14. Blood Collection for Laboratory Examination (2019). Available from http://www.nsplm.sk/dokumenty/odber%20krvi.pdf
15. Principles of Biological Material Sampling, Transport and Storage (2019). Available from http://www.synlab.sk/fileadmin/standortseiten/synlab_sk/pdf/ostatne/zasady-odberu-vzoriek.pdf

Method to Comparison of Cities

Michal Jerabek, Jan Kubat, and Vit Fabera

1 Introduction

Most people have competitive spirit, and they like to compare their results with others. This statement is especially true for municipal representatives. They would like to compare their town/village with other towns/villages. The output of comparison can be a measure of representative's fruitfulness as politicians. Many municipalities attempt to be a "smart city." One of the "smart city" definitions is introduced in [1]: "A smart city is a designation given to a city that incorporates information and communication technologies (ICT) to enhance the quality and performance of urban services such as energy, transportation and utilities in order to reduce resource consumption, wastage and overall costs. The overarching aim of a smart city is to enhance the quality of living for its citizens through smart technology." Essential characteristic is a focus on "to enhance the quality of living for its citizens". Not only citizens should be the target group, but the aim should be "to enhance the quality of staying for all people in the city." It means such people are considered who come to cities to work and of course visitors/tourists. Main attention should be paid to permanent residents, but especially in large cities, everyday commuters who have jobs or who travel for healthcare services cannot be ignored. But even properly set up tourism has a benefit for the city.

M. Jerabek (✉) · V. Fabera
Czech Technical University in Prague, Prague, Czech Republic
e-mail: jerabem1@fd.cvut.cz; fabera@fd.cvut.cz

J. Kubat
AMBIS, Prague, Czech Republic
e-mail: jan.kubat@wo.cz

L. Knapcikova et al. (eds.), *New Approaches in Management of Smart Manufacturing Systems*, EAI/Springer Innovations in Communication and Computing, https://doi.org/10.1007/978-3-030-40176-4_7

If we want to make a comparison of cities' "smartness," we have to:

- Find suitable parameters to compare. Section 2 discusses parameter selection.
- Select an appropriate method which is able to process values of found parameters; the DEA method is introduced in Sect. 3.

Section 4 describes the software solution.

2 Parameters of the "Smart City"

The main criterion on how successfully the concept of "smart city" is implemented is "to enhance the quality of staying for all people in the city." The set of evaluative parameters can be composed of quantitative indicators (e.g., the count of parking places), qualitative indicators (e.g., citizen's safety and e-help), or a combination of both. Parameters can be also computed using arithmetical operations applied to other parameters, typically sum, rate, and mean. It is possible to speculate about complexes or components—for example, the Smart Street is complex, but Smart Bench is a component [2]. Some examples of parameters are found below.

Typical parameters from the transportation point of view are:

- The length of communications
- Communication management and maintenance costs
- Traffic infrastructure investment
- The count of crossing with public transport preferences
- Average time to spend in congestion
- Public transport line length
- The count of employees in public transport
- The count of vehicles in public transport
- The count of transported persons
- Public transport costs

Parameters describing demographic aspects are:

- Number of the population
- Population structure
- The count of people travelling to work

Parameters describing cadastral territory are:

- The area
- The count of buildings
- The count of housing units

3 DEA Method

Charnes et al. [3] firstly introduced the data envelopment analysis (DEA) model in 1978. DEA estimates relative efficiency within the set of given decision-making units (DMUs); DMU is town/village in our application. It is a nonparametric method that combines given sets of inputs and outputs and calculates relative efficiency of each DMU. Since 1978, DEA becomes an important tool for measuring efficiency within various sectors. During the time, the original model got various extensions. Cooper et al. [4] came in 1999 with the imprecise data envelopment analyses (IDEA) and initiated the approach of fuzzy DEA [5].

The original DEA model [3] was called CCR model by its authors. It looks for the combination of optimal weights of given inputs and outputs so that the sum of weighted outputs divided by the sum of weighted inputs is maximal. The CCR model can be written as maximization problem for each DMU; see Eq. (1) [6]:

$$\begin{aligned} &\max E = \frac{v^T y_0}{u^T x_0} \\ &\mathit{Subject\ to} \\ &v^T y_j \le u^T x_j \\ &u \ge 0,\ v \ge 0 \end{aligned} \tag{1}$$

where u, v are the vector variables. They represent weights of inputs and outputs of DMU_0. Variables y_0 and x_0 are output and input vectors of the DMU_0. y_j and x_j are vectors of outputs and inputs of the DMU_j. The objective function E represents efficiency. If the objective function equals 1, the DMU is efficient. The original CCR model is input oriented and supposes constant returns to scale (CRS). The CCR model can be written as linear programming problem by adding a condition to have a sum of weighted inputs equal to 1, and so it is enough to look for maximum of the nominator only as in Eq. (2):

$$\begin{aligned} &\max E = v^T y_0 \\ &\text{Subject to} \\ &u^T x_0 = 1 \\ &v^T y_j \le u^T x_j \\ &u \ge 0,\ v \ge 0 \end{aligned} \tag{2}$$

For the interpretation purposes, it is convenient to construct the dual problem to the CCR problem in linear form (Eq. 3):

$$
\begin{aligned}
&\min F = \Theta \\
&\text{Subject to} \\
&y^{T}\lambda \geq y_0 \\
&x^{T}\lambda \leq \Theta x_0 \\
&\lambda \geq 0
\end{aligned}
\tag{3}
$$

where λ is the vector of variable that represents DMU weights. Θ is the scalar value that shows the necessary reduction of inputs to become efficient for the analyzed DMU (DMU_0). The vector λ also shows the peer DMUs to the analyzed DMU. Peer DMUs are DMUs that show the inefficient DMU the path on how to reach efficiency.

To construct the DEA model with variable returns to scale (VRS) instead of constant returns to scale, we add the constraint (Eq. 4) to the linear problem (Eq. 3):

$$
e^{T}\lambda = 1 \tag{4}
$$

where e is the vector of ones. Such model is called BCC model (Barnes, Charnes, Cooper).

By restricting λ to be the vector of binary numbers rather than nonnegative real numbers as in Eq. (3), we obtain the non-convex rather than convex set of feasible solutions. Such problem is called FDH (free disposal hull) [7].

We will illustrate the methods on the example of ten towns. We will compare their efficiency in the public expenses on smart road transport (Table 1).

The three mentioned methods of efficiency calculation are depicted in Fig. 1. It shows efficiency calculation example of one input (roads per employee) and one output (roads per citizen). The lines represent production possibility frontier. The DMUs on the line are efficient, while DMUs under the line are not. The constant

Table 1 Inputs and outputs

Town	Unit costs of maintenance (CZK/km)	Unit costs of service (CZK/km)	Unit costs of SSZ crossing (CZK/unit)	Roads per employee (km/employee)	SSZ crossings per crossing	Road length per citizen (km/citizen)	Parking places per citizen (unit/citizen)
A	737,563	448,802	527	2.67	0.42	0.0016	0.0250
B	764,358	418,945	519	1.31	0.59	0.0018	0.0428
C	180,937	249,126	185	2.90	0.56	0.0026	0.0328
D	418,650	4811	208	1.05	0.50	0.0028	0.0381
E	450,404	323,060	1241	1.60	0.52	0.0022	0.0338
F	221,471	108,701	909	1.86	0.17	0.0045	0.0188
G	367,781	313,417	2803	1.47	0.11	0.0029	0.0338
H	313,247	7016	2281	3.63	0.17	0.0031	0.0341
I	662,547	269,054	1689	2.44	0.12	0.0021	0.0317
J	430,861	91,407	16,064	1.67	0.08	0.0035	0.0257

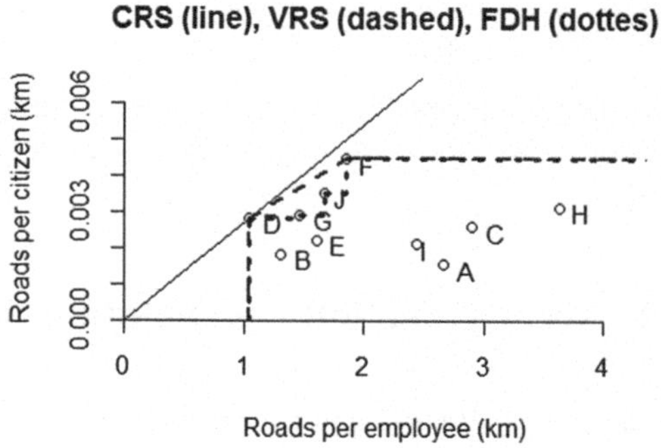

Fig. 1 Efficiency in input-oriented CRS (constant returns to scale), VRS (variable returns to scale), and FDH (free disposal hull) models for one input (roads per employee) and one output (roads per citizen) variables

Table 2 Efficiency of DEA model under VRS, CRS, and FDH options

Efficiency					Values of input variables to reach efficiency in VRS DEA model			
Town	CRS	VRS	FDH	Peer towns for inefficient towns in VRS model	Unit costs of maintenance (CZK/km)	Unit costs of service (CZK/km)	Unit costs of SSZ crossing (CZK/unit)	Roads per employee (km/employee)
A	0.42	0.50	0.57	C, D, F	372,289	43,918	266.1	1.35
B	0.95	1.00*	1.00*		764,358#	418,945#	518.6#	1.31#
C	1.00*	1.00*	1.00*		180,937#	249,126#	185.0#	2.90#
D	1.00*	1.00*	1.00*		418,650#	4811#	208.4#	1.05#
E	0.86	0.89	1.00*	B, C, D	401,736	86,711	229.9	1.43
F	1.00*	1.00*	1.00*		221,471#	108,701#	908.7#	1.86#
G	0.94	0.96	1.00*	C, D, F	353,955	52,252	340.8	1.41
H	1.00*	1.00*	1.00*		313,247#	7016#	2281.3#	3.63#
I	0.48	0.54	0.63	D, F	354,894	38,402	434.8	1.31
J	0.83	0.83	1.00*	D, F	335,277	48,738	504.5	1.39

returns to scale (CRS) showed only one efficient DMU (D); variable returns to scale (VRS), two efficient DMUs (D and F); and free disposal hull (FDH), four efficient DMUs (D, F, G, J).

If one constructs the DEA models for the all four input variables and three output variables as set in Table 1, one gets the results as in Table 2. The efficient towns have efficiency equal to 1 and are depicted with the mark $*$.

As one may see, there are four efficient towns in the CRS efficiency DEA model, five efficient towns in the VRS DEA model, and eight efficient towns in the FDH model. All efficient towns in CRS DEA were also efficient in VRS DEA and FDH models, and all efficient towns in VRS DEA model are also efficient in the FDH model. For the inefficient town, we may obtain the so-called peer towns. By product of result vector (matrix) λ and matrix of inputs x from Eq. (3), we can calculate matrix of input variable values to reach efficiency x^e:

$$x^{e} = \lambda^{T} x \tag{5}$$

For efficient DMUs, $x^e = x$. The values of VRS DEA problems are presented in Table 2. The already efficient towns have their x^e values marked with # in Table 2.

4 WebDEAr: Software Solution

The pilot software application WebDEAr is created to analyze municipality efficiency within project "Application of Nonparametric Methods (DEA, FDH) to Analyze and to Compare the Efficiency of Municipalities." This application should serve as a service application to municipalities to support their decision-making. Outputs of software should contribute to improve local politics. The application will be used by governing bodies, especially by the Ministry of the Interior of the Czech Republic that is application guarantee.

The application is designed as three-tier architecture (Fig. 2) having presentation, application, and database layers. The access to the application will be realized by web pages; import and export will be possible in Excel format.

The access to the actual version is authenticated. But some information is accessible without authentication. All operations and data about clients (IP address, browser, etc.) will be logged to be able to find clients in the case of troubles and improve the application.

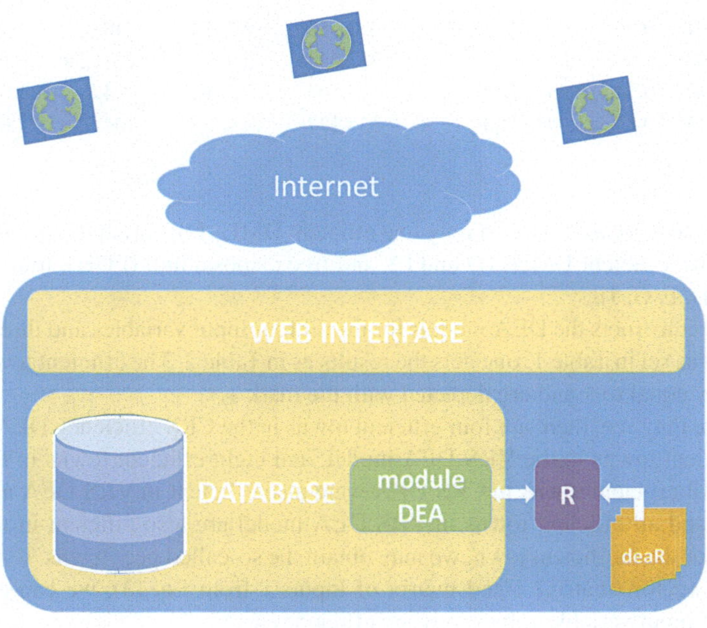

Fig. 2 The architecture of the WebDEAr application

The application will display as static data about towns/villages as dynamically calculated data. User will be able to change stored static data during session temporarily, and the efficiency will be recalculated.

Changes will be discarded after logout; persistent data will be only guaranteed one. Figure 2 shows the architecture of the WebDEAr application. The core is a database system to store data and DEA module programmed in PL/SQL. The module uses R programming language to calculate effectivity by the DEA method. The free package deaR will be applied which was created by Vicente Coll-Serrano, Rafael Benítez, and Vicente J. Bolós at the University of Valencia [8]. Current version 1.0 works with the following DEA models:

- Conventional DEA models
- Basic (radial) models (envelopment and multiplier forms)
- Directional distance function model
- (Weighted) additive model
- Superefficiency additive model
- Radial superefficiency model
- (Weighted) non-radial model
- Preference structure model
- (Weighted) slack-based model
- (Weighted) superefficiency slack-based model
- Cross-efficiency (crs12 and vrs13)
- Bootstrapping (Simar and Wilson algorithm)
- FDH model

Productivity

- Malmquist index

Fuzzy DEA models

- Kao and Liu model 14 [5]
- Possibilistic model
- Guo and Tanaka model
- Fuzzy cross-efficiency

The relational scheme of the database was designed based on requirement analysis. The scheme contains 23 relations (tables), and this is a kernel of data structures to store information (Fig. 3).

The main table is MUNICIPALITY containing basic information about towns/villages; the actual size is 6249 records. Categories of municipalities are stored in the CATEGORY table, and the category for each municipality is assigned by the relationship table CATEGORY_MUNICIPALITY.

Very important information is statistical and financial data. The list of statistical data is stored in the table LIST_STATISTICAL_DATA which contains reference to the table UNIT. The relationship of statistical data to municipalities is stored in the table MUNICIPALITY_STATISTICAL_DATA. Financial information is divided into PARAGRAPHS and ITEMS; their domains are saved in the

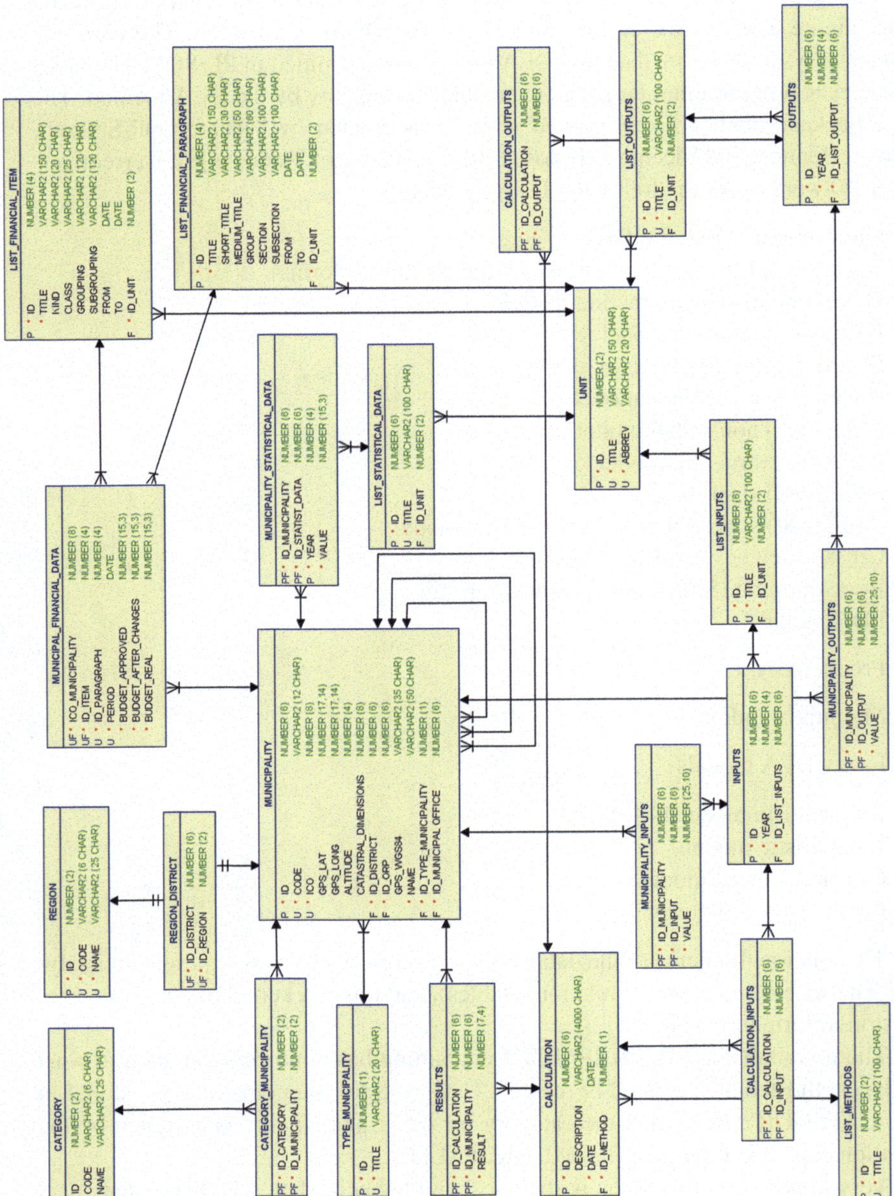

Fig. 3 Relational scheme of the database

LIST_FINANCIAL_PARAGRAPH and LIST_FINANCIAL_ITEM tables. Both tables have bindings to the UNIT table. Financial data about towns/villages are stored together with period in the table MUNICIPAL_FINANCIAL_DATA where three types of information are saved—budget approved, budget after changes, and real budget.

The table MUNICIPAL_FINANCIAL_DATA contains 5,263,845 records related to 524 records in the table LIST_FINANCIAL_PARAGRAPH and 527 records in the table LIST_FINANCIAL_ITEM. Financial data is collected from 2012 to 2017.

The DEA method to analyze the efficiency of municipalities uses several inputs and outputs from statistical and financial tables. Tables INPUTS and OUTPUTS serve as repository to store statistical and financial data which were used to compute efficiency. Tables reference to LIST_INPUTS and LIST_OUTPUTS. Used data (with regard to possibly later change) are saved in the tables MUNICIPALITY_INPUTS and MUNICIPALITY_OUTPUTS.

The database schema is taken into account to keep all results if calculations are repeated. Basic information about each calculation are stored in the table CALCULATION including reference to the used methods (the table LIST_METHODS). Results of the each calculation for the selected town/village are stored in the table RESULTS.

The originator of data was Czech Statistical Office (CZSO). There was an inconsistency in financial data among lists of municipalities, paragraphs, and items.

5 View Information About Municipalities

The list of municipalities contains basic information about each of them—name, identification number, municipalities with extended powers, district, region, and type of municipality. The municipalities can be sorted and filtered according to these criteria (Fig. 4). It is also possible to display detailed information about a

Fig. 4 List of municipalities

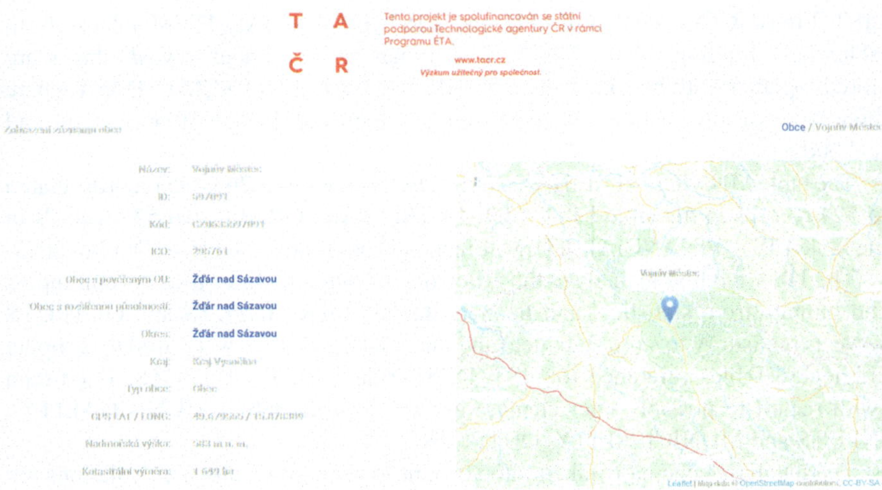

Fig. 5 Detail of municipality

specific municipality and thus GPS coordinates, altitude, cadastral area, and location on the map (Fig. 5).

In addition to these data, the database also includes financial data from 2012 to 2017 for:

- Housing, municipal services, and territorial development
- Social security benefits and benefits
- Culture, churches, and the media
- Physical education and leisure activities
- Healthcare
- Environmental protection
- Other public services and activities
- Defense
- Legal protection
- Industry, construction, trade, and services
- Water management
- Communication
- Fire protection and integrated rescue system
- Revenues
- Other research and development activities
- Security and public order
- Agriculture, forestry, and fisheries
- Transportation

Fig. 6 List of calculations

- General economic affairs and other economic functions
- Other service activities for population
- Employment policy
- Civilian preparedness for crisis situations
- State power, state administration, local government, and political parties
- Education and school services
- Social services and joint activities in social security and employment policy
- Financial operations
- Other activities

The database also include statistics - currently only Population for 2017.

The municipality details also include a list of all calculations with the results in which the municipality data were used (Fig. 6).

6 Calculation of Municipalities' Evaluation

The application contains several calculations of the evaluation of the municipalities, when the functionality of the application was verified. The first calculation was always made for each category of municipalities and for the input was used the financial item "Current expenditures" and for the outputs "Population" and "Cadastral Area". All data were used from 2017 and were calculated using the following R script:

```
#### paths
pathwd<- "C:/Users/JD31424/Documents/rkuos/dea/TACR2018/Pilot"
pathlib <- "L:/PROGR/R/r-3.1.2/library"
pathsrc <- "u:/Dokumenty/PROJECTS/R/src20131128"

#### path to installed packages
.libPaths(pathlib)
.libPaths()

library(Benchmarking)
library(readxl)
library(writexl)

setwd(pathwd)
DEAdata_o <- read_xlsx('Data_pro_pilot_R.xlsx')

## variable for group of municipation
skupiny<-aggregate(DEAdata[,'ico'],              list(Kategorie_
obce=DEAdata$Kategorie_Obce), length)

##
DEAdata <- subset(DEAdata_o,Bezne_vydaje_celkem > 0)
Vysl_dopln  <-  subset(DEAdata_o,  select  =  c(Kategorie_Obce,id,
Nazev_obce,Bezne_vydaje_celkem,Pocet_obyvatel,katastralni_vymera)
 ,Bezne_vydaje_celkem <= 0)

dopln_eff <- c(rep(1,length(Vysl_dopln)))
Vysl_dopln <- cbind(Vysl_dopln,dopln_eff)
colnames(Vysl_dopln)[7]<-'e$eff'

## calculation of DEA for every group of municipation
if(exists('Vysl')) rm(Vysl)
for(skup in skupiny$Kategorie_obce) {
# print(skup)
 skup_f  <-  subset(DEAdata,  select  =  c(Kategorie_Obce,id,  Nazev_
obce,     Bezne_vydaje_celkem,Pocet_obyvatel,katastralni_vymera),
Kategorie_Obce == skup)
   e   <-    dea(as.matrix(skup_f[,'Bezne_vydaje_celkem']),    as.
matrix(skup_f[,c('Pocet_obyvatel','katastralni_vymera')]),
RTS="vrs", ORIENTATION="in")
 if(exists('Vysl')) {
 Vysl <- rbind(Vysl,cbind(skup_f,e$eff))
 } else {
```

```
 Vysl <- cbind(skup_f,e$eff)
 }
}
Vysl<-rbind(Vysl,Vysl_dopln)
# Export do xls
write_xlsx(Vysl,'DEA_VRS_inputOriented.xlsx')
```

Another calculation (Fig. 7) was focused on efficiency in the field of librarianship for the year 2017, when they were used for the calculation:

Inputs
- The current expenditure on library activities
- The recalculated number of staff in the library
- The number of books at the beginning of the period

Outputs
- The number of registered readers
- The number of book loans
- The number of visitors to events organized by the library
- The increase in collection of books

The application allows to compare selected existing municipalities with another anonymous municipality, when it is necessary to enter the required data. This allows different data to be tested for impacts in the overall assessment. User can choose which municipalities to compare anonymously with filters, including distances by selecting a circle or rectangle on the map (Fig. 8).

It is necessary to enter the necessary parameters for the anonymous municipality (Fig. 9).

After the calculation, the results (Fig. 10) are shown, including the data used in the calculation. Since this function is used to verify the situation of "WHAT-IF," the results are not saved.

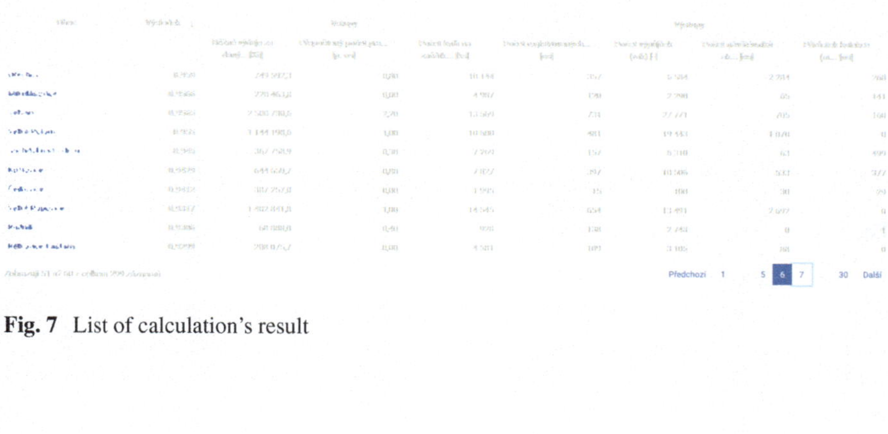

Fig. 7 List of calculation's result

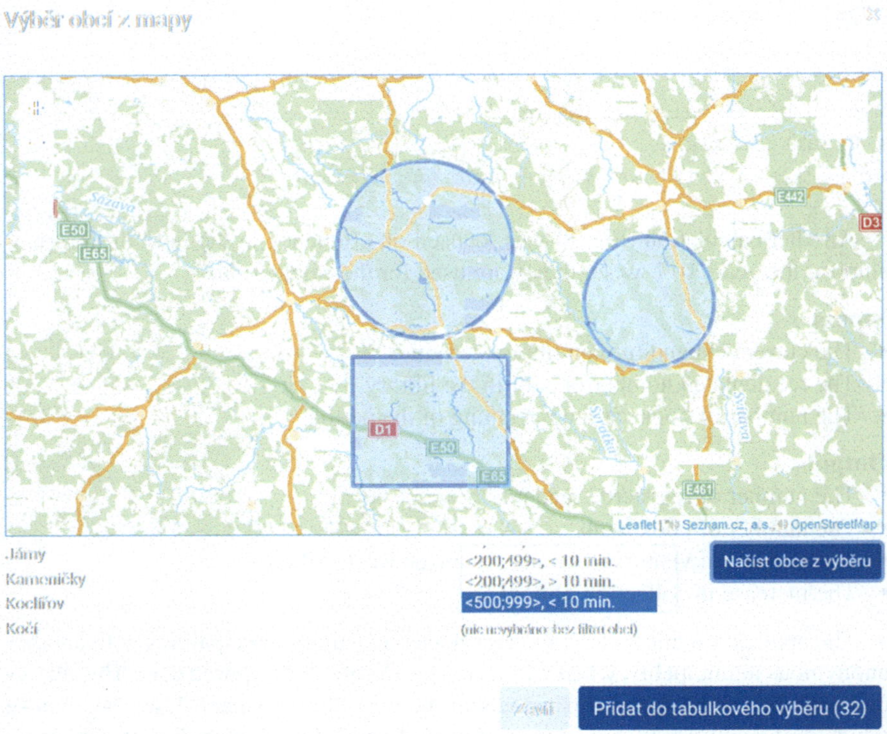

Fig. 8 Selection of municipation in map

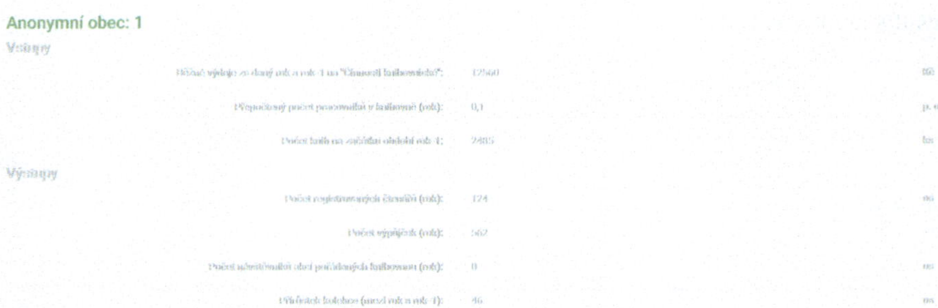

Fig. 9 Entering of the required data

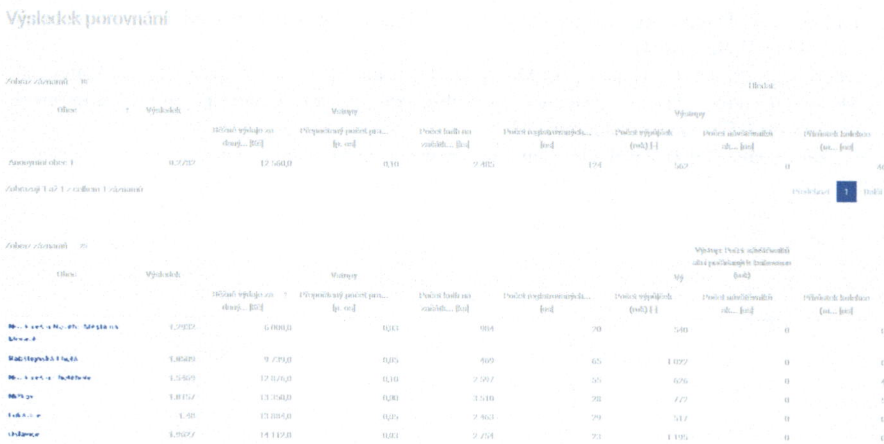

Fig. 10 Calculation result

7 Conclusion

The analyses of the efficiency of municipalities and their comparison are important tools for municipal politics and governments (ministry, region governments). The application of the nonparametric DEA method (data envelopment analysis) to analyze is one of the ways on how to transfer theory to the praxis and to contribute to the improvement of decision-making and control in public sphere. The paper concludes achieved results of project which solves application of DEA method in praxis. The designed software helps to put the method into practice.

Acknowledgment This work is solved within the project "Application of Nonparametric Methods (DEA) to Analyze and to Compare the Efficiency of Municipalities" that is supported by TAČR (Technology Agency of the Czech Republic), program Eta (project code TL01000463).

References

1. Definition - What Does Smart City mean? (2019). Available from https://www.techopedia.com/definition/31494/smart-city
2. Přibyl, P., Přibyl, O.: Definition of a Smart Street as Smart City's building element. In: Smart Cities Symposium Prague (2015). https://doi.org/10.1109/SCSP.2015.7181575. ISBN: 978-1-4673-6727-1
3. Charnes, A., Cooper, W.W., Rhodes, E.: Measuring the efficiency of decision making units. Eur. J. Operat. Res. **2**, 429–444 (1978)
4. Cooper, W.W., Park, K.S., Yu, G.: IDEA and AR-IDEA: models for dealing with imprecise data in DEA. Manag. Sci. **45**, 597–607 (1999)
5. Kao, C., Liu, S.T.: Fuzzy efficiency measures in data envelopment analysis. Fuzzy Set. Syst. **119**, 149–160 (2000)

6. Jablonský, J., Dlouhý, M.: Modely hodnocení efektivnosti produkčních jednotek. Professional Publishing, Praha (2004)
7. Deprins, D., Simar, L., Tulkens, H.: Measuring labor inefficiency in post offices. In: Marchand, M., et al. (eds.) The Performance of Public Enterprises: Concepts and Measurement, pp. 243–267. Elsevier, Amsterdam (1984)
8. Package deaR (2019). Available from https://www.uv.es/vcoll/deaR.html

The Proposal to Increase the Efficiency of the Production Process by a Software Module Tecnomatix Plant Simulation

Marek Kliment, Peter Trebuňa, and Miriam Pekarčíková

1 Introduction

The paper is focused on the production of components for the automotive industry. It is the production and assembly of components for car door handles. It describes the assembly process of the individual parts needed for a functional assembly unit used daily by people to open and close the car. In everyday life, we do not realize the difficult process and all the components needed to perform perfectly normal activity at the car door opening [1]. Every business, whether engaged in the production of products or the provision of services, aims to make everything work as efficiently as possible and to achieve the maximum with the least amount of resources. In improving each process, it is essential to use the tools that are available and achieve the highest degree of efficiency for both the business and the end customer, who often does not even realize what's behind the everyday consumption. Such means are also simulation software and various management methods [2]. We also use these resources to solve problems and shortcomings in the production of our contribution. The life cycle of the product and the means consumed affecting the final price need to be addressed. These are products and parts of products and human as well as material resources coming into production processes [3].

M. Kliment (✉) · P. Trebuňa · M. Pekarčíková
Faculty of Mechanical Engineering, Institute of Management, Industrial and Digital Engineering, Technical University in Kosice, Košice, Slovakia
e-mail: marek.klliment@tuke.sk; peter.trebuna@tuke.sk; miriam.pekarcikova@tuke.sk

L. Knapcikova et al. (eds.), *New Approaches in Management of Smart Manufacturing Systems*, EAI/Springer Innovations in Communication and Computing, https://doi.org/10.1007/978-3-030-40176-4_8

2 Production Process Characteristics

The manufacturing process we've been dealing with is the car door handle holder. It's part of the component which is located between the outer and inner car door handles (Fig. 1). Using the handle holder, the handles are mounted and held together. Handle holders are a very important component in cars because we can use them to move handles. The production line is capable of producing 1650 pieces for change. The time for making one handle holder is 110 s. Every 13 s, one finished product must come out of the line. On each workplace is operator. There are ten operators on this production line, and their number may change as needed. Each component has its specification, e.g., dimensions, appearance, material, and the like.

This assembly process consists of assembly and testing operations [4]. Assembly operations consist of eight locations where components are connected. The result is a finished product. For assembly operations on individual posts, it is necessary to distinguish:

- Left driver and passenger
- Right driver and passenger

The components of the entire handle holder mechanism are shown in Fig. 2.

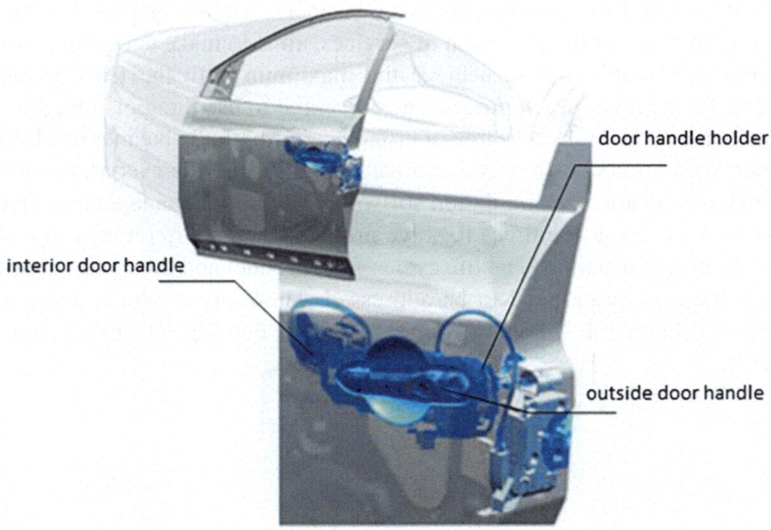

Fig. 1 A representation of a car door handle holder

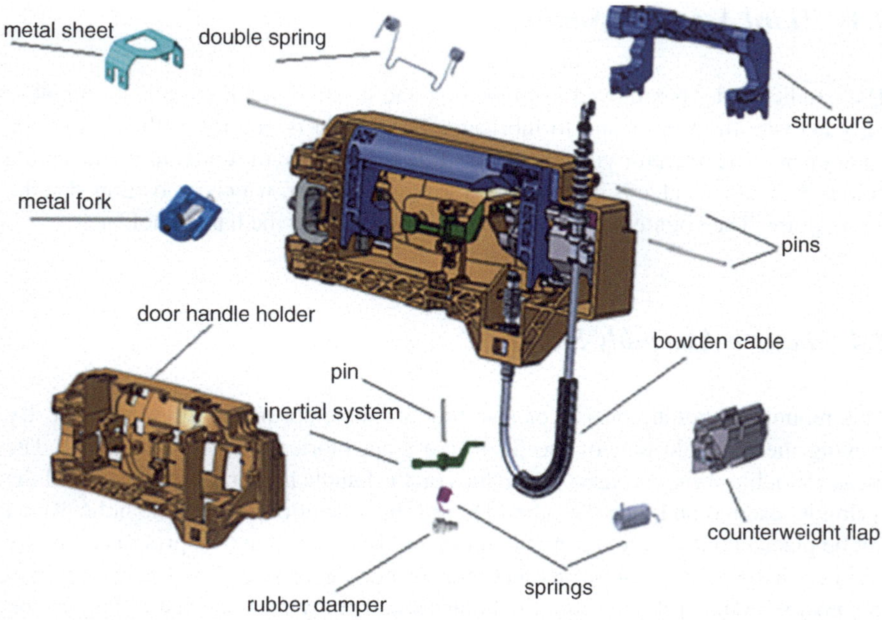

Fig. 2 Handle holder decomposition

2.1 *First Assembly Station*

On this position, rubber damper (used to reduce noise) is inserted in the counterweight flap. It is inserted using the specified preparation. Care must be taken to ensure that it is correctly installed, as it is inserted from above and incorrectly placed at the bottom counterweight flap. Finally, the flap is inserted into the lubrication seat. Pushing to the stop, lubrication component takes place [5]. After this operation, it is necessary to check that the lubricant is present in the spring groove, the cable, and the cable elastic. The lubrication seat is one designed for the left and right side of the component.

2.2 *Second Assembly Station*

In this position, a slider (a plastic component used to move the inertia latch during each door opening) is inserted into the counterweight flap, and a spring is then inserted. The assembly is inserted into the seat on the machine, and the start button is pressed. The spring coiling process is in progress. Before the process, the presence of a rubber damper and a slider is checked automatically, and after the process, the spring is correctly and sufficiently rotated on the counterweight flap.

2.3 Third Assembly Station

The handle holder is removed from the box and inserted into the lubrication seat to fully activate the sensor and to lubricate. The saddle is one for both sides of the component. The operator visually checks for the presence of lubricant in the handle holder. A Bowden cable is inserted into the handle holder, which is positioned at the fixing point. The counterweight flap is then inserted into the handle holder.

2.4 Fourth Assembly Station

This mounting station consists of inserting a double spring into the lubricant. By moving the movable part of the jig to the stop, lubrication will take place. The greased double spring is based on the pins on the handle holder. First, one side of the spring is inserted and then the other. During this operation, the handle holder must not be placed on the table during the process. Make sure that the spring is correctly oriented in the holder. The spring feet must rest on the bracket. The last operation in this post is to insert the pin into the holder and thereby fix it against falling out on further handling of the holder. Pin must be properly oriented. The finished components from the fourth assembly station are placed on a small rack, and the worker working on the other assembly station takes the components from it for further operation. There may be a maximum of six components on this intermediate storage.

2.5 Fifth Assembly Operation

On this position, a lubricated kinematic slider (a plastic component used to prevent the handle holder from noise) is inserted into the structure. The slider for the right side and the slider for the left side of the door differ not only in shape but also in color. The prepared structure is inserted into the lubrication seat as far as it will go, to automatically start lubrication. It is necessary to check that the lubricant is present on the component. The structure is inserted into the handle holder. The last operation is inserting one big pin. At this station, two lubrication seats for both sides of the component are designed so that it is not possible to lubricate the structure without a slider.

2.6 Sixth Assembly Stage

Another operation is to insert a second pin into the door handle holder. The bracket prepared in this way is inserted into the replaceable seat; the Bowden cable is inserted into the holder. The start button is pressed to initiate the pin pushing process into the handle holder. It is a process in which the machine pushes the pin into the door handle holder so that the components that are inside it are not released. Another component is being prepared during this process with holder. After its completion, a rubber damper is inserted into the manufactured handle holder, which is still in the machine seat. The cable is released from the bracket followed by the presence and position of the pins. Pin must be pushed; other operations cannot be performed in the case of pin not pinned.

2.7 Seventh Assembly Stage

The penultimate assembly operation is the insertion of the inertial system into the lubrication seat. The inertial system is pushed into the saddle as far as it will go to start the lubrication process. The saddle is one for both sides of the component where the component side is detected. Subsequently, the spring is manually inserted into the inertial system. The spring foot must be fixed in an inertial system. The inertial system is then inserted into the door handle holder. The spring foot must be in the holder groove to ensure proper functioning of the component. The middle part of the inertial system fits into the grooves on the bracket.

2.8 Eighth Assembly Operation

The final assembly operation is to push the pin into an inertial system. The holder is inserted into a replaceable seat, and a small pin is inserted into the groove. The start button is pressed to start the push-in process. The sheet metal component is then inserted into the handle holder. With both fingers, press the plate onto the handle holder until it stops so that all three latches engage.

2.9 Test Stations

The manufacturing process does not end with a complete folding of the car door handle holder. Every assembled product must be tested at test stations.

The test stations consist of two identical stations where the final product testing takes place. By testing, they can detect errors and deficiencies on the product. This prevents complaints that are undesirable for every business.

A fork is inserted into the handle holder before testing the product. Subsequently, the holder is inserted into the tester, where it must be properly fixed. Here are also removable palettes by page. The last operation is to press the start button to start the test cycle. Another product is prepared during testing. If the test has been completed without complications, the printer will print a small label with the date, time, and product designation on the barcode that will be affixed to the designated location. The label is then scanned with a small scanner. After loading the piece, the product is placed in the box.

2.10 Production Process Simulation

The entire production process was reflected in the simulation using the Tecnomatix Plant Simulation software module. Simulation (Fig. 3) also shows the schematic ground plan of the entire production process and the method of disposition of workplaces [6].

On the production layout, we can see 12 jobs at each job position. In the chapters above, it has been mentioned that ten workers serve the line. In the simulation, it is possible to see the pathway after which operators come to the line but also two shorter paths directly at the line (Fig. 4). After this, two operators move between jobs to lines and hold two posts in the production process [7, 8].

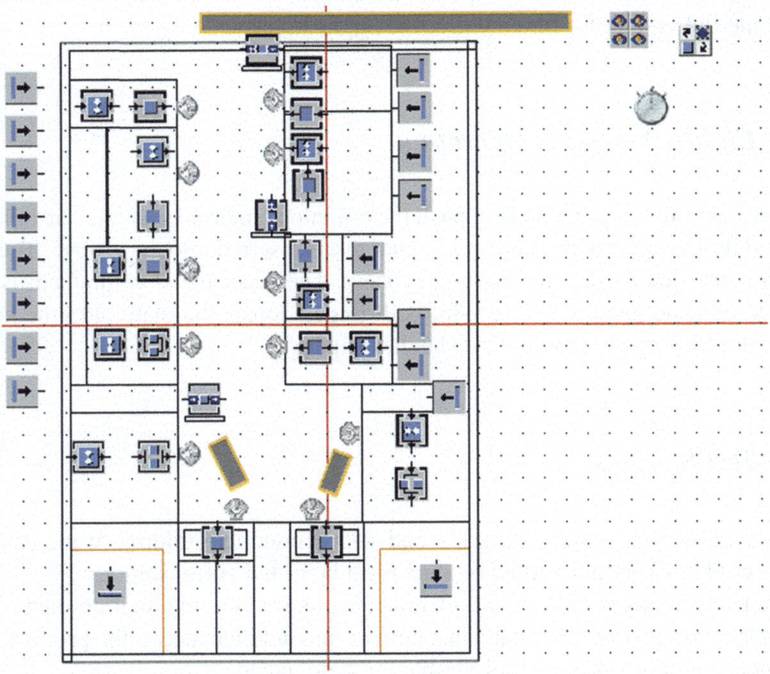

Fig. 3 Simulation of manufacturing process of car door handle holder

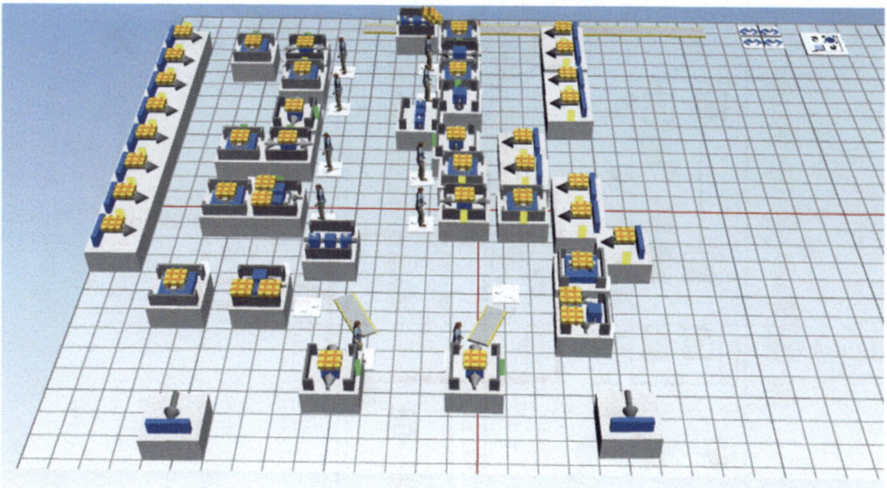

Fig. 4 Simulation in 3D showing and in the moving

3 Proposals to Increase the Efficiency of the Production Process

Based on the analysis and research of the production process, we try to find the most effective options that can help to increase efficiency in the future.

Among the suggestions for improvement, the following are included:

- Machine modification
- Shortening production time

3.1 Machine Modification

The first proposed variant of the production process is to modify the machine. The term machine modification means that we will modify certain parameters of the machine and we add changes to the manufacturing process that the machine will perform. Modifying the machine would be done by the company that supplied the particular machine. Our goal is to save time on assembly work. We can save time if we merge operations performed on the first and second production station [9]. This proposal can eliminate one operator. The suggestion is that inserting the rubber damper into the counterweight flap will be automatically inserted into a machine. The operator on the second production post lubricates the counterweight flap, inserts a spring and slider, and then inserts it into the machine. The rubber dampers will be automatically inserted into the machine during fixation of the counterweight flap in

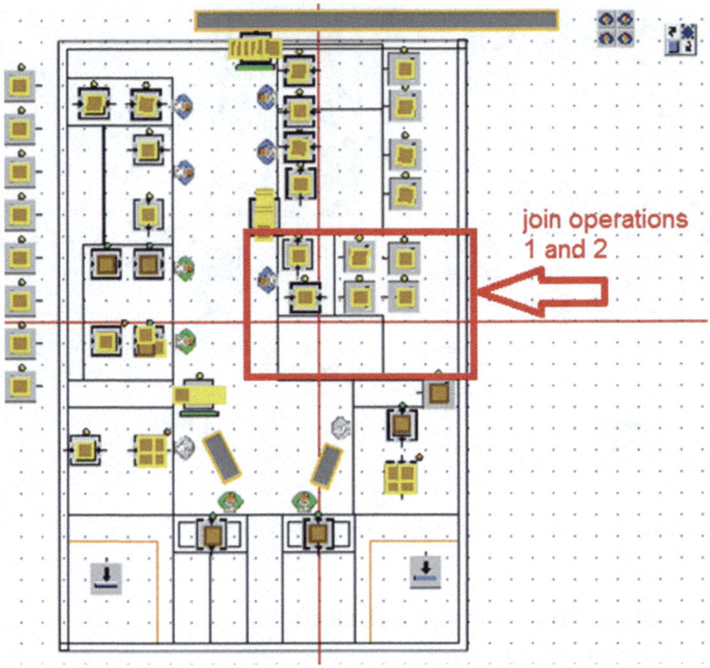

Fig. 5 Proposal of machine modification in simulation

the machine. Detecting the presence of a rubber damper will be the same as now. The machine's time and dimensions will be unchanged.

When inserting with the help of a machine, there is a risk of misalignment of the rubber muffler. The operator stayed in this undemanding operation for over 10 s.

While observing the manufacturing process and the time measurements, we found that the time required to perform the operation on the first station is 10.6 s and the time on the second station is 15.3 s. By combining these two processes with modification, we can do two operations in 12.3 s. We save 13.6 s.

The proposal can be seen reflected in the simulation in Fig. 5.

3.2 Shortening Production Operations Using the Hoshin Method

The second option to increase the efficiency of the production process is to shorten production operations. In this variant, we used the Hoshin method and the MTM method [10]. We made an analysis on the production line, which consisted of measuring the times. On the production line, we made 20 measurements during one change. We measured from component grip to product completion [5].

Table 1 Production operation times

Nr. of meas.	Pos. nr. 1	Pos. nr. 2	Pos. nr. 3	Pos. nr. 4	Pos. nr. 5	Pos. nr. 6	Pos. nr. 7	Pos. nr. 8	Pos. nr. 9
1	7.4	14.0	13.8	12.5	16.2	13.0	12.1	22.5	26.9
2	7.8	14.1	12.3	12.3	16.1	13.3	13.5	23.6	24.7
3	8.6	13.3	11.6	12.5	16.3	14.3	13.5	23.7	24.3
4	8.7	13.1	12.4	13.5	16.8	13.7	13.2	22.6	24.7
5	8.8	13.1	12.1	11.8	16.2	13.9	12.5	24.4	25.1
6	8.1	13.1	12.9	12.2	15.3	14.9	12.6	23.5	23.1
7	8.7	13.2	14.1	12.0	15.7	12.8	12.1	22.1	24.4
8	8.9	12.7	13.2	11.3	16.3	13.3	12.1	22.1	24.8
9	8.8	13.7	14.1	10.8	15.7	13.5	11.4	22.4	24.7
10	8.8	14.1	13.1	11.9	16.4	13.7	11.8	23.0	24.3
11	8.1	13.5	12.1	11.3	15.2	13.6	13.1	22.1	24.9
12	8.5	13.3	13.1	12.1	17.4	13.8	11.6	22.7	24.4
13	7.8	13.1	13.3	12.1	16.3	13.9	12.4	22.5	26.8
14	8.1	13.3	13.2	12.7	15.9	12.6	12.1	22.6	25.4
15	8.9	13.1	12.5	12.5	16.9	12.6	11.4	22.1	25.1
16	8.1	13.7	12.1	13.2	17.4	12.3	11.7	23.0	25.8
17	8.0	13.6	12.5	12.6	17.7	12.9	11.4	22.7	24.6
18	7.8	12.9	12.5	11.5	15.9	13.1	11.2	22.5	24.4
19	8.8	13.8	12.6	11.2	17.3	13.5	12.5	23.1	25.3
20	8.1	12.8	13.7	11.1	17.1	15.4	11.9	22.5	24.6
Overall	166.8	267.4	257.1	241.1	328.0	269.9	244.0	455.5	498.0
Average	8.3	13.4	12.9	12.1	16.4	13.5	12.2	22.8	24.9
Minimum	7.4	12.7	11.6	10.8	15.2	12.3	11.2	22.1	23.1
In %	13%	5%	11%	11%	8%	10%	9%	3%	8%

Table 1 shows measurements on a production line. Position 9 includes one test station and scan labels on the product. The times of individual operations are not the same, but there are small deviations.

In the graph (Fig. 6), we can see how the individual operations on the posts move unevenly. Our goal is to break down individual operations into smaller operations. We choose the optimum time for operations to be at about the same time. Small deviations are permitted [11].

We use the Hoshin method for this improvement design. We did a lot of measurements. Our goal was to combine some operations. We increase efficiency, and we go in depth. We split the operations into smaller operations using the MTM method (Table 2) [10].

The operations performed on the production line were divided into small and more detailed operations. In this improvement proposal, we also envisaged variant no. 1 thus with the modification of the machine [2].

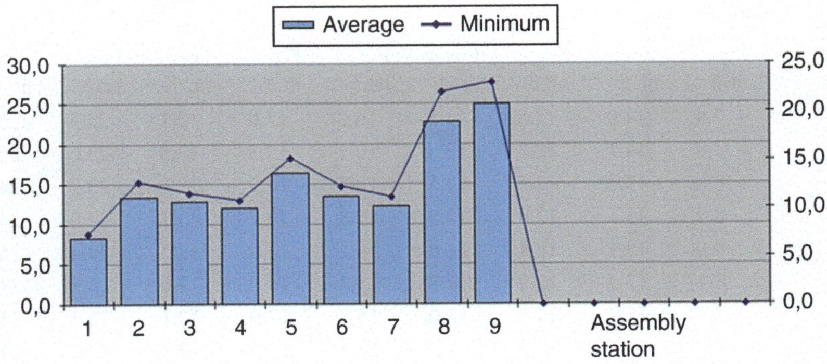

Fig. 6 Production operation times

Table 2 Separation of operations

Position	Operation type	Time in second
Pos. Nr. 1	Lubricating the counterweight flap	12.3
	Slider insertion	
	Inserting the spring	
	Inserting rubber damper	
Pos. Nr. 2	Lubrication of the handle holder	4.1
	Cable fixation	3.2
	Inserting the lever	5.6
Pos. Nr. 3	Spring lubrication	4.1
	Inserting the spring	4.8
	Inserting the pin into the lever	3.2
Pos. Nr. 4	Insert slider into internal structure and erase	7
	Insert internal structure and pin	9.4
Pos. Nr. 5	Inserting a pin	3.6
	Pressing pins	9.9
Pos. Nr. 6	Lubrication of inertial system	3.7
	Inserting a spring into an inertial system	3.9
	Inserting an inertial system into the handle holder	4.6
Pos. Nr. 7	Pushing pins into an inertial system	18.7
	Inserting sheet metal	4.1
Pos. Nr. 8	Insert the metal fork into the handle holder	4.0
	Testing the handle holder	13.6
Pos. Nr. 9	Car door handle holder pack	7.3

In this improvement proposal, we have tried to align the operations so that there are no time jumps on each production post. Our goal was to improve production times and reduce the number of operators on the production line.

In Table 3, we have aligned production operations using the Hoshin and MTM methods. Of the nine stations, we arrived at seven stations [10, 11].

Table 3 Alignment of manufacturing operations

Position	Operation type	Time in second	
Pos. Nr. 1	Lubricating the counterweight flap	12.3	16.4
	Slider insertion		
	Inserting the spring		
	Inserting rubber damper		
	Lubrication of the handle holder	4.1	
Pos. Nr 2	Cable fixation	3.2	17.7
	Inserting the lever	5.6	
	Spring lubrication	4.1	
	Inserting the spring	4.8	
Pos. Nr. 3	Inserting the pin into the lever	3.2	19.6
	Insert slider into internal structure and erase	7	
	Insert internal structure and pin	9.4	
Pos. Nr. 4	Inserting a pin	3.6	17.2
	Pressing pins	9.9	
	Lubrication of inertial system	3.7	
Pos. Nr. 5	Inserting a spring into an inertial system	3.9	16.6
	Inserting an inertial system into the handle holder	4.6	
	Inserting sheet metal	4.1	
	Insert the metal fork into the handle holder	4.0	
Pos. Nr. 6	Pushing pins into an inertial system	18.7	18.7
Pos. Nr. 7	Testing the handle holder	13.6	20.9
	Car door handle holder pack	7.3	

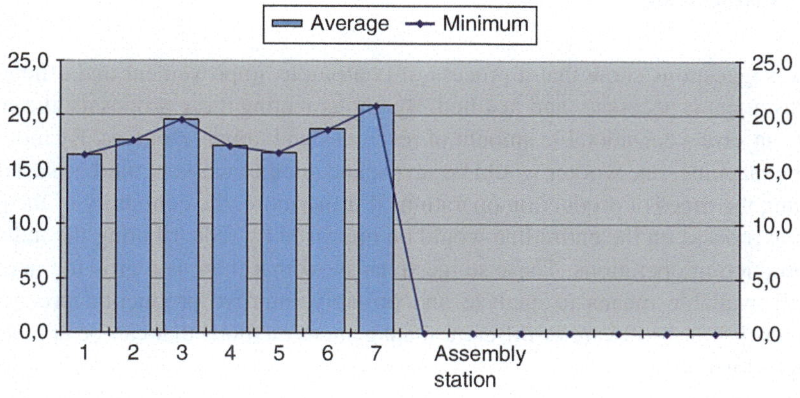

Fig. 7 Graph of manufacturing operations after alignment

In the graph (Fig. 7), we can see the movement of times on individual posts after the times in production operations were aligned. It is obvious that the movement of times is much more uniform than in the first case.

The proposal in alignment operations we have spilled over into the simulation. As we can see on the line, there is another operator missing from this proposal, and it is visible from Fig. 7 that the flow of the production process has also improved (Fig. 8).

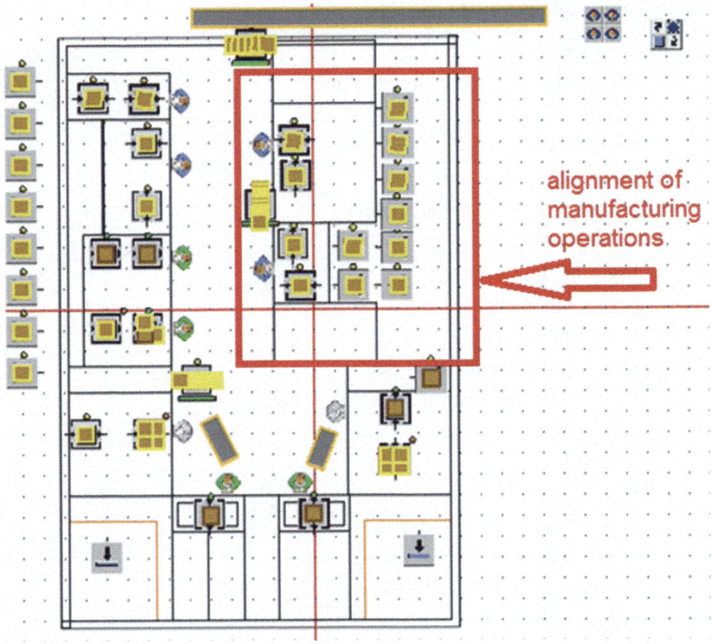

Fig. 8 Simulation design of the reconciliation operation

4 Conclusion

These suggestions show that a process of continuous improvement and efficiency improvement is necessary and justified. By implementing these proposals, the company can save a considerable amount of resources on human resources. By modifying the machine, one worker would be saved, and another worker could be saved by aligning the times of production operations. Furthermore, the continuity of the production process on the entire line would be improved by redistributing the individual production operations. These suggestions show that there is a growing need to use all available means to analyze and possibly improve production processes, whether they are software or different management methods that can be applied in a given situation.

Acknowledgment This article was created by implementation of the grant project VEGA 1/0708/16 "Development of a new research methods for simulation, assessment, evaluation, and quantification of advanced methods of production" and APVV-17-0258 "Digital engineering elements application in innovation and optimization of production flows."

References

1. Poór, P., Šimon, M., Karková, M.: CMMS as an effective solution for company maintenance costs reduction. In: Production Management and Engineering Sciences, pp. 241–246. CRC Press, Boca Raton, FL (2015)
2. Wicher, P., Staš, D., Karkula, M., Lenort, R., Besta, P.: A computer simulation-based analysis of supply chains resilience in industrial environment. Metalurgija. **54**(4), 703 (2015)
3. Straka, M., Lenort, R., Khouri, S., Feliks, J.: Design of large-scale logistics systems using computer simulation hierarchic structure. Int. J. Simulat. Model. **17**(1), 105–118 (2018)
4. Edl, M., Lerher, T., Rosi, B.: Energy efficiency model for the mini-load automated storage and retrieval systems. Int. J. Adv. Manuf. Technol. **70**, 97–115 (2014). ISSN: 0268-3768
5. Rosova, A., Malindzakova, M.: Material flow - starting point for recovery of inputs in the production company. In: International Multidisciplinary Scientific GeoConference Surveying Geology and Mining Ecology Management, SGEM 3(5), 14th International Multidisciplinary Scientific Geoconference and EXPO, SGEM Albena, Bulgaria (2014)
6. Klos, S., Ptalas-Maliszewska, J.: Badania kluczowych problemów w zarządzaniu utrzymaniem ruchu. In: Knosala, R. (ed.) Innowacje w zarządzaniu i inżynierii produkcji, vol. 2, pp. 678–687. Oficyna Wydaw. Polskiego Towarzystwa Zarządzania Produkcją, Opole (2014). ISBN: 978-83-930399-6-8
7. Krajčovič, M., Grznár, P., Fusko, M., Skokan, R.: Intelligent logistics for intelligent production systems. Sci. Lett. Univ. Žilina. **20**(4), 16–23 (2018). ISSN 1335-4205
8. Knapčíková, L., Balog, M., Ruggiero, A., Husár, J.: Composites manufacturing: a new approaches to simulation. In: Smart Technology Trends in Industrial and Business Management, pp. 319–327. Springer International Publishing AG, Cham (2019). ISBN 978-3-319-76997-4
9. Bangsow, S.: Tecnomatix Plant Simulation. Modeling and Programming by Means of Examples. Springer, New York, NY (2015). ISBN 978-3-319-19502-5
10. Kesterson, R.K.: The Basic of Hoshin Kanri. CRC Press, Boca Raton, FL (2014). 165. ISBN 978-1-4822-1869-5
11. Saniuk, S., Samolejova, A., Saniuk, A., Lenort, R.: Benefits and barriers of participation in production networks in a metallurgical cluster - research results. Metalurgija. **54**(3), 567–570 (2015)

Different Approaches in Building Digitization Through the Use of 3D Laser Scanning

Katarína Krajníková, Jana Smetanková, Peter Mésároš, and Marcel Behún

1 Introduction

There exist a lot of methods and techniques for building recording. Naturally it starts at simplest manual recording up to using modern technologies that provide the possibility of virtual presentations of parts of the building and its whole or of old historic buildings.

Building Information Modelling (BIM) is now rapidly penetrating the Architecture, Construction, Engineering and Facilities Management sectors. However, it is notable that this increase in use has been predominantly for new builds, as well as on the design stage of those new constructions [1]. The terrestrial laser scanners (TLS) produce a 3D point cloud by measuring the range and two angles (azimuth and elevation) to points on the surfaces of objects in a scene. Establishing metrological traceability of TLS measurements is a challenge. These data are useful for many applications, such as 3D landscape modelling for visualization in planning, simulation for environmental management and navigation for robot or vehicle. In many instances, the surveyed scenes are usually very large. Therefore, in order to generate an entire 3D scene, the multiple scanning point cloud data from different positions are needed to be transformed into a unified coordinate reference system [2, 3].

Wang and Cho in [4] state that to recognize objects and extract useful object information from point clouds, object recognition techniques have frequently been

K. Krajníková · J. Smetanková · P. Mésároš
Faculty of Civil Engineering, Technical University of Košice, Košice, Slovakia
e-mail: katarina.krajnikova@tuke.sk; jana.smetankova@tuke.sk; peter.mesaros@tuke.sk

M. Behún (✉)
Institute of Earth Resources, Faculty of Mining, Ecology, Process Control and Geotechnologies, Technical University of Košice, Košice, Slovakia
e-mail: marcel.behun@tuke.sk

L. Knapcikova et al. (eds.), *New Approaches in Management of Smart Manufacturing Systems*, EAI/Springer Innovations in Communication and Computing, https://doi.org/10.1007/978-3-030-40176-4_9

applied in recent studies in the AEC/FM domain. One method is the method of extracting geometric information items of a building from point cloud data collected from a laser scanner, for building management. Advanced techniques and improvements in devices have resulted in textured point cloud data becoming available.

Terrestrial laser scanning (TLS) can capture point clouds of the damaged state of building facades rapidly, remotely and accurately [5]. The Scan-to-BIM process has been the subject of numerous studies. Tang et al. in [6] surveyed techniques developed in civil engineering and computer science that can be utilized to automate the process of creating as-built BIMs. The authors subdivided the overall process into three core operations: geometric modelling, object recognition and object relationship modelling. They surveyed the state-of-the-art methods for each operation and discussed their potential application to automated as-built BIM creation. They also outlined the main methods used by these algorithms for representing knowledge about shape, identity and relationships. Bosché et al. [7] explored the opportunity for frequent, detailed and semantically rich assessment of as-built status of MEP works in construction projects by joining 3D laser scanning and 3D/4D BIM models, and Sacks et al. [8] proposed an approach for semantic enrichment that supports object recognition and object relationship modelling.

In the case when an as-designed BIM model of the works to be tracked is available, the prior information contained in the model can be leveraged to not only detect and recognize the objects contained in the model but also identify them [9]. Bosché and Haas [10, 11] proposed such an approach and refer to it as "Scan-vs-BIM". In the Scan-vs-BIM approach, 3D laser-scanned point clouds are first aligned in the coordinate system of the 3D model. This can be done using site benchmarks or using automated or semi-automated registration techniques [12].

Unmanned aerial vehicles (UAVs) have become popular platforms for collecting various types of geospatial data for various mapping, monitoring and modelling applications. With the advancement of imaging and computing technologies, a vast variety of photogrammetric, computer vision and, nowadays, end-to-end learning workflows are introduced to produce three-dimensional (3D) information in a form of digital surface and terrain models, textured meshes, rectified mosaics, CAD models, etc. These 3D products might be used in applications where accuracy and precision play a vital role, e.g. structural health monitoring. Therefore, extensive tests against data with relevant characteristics and reliable ground truth are required to assess and ensure the performance of 3D modelling workflows [13].

UAVs can be divided into devices with moving on surfaces producing buoyancy, multi-copters or planes. In practice, the use and selection of the composition are mainly based on the purpose. While multi-copters have a small range, they have positive features such as hovering, take-off and landing from one point or the ability to scan the territory at low speeds, which is already used by LIDAR scanning of terrain and objects. In [14], aircrafts, on the other hand, have a long-range performance and a higher speed with the ability to scan a larger area.

Recent years have also seen extensive national and international civil applications of unmanned aerial vehicles in aerial photography, geological surveys, high-voltage line inspection, oilfield pipeline inspection, highway administration, forest fire patrol, toxic gas inspection, drug detection, emergency rescue, ambulance, agriculture and so on [15]. Unmanned aerial vehicles have also been applied in the physical inspection of civil engineering. This is not surprising as using GPS-enabled UAVs for aerial surveying is very cost-effective in comparison to hiring an aircraft with photogrammetry equipment. With GPS-equipped drones, digital cameras and powerful computers, surveys have an accuracy right down to 1 cm [16].

The manual measurements are more time-consuming with respect to the automated methods. Therefore, the automated methods are preferred instead of the manual ones. In that the resolution of the laser is higher, single structures like stones could be seen in the laser elevation model. On the other hand, in both data sets, blunders occur. For the UAV, the main difficulties were on walls and structures with vertical surfaces, which were not covered in different image strips. The laser could not acquire points in the holes (mining entrances); therefore the UAVs fit better in these areas. Using the advantages of both data sets, in future work, we can combine them to a more precise elevation model [17]. The laser scanner can be used for interpretation of the architecture of single walls and rooms due to the high point density.

Gruszczyński et al. in [18] state clearly that a considerable advantage is shown in UAV photogrammetry over TLS. Applying TLS to the surveying of civil engineering sites and the high accuracy of point positioning offered by this method suggest it has an advantage over UAV in terms of accuracy. The spatial distribution of UAV data was more even and thus more favourable than that of TLS data. UAV photogrammetry proved to be more accurate than TLS when applied to terrain mapping. Furthermore, with equipment costs and surveying time taken into account, UAV turned out to be the more efficient method in this case [18].

The selection of the method that will be used in 3D modelling study depends on the scale and shape of the object and also the applicability of the method [19]. Laser scanners are high-cost instruments which collect high-density spatial data in a short time from the measurement area. On the other hand, image-based 3D measurement uses images to create 3D point cloud data. The image matching and the creation of the point cloud can be done automatically. The measurement to all details with particular spaces by the UAV require more time and labour. For an application of photogrammetry by UAV, at least two photographs of the same building are required as well as the exact distance between two points visible in at least one photograph [20].

2 Used Technical Equipment and Methodology

The Faculty of Civil Engineering has a simple layout in the L-shape with a built-up area of 3150 m^2, a height of 20 m and a roof angle of 24°. The main aim of digital measurement was to measure the exterior of the building completely; it means the facade with the roof. This building can be measured by three methods, and each of them has its own specifics, benefits but also limitations and disadvantages.

Digital surveying of the Faculty of Civil Engineering was divided into three methods:

(a) **Terrestrial laser scanning** (facade of the building) + **Aerial photogrammetry** (roof of the building)
(b) **Terrestrial laser scanning** (facade and roof of the building)
(c) **Aerial photogrammetry** (facade and roof of the building)

2.1 Using the Terrestrial Laser Scanning Combined with Aerial Photogrammetry

For the façade of the building, the laser scanner FARO Focus 3D X130 was used. On Fig. 1, the real positions of the laser scanner (*blue points*), start and end of scanning (*red points*), the direction of scanning around the entire building and the real photos of build of the faculty can be seen.

1. The overall time of measurement was approximately 12 h.
2. Created 48 scanner positions.

For creating aerial photographs from the building's roof, the DJI MAVICO PRO drone classified as a quadcopter (with four rotors) was used. Furthermore, a remote control, a DJI GO application and a free online application DroneDeploy were used.

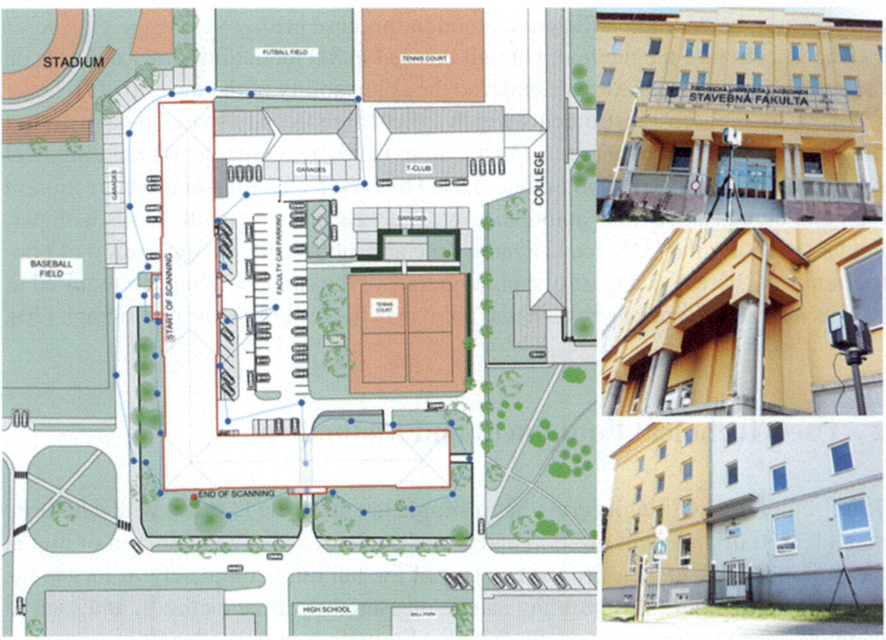

Fig. 1 Technological process of terrestrial laser scanning/focused on the facade from the building (source: authors)

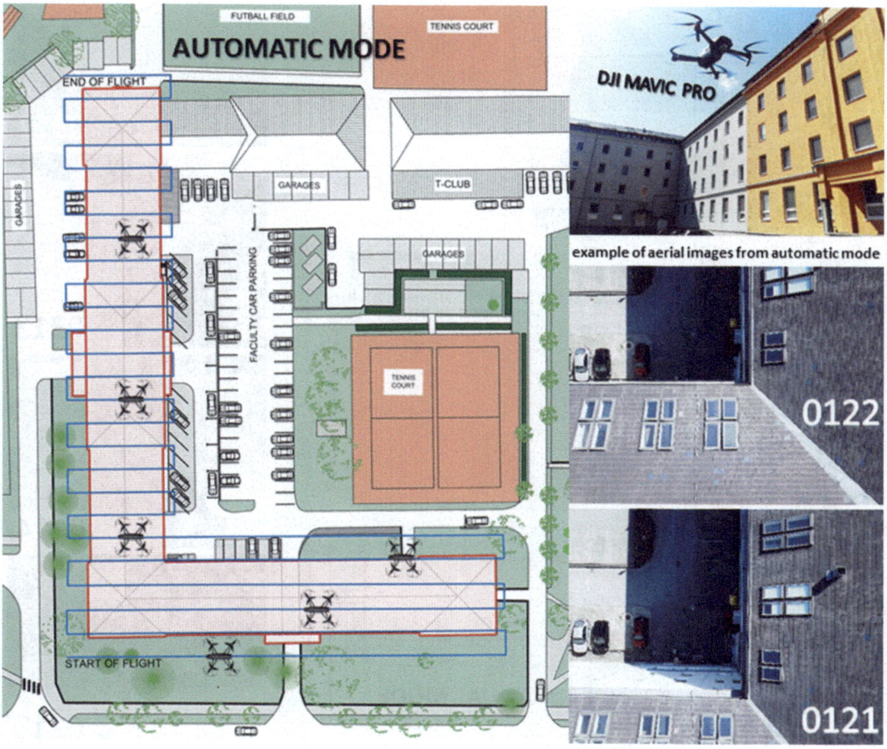

Fig. 2 Aerial photogrammetry/automatic mode (source: authors)

At first, a drone flight plan had to be set. This plan may be either automatic (Fig. 2) or manual (Fig. 3). Both of these modes were tested.

1. Automatic mode: 195 aerial images for 5 min
2. Manual mode: 107 aerial images for 20 min (remote control manual flight requires experienced pilot)

The phase of data processing was divided into three stages:

1. **Getting two-point clouds from two technologies:**
   (a) For raw data processing from scanner, the Faro SCENE software was used—registration, deleting of redundant points, colour application based on panoramic photos and export of the point cloud in ∗E57 format to the Autodesk Recap software were performed.
   (b) For raw data processing from aerial surveying, the Agisoft Photoscan Professional software was used—sparse point cloud in the highest-quality dense point cloud in ultrahigh quality. The overall time for creating the point cloud from 302 aerial photographs (from both modes) was 31 h. In general, lesser quality means less time. In this process, also the computer's parameters affect the overall time.

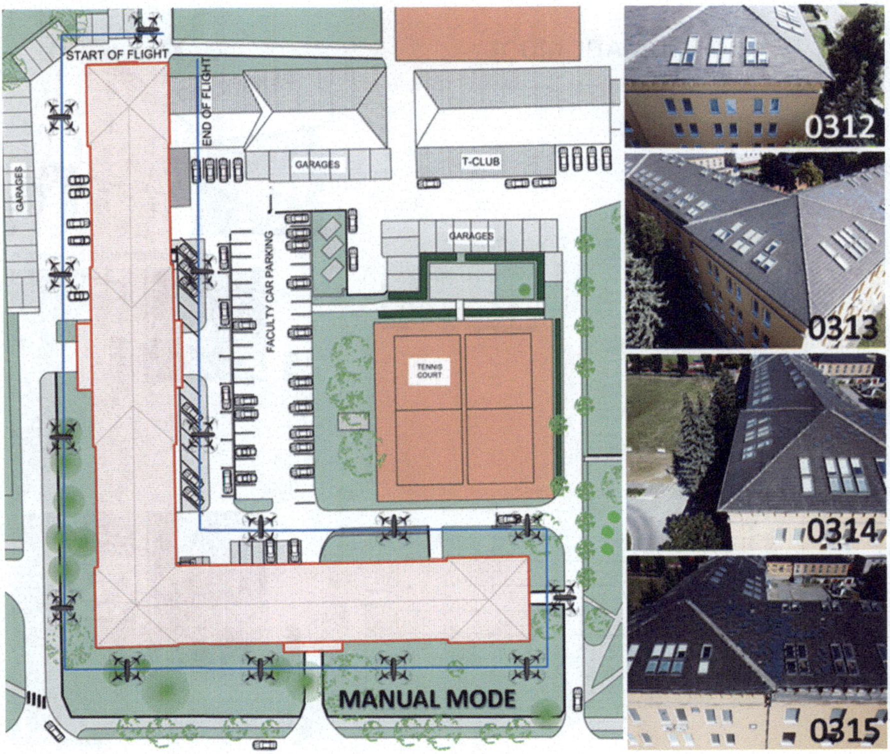

Fig. 3 Aerial photogrammetry/manual mode (source: authors)

(c) The overall time for creating the point cloud from 302 aerial photographs (from both modes) and from the scanner was 32 h (31 + 1 h).

2. **Creating a common coordinate system:**

(a) The coordinate system of the point cloud from aerial photographs (from the software Agisoft) was adapted to the coordinate system of the point cloud from laser scans (Fig. 4).
(b) The overall time for these processes was 1 h.

3. **Connection of two-point clouds into one:**

(a) Connecting the point cloud from the facade with the point cloud from the roof construction. The CloudCompare software was selected.
(b) The overall time in connecting two-point clouds was only a few minutes.

It should be noticed that using the UAV technology in conditions of Slovakia requires to obtain permission for creating aerial images from the Ministry of Defence of the Slovak Republic. The waiting period for this permission lasts about 2 or 3 weeks. After creating aerial images in the terrain, it is necessary to send obtained images for assessment again to the Ministry of Defence of the Slovak

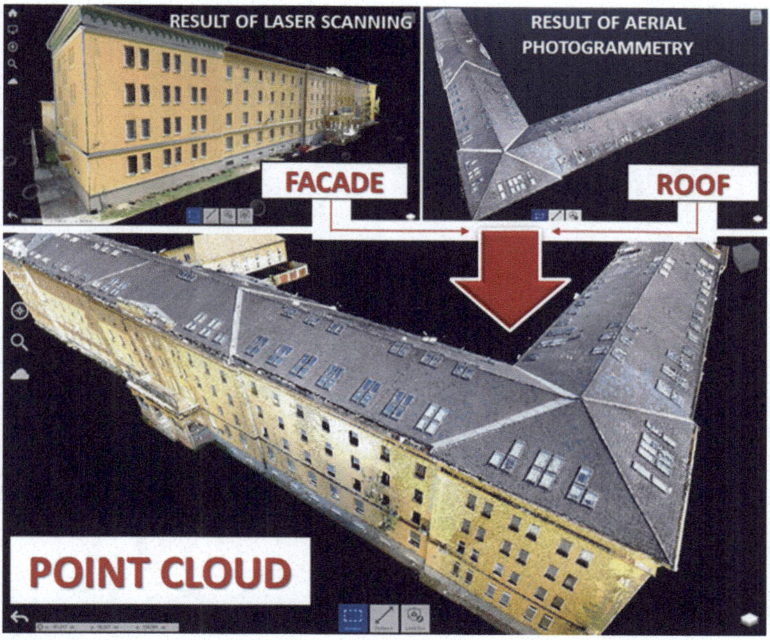

Fig. 4 Result of connecting two-point clouds from two technologies (source: authors)

Republic. The waiting period for assessment lasts again about 2 or 3 weeks. During deciding which of the technologies to use for digital as-built surveying, these waiting periods can be considered as one of the disadvantages for using UAV technology.

2.2 *Using the Terrestrial Laser Scanning Only*

Because of extension laser scanning also on roof of the faculty, the new scan positions of scanner had to be added. In Fig. 5, it can be seen the first positions of the laser scanner (blue points from the first using in the case A.) but also new additional scan positions (red points far away from scanned building).

These extra scans would have to be created at a distance of 50–70 m from the building. The range of scanning for the Faro Focus is up to 130 m; therefore it would be possible to measure the largest surface area of the roof structure, and this distance was not a problem for the Faro Focus laser scanner. The overall data collection time would be 2 days in this case. Expected time for the overall data processing time would be approximately 5 h. The quality of the resulting point cloud from the roof surface would certainly be slightly worse than the quality of aerial photogrammetry using an unmanned aircraft. For this reason, we consider this solution realistic but not very effective in measuring the roof of a building.

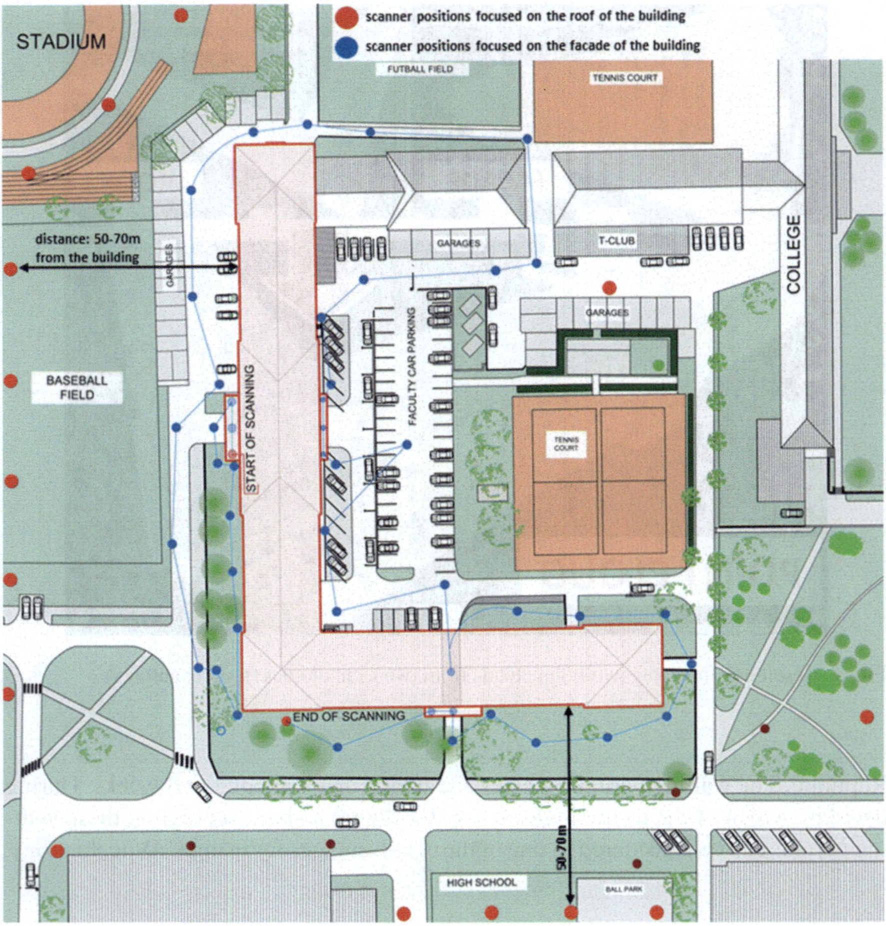

Fig. 5 A conceptual design of scanning using only the terrestrial laser scanner (source: authors)

2.3 Using the Aerial Photogrammetry Only

This method was used to measure not only the roof structure of the building but also the facade of the building completely. For creating of aerial images from the building, the drone Falcon 8 was chosen. This type of drone is possible to classify an octocopter (number of rotors 8). The disadvantage of this drone is its take-off weight (2.8 kg), which has a major impact on battery life. The slightly cloudy weather could not be influenced, so there were many shadows on the facade of the building. These effects meant that one battery lasted only 7 min. The total flight time was 2 h, so ten batteries had to be replaced during the flight. Aerial photos were also created in automatic and manual mode. Totally, 1050 aerial photographs were taken with a camera resolution of 36 megapixels. As mentioned above, using the UAV technol-

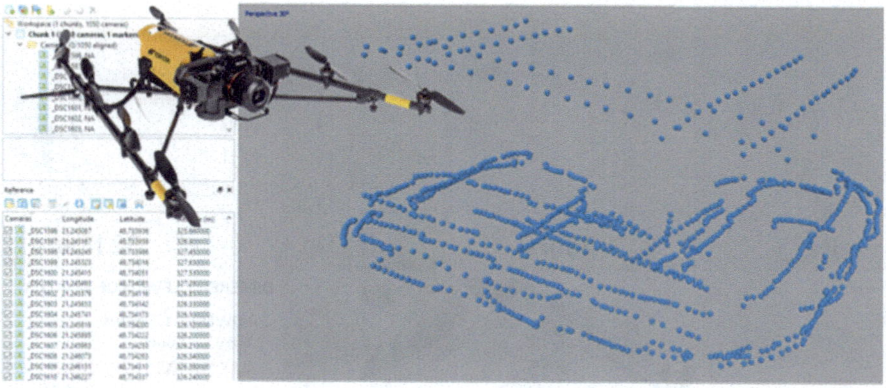

Fig. 6 Data collection of aerial photogrammetry (source: authors)

ogy in conditions of Slovakia requires to obtain permission for creating aerial images from the Ministry of Defence of the Slovak Republic (Fig. 6).

The process of data processing aerial images:

- In automatic mode 100 images, created in 3 min.
- In manual mode 950 images, created at different heights to capture all the necessary details on the building's facade.
- Images processed by the software Agisoft (used the same computer as in the previous case).
- The overall data processing time in the medium quality of the point cloud was 33 h (high- and ultrahigh-quality point cloud has not been tested, the estimated time of ultrahigh quality for 1050 aerial images could be up to 1 week).

In Fig. 7 the final quality of the point cloud from the building's facade is shown. It could be said that this final quality is sufficient for scanning needs.

For determining the weights and the resulting model of which type of scanning can be voted as the best, the Analytic Hierarchy Process (AHP) method developed by Saaty was used. None of the mentioned methods (a–c) was preferred; they were equally weighted as the alternatives:

- A1: **Terrestrial laser scanning + Aerial photogrammetry**
- A2: **Terrestrial laser scanning**
- A3: **Aerial photogrammetry**

The criteria for this AHP method were set, which could be represented by numbers, called quantitative criteria:

- K1: **Total time of scanning building** (manual working with devices)
- K2: **Costs** (for installation and operating costs, i.e. purchase devices, buying compatible software)

Fig. 7 The results of aerial photogrammetry (source: authors)

- K3: **Data processing—time of processing the results and format them to the required format, if it is needed** (and which is compatible with the software supporting BIM, i.e. ArchiCAD, AutoCAD, Revit and so on)

The basis for decision-making is empirical decision-making criteria, and using AHP is to outline the whole decision problem as a hierarchical structure [21]. AHP is based on the value of the information obtained and derives ratio scales from paired comparisons of criteria to discover and correct logical contradictions. The decision tree which represents the best alternative according criteria is the main created model and result. AHP helped us to make decisions in a more rational way and to make them more transparent and better understandable [22]. Using AHP as a supporting tool for decision-making helps to gain a better insight in complex decision problems. To consider about the structure of decision tree force the researcher to think through the problem as about a complex problem, consider possible alternatives (decision criteria) and select more significant criteria pairwise with respect to the decision objective.

The criterial matrix for evaluation alternatives according to criteria is represented by $Y = (y_{ij})$, where rows are alternatives and columns are criteria. For every criterion, it was necessary to calculate its weight (number from 0 to 1). The more important criterion, the higher number of its weight (denoted v_j for criterion K_j, $j = 1 \ldots, n$). There exists more methods to gain weights of criteria, for example, method of entropy (no preferences, weights of criteria are equal ($w_j = 1/n$)), method of order and Fuller's method (ordinal information about criteria are known) and, the most important for AHP method, Saaty's method (ordinal information and distances between criteria are known). Scale of points for criteria are 1, 3, 5, 7, 9, where 1 means criteria are equal; 3, the first criterion is more significant than the second one; and 9, the first criterion is absolutely more significant than the second one.

Mathematically the AHP method is based on the solution of an eigenvector problem. The process of calculating of eigenvectors is one of the basic principles in linear algebra [23]. The results of the pairwise comparisons are arranged in a Saaty's matrix. The four matrices are gained, one matrix from criteria and three matrices from comparison alternatives based on every criterion and for every of them which has calculated its eigenvector. Saaty's matrix is squared and consists of estimated elements $s_{ij} \oplus \frac{w_i}{w_j}$ which are ratios of weights i-th and j-th criterion. Finally, the criteria preferences matrix and ranking matrix are created. The results from ranking matrix were arranged in the decision tree.

3 Results

The results of these three methods were compared, based on their main properties.

The main observed advantages of terrestrial laser scanning technology are:

- Measuring all important spatial information from the exterior of the Faculty of Civil Engineering—the facade of the building with high spatial details and the roof of the building with low quality
- Doesn't require permission from the Ministry of Defence of the Slovak Republic
- Any weather conditions
- Fast data-processing phase of laser scans

The main observed disadvantages of terrestrial laser scanning technology are:

- Long overall data collection time
- Price of the laser scanner (approx. 30,000 €) + additional hardware devices and software Faro Scene (approx. 5000 €)
- Most roof constructions represent a limitation for terrestrial laser scanning
- Difficult handling with a terrestrial laser scanner

The main observed advantages of aerial photogrammetry technology are:

- Measuring all important spatial information from the exterior of the Faculty of Civil Engineering—the roof and the facade of the building were measured in high spatial details (resolution of the digital camera (36 megapixels) is enough for high-dense point cloud).
- Speed of the data collection (approx. 90% faster than laser scanning).
- An automatic drone flight mode with automatic aerial image creation—only if the camera is rotated vertically to the ground.
- Almost easy with drone handling.
- Measuring places that may be dangerous for the scanner operator.

The main observed disadvantages of aerial photogrammetry technology are:

- Weather conditions—wind speed, low temperature, cloud amount, shadows and dazzling sunshine.
- Danger of attacking the drone by birds.
- Low battery life (approx. 7 min)—ten batteries were required for the overall data collection over 2 h.
- Demanding data processing (the number and quality of aerial images, software used, parameters of computer).
- In a certain type of building, the principle of aerial photogrammetry can replace technology of terrestrial laser scanning completely (if we are only talking about measuring of an exterior of some building).
- Price of the drone (approx. 40,000 €) + additional software Agisoft Photoscan Professional (price approx. 3500 €)
- Permission of the Ministry of Defence. In the Slovak Republic, all aerial images/photos that have been created from unmanned aerial vehicle (UAV) and which contain the metadata (GPS coordinates, longitude, latitude, height, yaw, pitch, roll, etc.) require permission from the Ministry of Defence. Waiting period is approx. 3 weeks. After taking aerial images, it was necessary to send the aerial images back to the Ministry of Defence and wait 2 or more weeks to overall reviewing of aerial images. The overall waiting period can be approx. 1 month or more, and this waiting period is a disadvantage for this technology.

The results of AHP method to find out which method is the most effective according to established criteria were obtained; criteria preferences matrix (Table 1) and ranking matrix (Table 2) were calculated. In the consequential decision tree which can be seen in Fig. 8, using aerial photogrammetry only is the best method with the weight 0.622.

Table 1 Preferences matrix

Preferences matrix	K1	K2	K3
A1	0.087	0.055	0.067
A2	0.149	0.29	0.467
A3	0.764	0.655	0.467

Table 2 Ranking matrix

Ranking matrix	K1	K2	K3
A1	0.029	0.018	0.022
A2	0.049	0.096	0.154
A3	0.252	0.216	0.154

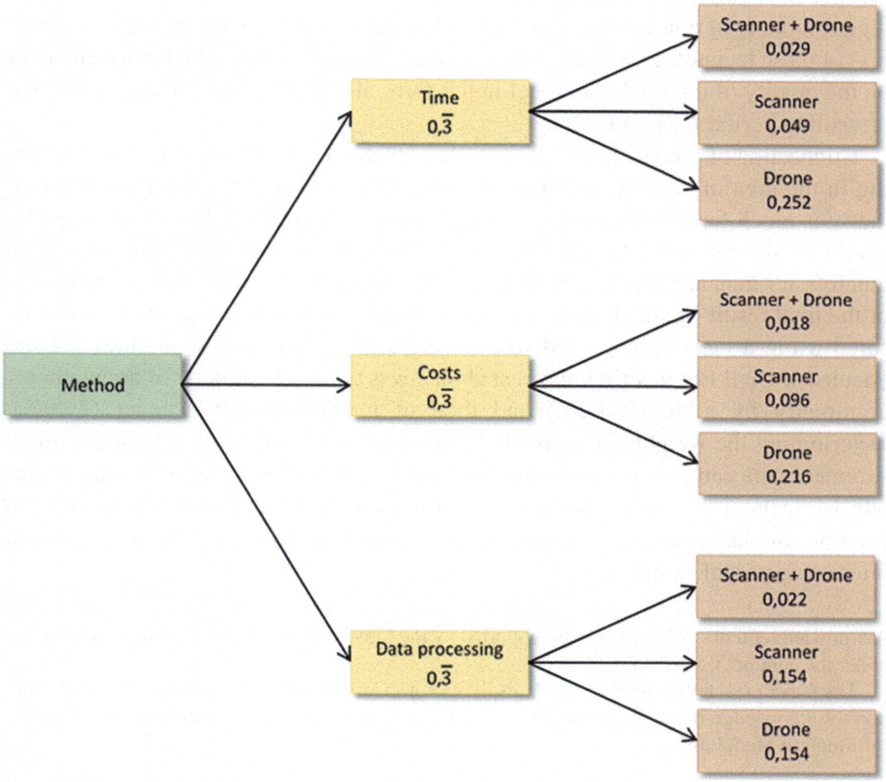

Fig. 8 The decision tree—the result of AHP method

4 Conclusion

Modern methods of building scanning for creating a 3D model are currently used frequently. Creating a 3D virtual model for the recording of building architecture is a very efficient and high-quality presentation method that is made possible by the availability of various modern technologies. The terrestrial laser scanner (TLS) is an appropriate choice for measurements that are sufficient to be performed from the ground. TLS gathers high-density 3D spatial data from the measurement area in a short time. With this technology, however, it is a problem to locate some parts of the roof structures of the building or some other complicated details. In this situation, the final product can be a problem because the point cloud doesn't contain all spatial information. Therefore, other supportive technologies for efficient recording of these details have been discovered. Using aerial photogrammetry has become a very good alternative to a terrestrial laser scanning. The output of both technologies is the point clouds, which can then be combined into one point cloud to create a full 3D model of building. By using these methods, it is possible to create very accurate digital models of constructions and spatial entities. Despite the considerable progress of these technologies, there always exist some limitations, which have an effect on the quality of the final 3D model in the form of a point cloud, for example, time or waiting period for permission.

Connecting of both technologies together creates a very powerful tool for digitizing in the area of building construction. It can be seen that the aerial photogrammetry is much faster than laser scanning (approx. about 90%). But, if we compare results of the overall data processing time, we can see that the laser scanning is much faster than aerial photogrammetry. Sometimes, the overall quality and details of the final result (point cloud) are more important. Some buildings are better measured using a laser scanner and some using a drone. For this case study for the Faculty of Civil Engineering, the best choice was to use technology of aerial photogrammetry by a drone (62%) and then of terrestrial laser scanner (29.9%). Determining the weights of individual criteria more sensitively based on the measurement data can help to obtain different results for the choice of individual methods. More precisely finding the boundary, for example, when it will be preferable to use only the laser scanning method before another method, etc., would also be an interesting partial result.

Acknowledgements This work was supported by the Slovak Research and Development Agency under the contract no. APVV-17-0549.

The paper presents a partial research result of project VEGA 1/0828/17 "Research and application of knowledge-based systems for modeling cost and economic parameters in Building Information Modeling".

References

1. Adán, A., Quintana, B., Prieto, S.A., Bosché, F.: Scan-to-BIM for 'secondary' building components. Adv. Eng. Informat. **37**, 119–138 (2018). https://doi.org/10.1016/j.aei.2018.05.001
2. Muralikrishnan, B., Rachakonda, P., Lee, V., Shilling, M., Sawyer, D., Cheok, G., Cournoyer, L.: Relative range error evaluation of terrestrial laser scanners using a plate, a sphere, and a novel dual-sphere-plate target. Measurement. **111**, 60–68 (2017). https://doi.org/10.1016/j.measurement.2017.07.027
3. Pejić, M., Ogrizović, V., Božić, B., Milovanović, B., Marošan, S.: A simplified procedure of metrological testing of the terrestrial laser scanners. Measurement. **53**, 260–269 (2014). https://doi.org/10.1016/j.measurement.2014.03.024
4. Wang, C., Cho, Y.K.: Performance evaluation of automatically generated BIM from laser scanner data for sustainability analyses. Proc. Eng. **118**, 918–925 (2015). https://doi.org/10.1016/j.proeng.2015.08.531

5. Shini, R.Z., Sacks, R., Ma, L., Filin, S.: Towards generation of as-damaged BIM models using laser scanning and as-built BIM First estimate of as-damaged locations of reinforced concrete frame members in masonry infill structures. ISPRS J. Photogram. Rem. Sens. **126**, 168–179 (2016). https://doi.org/10.1016/j.aei.2016.04.001
6. Tang, P., Huber, D., Akinci, B., Lipman, R., Lytle, A.: Automatic reconstruction of as-built building information models from laser-scanned point clouds: a review of related techniques. Auto. Constr. **19**(7), 829–843 (2010). https://doi.org/10.1016/j.autcon.2010.06.007
7. Bosché, F., Guillemet, A., Turkan, Y., Haas, C., Haas, R.: Tracking the built status of MEP works: assessing the value of a scan-vs-BIM system. J. Comput. Civ. Eng. **28**(4), 1 (2014). https://doi.org/10.1061/(asce)cp.1943-5487.0000343
8. Belsky, M., Sacks, R., Brilakis, I.: Semantic enrichment for building information modelling. Comput. Aided Civ. Inf. Eng. **31**(4), 261 (2015). https://doi.org/10.1111/mice.12128
9. Kim, C., Son, H.: Automated construction progress measurement using a 4D building information model and 3D data. Auto. Constr. **31**, 75–82 (2012). https://doi.org/10.1016/j.autcon.2012.11.041
10. Bosche, F., Ahmed, M., Turkan, Y., Haas, C.T., Haas, R.: Automated retrieval of 3D CAD model objects in construction range images. Auto. Constr. **17**(4), 499–512 (2015). https://doi.org/10.1016/j.autcon.2007.09.001
11. Bosche, F.: Automated recognition of 3D CAD model objects in laser scans and calculation of as-built dimensions for dimensional compliance control in construction. Adv. Eng. Informat. **24**, 107–118 (2010). https://doi.org/10.1016/j.aei.2009.08.006
12. Bosche, F., Ahmed, M., Turkan, Y., Haas, C.T., Haas, R.: The value of integrating Scan-to-BIM and Scan-vs-BIM techniques for construction monitoring using laser scanning and BIM: the case of cylindrical MEP components. Auto. Constr. **49**(B), 201–213 (2015). https://doi.org/10.1016/j.autcon.2014.05.014
13. Shahbazi, M., Menard, P., Sohn, G., Theau, J.: Unmanned aerial image dataset ready for 3D reconstruction. Data Brief. **25**, 103962 (2019). https://doi.org/10.1016/j.dib.2019.103962
14. Pecho, P., Magdolenova, P., Bugaj, M.: Unmanned aerial vehicle technology in the process of early fire localization of buildings. Transport. Res. Proc. **40**, 461–468 (2019). https://doi.org/10.1016/j.trpro.2019.07.067
15. Zhong, X., Peng, X., Yan, S., Shen, M., Zhai, Y.: Assessment of the feasibility of detecting concrete cracks in images acquired by unmanned aerial vehicles. Auto. Constr. **89**, 49–57 (2018). https://doi.org/10.1016/j.autcon.2018.01.005
16. Corrigan, F.: Introduction to UAV Photogrammetry and Lidar Mapping Basics. DroneZon (2018). Available from https://www.dronezon.com/learn-about-drones-quadcopters/introduction-to-uav-photogrammetry-and-lidar-mapping-basics/
17. Eisenbeiss, H., Zhang, L.: Comparison of DSMS generated from mini UAV imagery and terrestrial laser scanner in a cultural heritage application. In: ISPRS Commission V Symposium Image Engineering and Vision Metrology, pp. 1–7 (2006)
18. Gruszczyński, W., Matwij, W., Ćwiąkała, P.: Comparison of low-altitude UAV photogrammetry with terrestrial laser scanning as data-source methods for terrain covered in low vegetation. ISPRS J. Photogram. Rem. Sens. **126**, 168–179 (2017). https://doi.org/10.1016/j.isprsjprs.2017.02.015
19. Altuntas, C.: Integration of point clouds originated from laser scanner and photogrammetric images for visualization of complex details of historical buildings. Int. Arch. Photogram. Rem. Sens. Spat. Inf. Sci. **XL-5**(W4), 431–435 (2015). Available from https://www.semanticscholar.org/paper/INTEGRATION-OF-POINT-CLOUDS-ORIGINATED-FROM-LASER-Altuntas/3a5a22af37ddc29da529aabce73bf4cf27804591/
20. Pejic, P., Krasic, S.: Modern methods of three-dimensional presentations of the building heritage. In: IV International Symposium for Students of Doctoral Studies in the Fields of Civil Engineering, pp. 111–119 (2012). https://doi.org/10.5194/isprsarchives-XL-5-W7-207-2015

21. Saaty, H., et al.: Group Decision Making: Drawing Out and Reconciling Differences. RWS Publications, Pittsburgh, PA (2008)
22. Goepel, K.D.: Implementation of an online software tool for the analytic hierarchy process (AHP-OS). Int. J. Anal. Hierar. Proc. **10**(3), 469–487 (2018). https://doi.org/10.13033/ijahp.v10i3.590
23. Cherney, D., Denton, T., Rohit, T., Waldron, A.: Linear Algebra (2013). Available from https://www.math.ucdavis.edu/~linear/linear-guest.pdf

Development of Digital and Managerial Competencies and BIM Technology Skills in Construction Project Management

Tomáš Mandičák, Peter Mésároš, Marcel Behún, and Annamária Behúnová

1 Introduction

The current digital era requires different requirements than the last century. Architecture, engineering, and construction (AEC) has changed significantly over the past few years. The requirements for individual experts have changed. Every day, project managers take a number of decisions that need to be supported by their competencies and skills. Therefore, project managers try to use their competencies and skills for project success. However, what was a competitive advantage yesterday and guarantees the success of the construction project is now a necessary necessity for the project to be viable at all. Generally, last decades international and local communities have recognized the construction field and construction sector as vital in encouraging societal change toward sustainable development in a global context [1]. According to the United Nations Earth Summit Agenda from 2009, the action blueprint for "promoting sustainable human settlement development" advocates the promotion of sustainable construction industry activities and working together to take action toward achieving sustainable human resource development and capacity-building for human settlement development. In this regard, project managers and

T. Mandičák · P. Mésároš
Faculty of Civil Engineering, Technical University of Košice, Košice, Slovakia
e-mail: tomas.mandicak@tuke.sk; peter.mesaros@tuke.sk

M. Behún
Institute of Earth Resources, Faculty of Mining, Ecology, Process Control and Geotechnologies, Technical University of Košice, Košice, Slovakia
e-mail: marcel.behun@tuke.sk

A. Behúnová (✉)
Department of Industrial Engineering and Informatics, Technical University of Kosice, Prešov, Slovakia
e-mail: annamaria.behunova@tuke.sk

L. Knapcikova et al. (eds.), *New Approaches in Management of Smart Manufacturing Systems*, EAI/Springer Innovations in Communication and Computing, https://doi.org/10.1007/978-3-030-40176-4_10

leaders in the construction industry by transforming subordinates as well as influencing their development [2]. On the other hand, sustainable performance in construction projects starts with systematic project managing and increasing leaders' competencies and skills. Subordinates can also transform the project toward sustainability and achieve better productivity [3]. Architecture, engineering, and construction has become both a vehicle for improving first of all especially the quality of construction project and an entity that can determine the environmental and social sustainability of development efforts [4]. The concept of knowledge-based society is closely linked to knowledge-based economy characteristic of globalization and extraordinary utilization of human thinking and knowledge in a specific way embedded in state-of-the-art technologies. One of the consequences of knowledge economy emergence lies in its new and strict demands on managerial competencies on all managerial levels in the organizations that want to be successful under these new conditions [5]. Even though the managerial role remains managerial, there are many new conditions influencing the current managerial role.

2 Problem Statement and Theoretical Background

A lot of studies discuss transformational leadership that have discussed other aspects of measuring the leaders' and managers' quality [6–9]. Based on this, there are developed questions to assess the quality of transformational leadership in two dimensions. First of all, it is so-called develop followers into leaders [8]. Last, it's inspire followers to go beyond their own interest.

In another point of view, preceding research on leadership outlined that the leader's ability to foster cooperative aims and motivate participants of construction projects and project members to attain such aims highly influences project performance. Effective managers are those who can change the style of task requirements and the subordinates' needs, even in the middle of a project. Consequently, different conditions may affect the leadership behavior of a leader. Nevertheless, there has not been enough research on the effect of leadership competencies and transformational leadership qualities on success criteria in sustainable building projects [10].

Sajgalikova presents in her research, most of the respondents present their motivation to apply the surveyed competencies at work [11]. Language skills account for the only exception; only one third of the respondents are motivated to apply them. Flexibility, analytical skills, and respect for the others are those that managers feel motivated most to utilize. The size of an organization exercises an impact on the extent to which managers are motivated to apply the identified competencies: the larger the organization, the more the managers are motivated to apply them. The results show that, in general, foreign managers are motivated to utilize the new managerial competencies more than their Slovak peers. It can be concluded that the high percentage of managers motivated to apply new managerial competencies in knowledge-based economies establishes good prospects. It indicates the possibility for managers to apply them more intensively. The finding is based on statistically relevant relationship between the motivation to apply the competencies and frequency of their application. It shows that if motivation for competency application exists, frequency of their utilization grows.

In the digital age, the managerial competencies are closely connected with digital competencies. That's not only about using ICTs, but it's about developing competencies too. According to European Union documents, the digital competency involves the confident, critical, and responsible use of, and engagement with, digital technologies for learning, at work, and for participation in society [12]. The digital competency is something bigger than just use of ICT tools. The most important part of digital competencies is focusing on cognitive skills involving active and creative use of ICT and considering their use in the workplace for lifelong learning and also to become integral digital citizens [13].

Architecture, engineering and construction are projects where there are sources of income for companies. Therefore, construction industry is probably the largest project-based sector. A lot of businesses within the industry are interested in establishing the competencies of project managers since the project's success is highly dependent on its managerial and digital skills [14].

Construction projects usually follow a predefined life cycle. Regardless of size and complexity of construction project, it passes through sequential stages from starting, designing, and preparing, carrying out the project work, to project closing [15]. The requirements on time, cost, and quality present the several problems that exist in the implementation of construction projects or in the last stage of project but are frequently ignored, and only a small proportion of discussions in literature technically focus on the ending phase. They normally regard project ending as natural and uncomplicated.

The construction project manager is a core professional who connects all participants of construction projects in the last project stage and has to collaborate with senior managers, communicate with customers, formulate and execute the strategy related to the last stage of project, and so on (strategize contingency and collect and archive knowledge as organizational process assets). Their asset to the project in the last stage is paramount. Project managers face a large amount of stressful issues in the ending phase. Other researcher [16] demonstrated that work-related issues would easily convert to stressors in the last stage of project, and initially motivating and challenging issues might become a pointless burden if the project is about to shut down abruptly.

Another study [17] confirmed the industry's growing awareness of the relationship between achieving project success and constructing project management competencies and skills. Successful construction businesses actually focus on ensuring that project managers acquire the core competencies required to be successful in their projects. According to study [18], the project manager has a direct impact about 34–47% on project success. Based on these previous studies, it is clear that construction project managers play an important role in determining the success of a project.

The Project Management Institute (PMI) first documented its nine knowledge areas (integration, time, cost, procurement, quality, communication, human

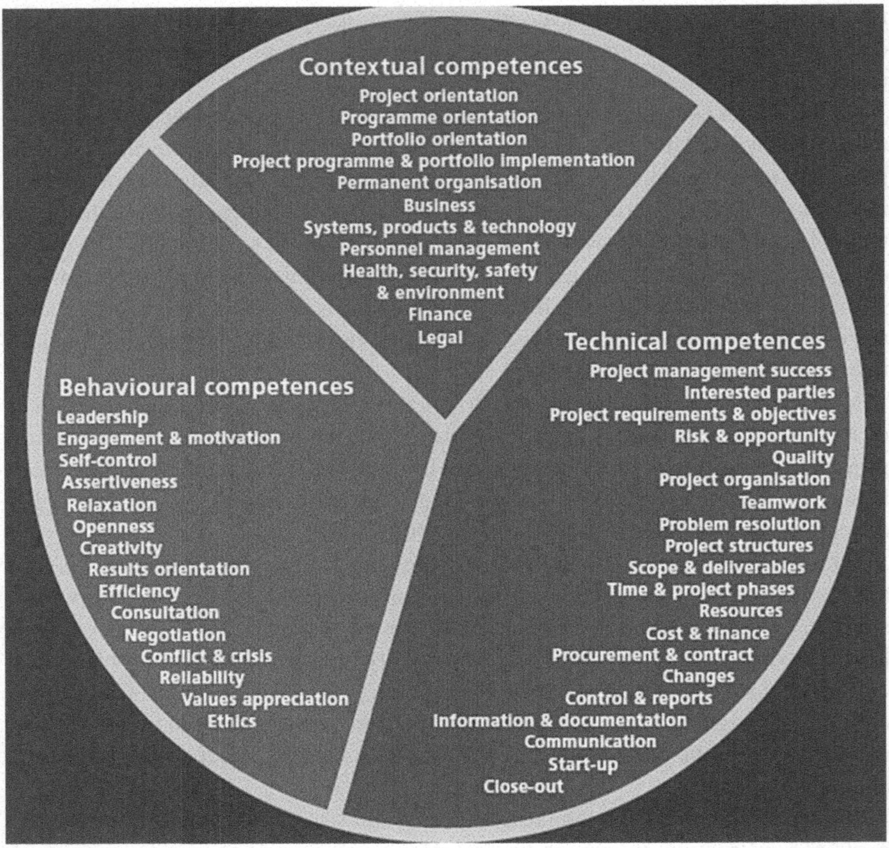

Fig. 1 Competencies of project managers according to IPMA

resource, scope, and risk) in the *A Guide to the Project Management Body of Knowledge* (*PMBOK Guide*) in 1987 in an attempt to document and standardize generally accepted project management information [19]. Other sources describe managerial competencies of project managers. There are three sectors of competencies in Fig. 1.

Each of the nine knowledge areas contains processes that need to be accomplished within its discipline in order to achieve an effective project management program. For instance, project cost management encompasses processes that are required to ensure the project is completed within the approved budget and consists of resource planning, cost estimating, cost budgeting, and cost control.

Considering all the challenges of construction project managers are facing, it is evident that one area where construction project managers need to make big improvements is in the field of construction project management. Skills on their own, including their applications, do not make an effective people project manager. Behaviors drive outcomes. Specific behaviors for each skill need to be applied by project managers to make these skills truly effective. It is the application of these that is of paramount importance and is the catalyst to being an effective people project manager [20, 21] (Table 1).

Competencies of project managers are closely related to ICT competencies and especially in construction project management with building information modeling (BIM) skills. The implementation of building information modeling (BIM) has resulted in the development of new work practices, roles, and responsibilities related to the new technology. The literature has identified BIM actor roles as relevant for successful BIM project implementation, and these roles include not only technical competencies but also skills related to process change and management. In a recent study, Uhm et al. [34] analyzed job advertisements of BIM roles and found that 35 different job titles were used with different responsibilities and tasks. Another similar study by Davies et al. [35] focused on how the roles and responsibilities of BIM actors were defined in documents and national/international BIM guidelines and handbooks. Both studies discuss that there is a development of a standardized BIM practice but that the development of the role is uncoordinated and that the role and its responsibilities would require further attention for the development of BIM roles. Previous studies have discussed the emerging new role of the BIM actor on a conceptual level or as a review, but there is a lack of empirical studies on this new role and its characteristics, tasks, perceived responsibilities, and development as an established professional role in the architecture, engineering, and construction industry [36].

BIM managers are responsible for managing people, information, and the model itself during projects. Thus, BIM managers require skills that might be similar to the skills of project managers, such as teamwork and communication skills, while also requiring deeper skills in BIM such as understanding BIM standards and BIM workflow. However, BIM managers have limited opportunity in their career advancements because the industry has poor awareness on the skills and expertise that a BIM manager provides. As a result, some BIM managers are considering leaving their current organization to pursue consulting positions specializing in BIM. Hence, comparing the skills between BIM managers and project managers can help employers understand the value that BIM managers bring while assisting the education community to personalize BIM education [37].

Rahman et al. [37] discuss about the shared skills and unique skills between project managers and BIM managers based on the BIM skills and their correlated skills identified in previous sections. Eighteen skills are shared together by project managers and BIM managers. Project managers have five unique skills, while BIM managers have 11 unique skills. This illustrates that project managers and BIM managers have more shared skills than unique skills between them. There are a number of possible reasons for this similarity between the positions. Potentially, there is not yet a consistent set of BIM-related responsibilities for different project participants on different projects. In other words, it is possible that on some projects when a dedicated BIM manager is present, the act of managing steel detailing is performed by that individual, whereas on other projects where a dedicated BIM manager is not staffed, a project manager may assume this responsibility. Another

Table 1 Managerial and digital competencies of construction project managers

Yao and Luu [22]	Professional ethics Teamwork potential Attitudes toward consultancy Management and ownership Creativity
Wang [23]	Interpersonal communication skills Technology skills Information skills Application skills Learning skills Organization skills Strategic perspective Resource management Critical analysis
Dulewicz and Higgs [24]	Communication Empowering Developing Achieving
Daft and Pirola-Merlo [8]	Develops followers into leaders Inspire followers Put together the budget and negotiate cost estimates Choose the most efficient construction method and strategies Stay in touch with the clients for works or budget-related issues Arrange the work timetables Discuss about technical and contract details with workers and other professional parties Cooperate with building and construction specialist Keep an eye on the personnel in construction onsite
Liikamaa [25]	Emotional awareness Self-confidence Self-assessment Trustworthiness Maintaining order Flexibility Innovation Responsibility Seeking information Production efficiency Decision quality Stress management Analytical thinking Conceptual thinking Language proficiency Achievement drive Commitment Initiative Optimism Understanding others Developing other people Leveraging diversity Organizational savvy Communications Conflict management Management Leadership Relationship building Collaboration

(continued)

Table 1 (continued)

Giudici and Filimonau [26]	Able to listen Able to delegate authority Increasing a good overall relationship with team Be efficient Be confident Make the right decision Motivating team members Be helpful Understanding Be authoritative
Shi et al. [27]	Construction cost planning and control Civil engineering construction Engineering contract and law Construction project management Construction regulations Civil engineering drawing Building structure Building architecture Engineering economics Construction materials Financial management Construction equipment Administrative regulation Management science Economic law Civil engineering surveying Micro- or macro-economics Urban planning Operational science Accounting Statistics Banking and insurance
El-Baz and El-Sayegh [28]	Technical competencies (statistical analysis, decision analysis, resource optimization, information technology control, SCM) Management competencies (planning and managerial competencies as strategic and operation planning, human resource planning, change management) Leadership management (ability of managers to have effective communication, vision, responsibility, and motivation) Financial competencies (cost management ability, financial analysis, accounting)
Kerzner [29]	Leadership abilities Creativity Ability to make decisions Ability to identify problem Ability to organize work to subordinates Effectiveness Ability to persuade and having ambition Vision and leadership abilities
Project Management Institute (PMI)	Knowledge competencies Performance competencies and personal competencies

(continued)

Table 1 (continued)

IPMA	**Technical competencies:** Project management success Interested parties Project requirements and objectives Risk and opportunity Quality Project organization Teamwork Problem resolution Project structures Scope and deliverables Time and project phases Resources Cost and finance Procurement and contract Changes Control and reports Information and documentation Communication Start-up Close-out
	Behavioral competencies: Leadership Engagement and motivation Self-control Assertiveness Relaxation Openness Creativity Results orientation Efficiency Consultation Negotiation Conflict and crisis Reliability Values appreciation Ethics
	Contextual competencies: Project orientation Programmed orientation Portfolio orientation Project program and portfolio implementation Permanent organization Business Systems, products, and technology Personnel management Health, security, safety, and environment Finance Legal
Balula [12]	Confident Be critical and responsible Engagement Participation in society

(continued)

Table 1 (continued)

The European Commission's science and knowledge service [30]	Information and data literacy Communication and collaboration Digital content creation Safety Problem solving
Lichtarski [31]	Creating trust Ability to persuade
Dziekoński [14]	Effectiveness Ability to make a decision
ASCE [32]	Ability to organize work Ability to formulate goals
The Chartered Institute of Building [33]	Ability to motivate employees Ability to communicate and data exchange Ability to solve problem Ability to resolve conflicts Be flexible in management Ability to negotiate Focus on the aims

Source: own processing based on literature review [8, 12, 14, 22–24, 28, 29, 31–33]

potential explanation for this similarity could be related to when these skills were endorsed. It is possible that project managers had received similar skill endorsements to BIM managers from earlier in their careers when they were doing tasks that were more closely aligned to BIM management. LinkedIn does not provide information on the time frame of those endorsements. Hence, this article could not validate whether the skills were endorsed as a project manager or before becoming a project manager. While many of the correlated skills are shared between BIM managers and project managers, there are also a number of unique skills that are only correlated with one of the positions. Project managers have five unique skills, while BIM managers have 11 unique skills that are correlated with their promotion to their respective positions. The differences in the unique skill sets between project managers and BIM managers indicate some unique responsibilities between them. This may illustrate the values that the BIM managers can provide into construction projects, whereas there are certain skills that are not possessed by project managers. Thus, employers might need to reconsider the importance of BIM managers in construction projects and to be cautious in combining the responsibilities of BIM managers into project managers. The construction field presents a number of relative processes. The success of the construction project is primarily influenced by the order of these processes. The primary goal is to execute individual processes in the right order and in the right way. The construction project has a temporary character, and the result of these processes is a specific product. The construction project is a one-time organized process of construction, renovation, etc. An important factor in the project implementation process is the project team from which the success of the project depends [38]. Construction projects are strictly defined by result

Table 2 BIM skills of project managers in construction project management [39]

Competency	Partial competency
Managerial	Leadership Strategic planning
Functional	Finance Accounting Budgeting Human resources Collaboration
Administration	Organizational skills Policies Procedures
Operation	Designing Conceptualizing
Technical	Analyzing Simulating Quantifying and estimating Modeling and drafting Documentation and detailing
Implementation	Model management Component development Training
Support	ICT support
Research and development	Software developments General R&D Coaching Teaching Knowledge sharing Industry engagement

requirements, the cost, and time constraints and are bounded by the environment in which they are implemented. Table 2 brings BIM skills of project managers in construction project management.

Recent BIM-related publications, including the US National Building Information Modeling Standards (NBIMS), have highlighted the importance of collaboration in a project. However, in BIM education facilities or training centers, the prevailing perception is still that BIM software skills are much more important for conducting BIM well than are other aspects of BIM competencies, such as collaboration and coordination skills [34]. A survey of the BIM education track for undergraduate students and employees of the BIM industry performed indicated that the respondents perceived BIM software application skills to be the most important learning outcomes of BIM education programs, followed by knowledge of BIM concepts and literature, understanding BIM standards and interoperability issues, BIM internship and work experience, and network-based BIM model management knowledge. Along the same line, a survey by Zuppa et al. [40] showed a pervasive perception in the architecture, engineering, and construction (AEC) industry that BIM was a tool for visualization and coordination that helped to avoid the errors and omissions.

On the other hand, quite a few studies have identified types of competencies other than software skills as core competencies for BIM professionals. An analysis of 31 online BIM-related job advertisements in the USA by Barison and Santos [41] classified BIM jobs into eight types. The eight types of BIM jobs were BIM manager, BIM modeler, BIM trainer, BIM director, BIM technician, BIM consultant, BIM marketing manager, and BIM software support engineer. However, since most of the collected jobs (70% of the samples, 22 cases) were BIM manager positions, and only two cases of other job types were considered in their paper, they focused on the BIM manager and analyzed six competencies for BIM managers: aptitude, education, experience, skill and ability, knowledge, and attitude [42]. Barison and Santos concluded that the competencies required for a BIM manager include the ability to work with a computer, a Bachelor of Science degree in AEC, industrial experience from 3 to 7 years, oral and written communication skills, presentation skills, systemic and critical thinking, teamwork, programming language skills, and interpersonal skills; however, the sample size of this study was too small to draw any statistically meaningful conclusion. Naturally, the research method was exploratory rather than descriptive [34].

Several other studies also proposed a classification for BIM jobs. For instance, Joseph [43] classified BIM jobs into seven types based on his experience: BIM managing director, BIM manager, BIM job captain (model manager), BIM coordinator, BIM technician, and BIM application specialists I and II. Abdulkader [42] also proposed eight types of BIM jobs based on his observations: BIM manager, BIM modeler, BIM analyst, BIM application developer (BIM software developer), BIM modeling specialist, BIM facilitator, BIM consultant, and BIM researcher. Nevertheless, these classifications were all developed based on observations or experience, without a sound theoretical foundation [34].

In addition to recent research that has revealed BIM job types and competencies, Succar et al. [39] proposed the only competency framework for BIM through a job advertisement analysis using taxonomy-based clustering. This framework classified competency into three groups—core competency, domain competency, and execution competency—and matched each group. According to this framework, for example, when a core competency group shows creativity, the core activity is related to the activity of design conceptualization in a domain competency, and the operation skill of the "ArchiCAD" software is related to the execution competency group. Based on all literatures, it was set the most frequently a thematic similar competency, which is in more detail described in Table 3.

Based on these facts, BIM is a tool that helps eliminate costs and time load across the building's life cycle, and probably, it has a positive effect on the development of digital and managerial competencies. Of course, on productivity and economic parameters impact other factors like materials and other aspects with them (e.g., waste management and so on) [44, 45]. In spite of these facts, BIM technology includes many points of view. Currently, the industry is undergoing rapid changes that will also affect the population, and one of the reasons is BIM technology [46].

Table 3 Groups of competencies and skills of project managers

Technical competencies	Behavioral and managerial competencies	BIM skills
• Ability to organize work to subordinates • Ability to make decisions • Ability to formulate goals • Ability to use project management software • Ability to design project	• Ability to communicate • Ability to motivate team members • Help in solving problems • Ability to resolve conflicts • Ability to human resource planning • Ability to motivate • Ability to take responsibility	• Designing • Drawing • Rendering • Share information • Understand complexity of construction project • Updating project data

Source: own processing based on literature review

3 Methodology

The data collection was done by questionnaire survey in digital form. The questionnaire includes questions about digital and managerial competencies of project managers and their BIM skills in construction project management. It's perception of experts (project managers). The main important issue of research was about analyzing digital and managerial competencies of construction project management and BIM skill development by BIM technology. Project managers used the Likert scale for the quantification of impact in the main research questions (1, low impact on development of managerial and digital competencies and BIM skills, 5, high impact on development of managerial and digital competencies and BIM skills).

The research sample includes participants of construction projects (project managers in Slovakia). Return rate was 4.89%. Research sample incudes 2% of project managers of large companies, 19% of project managers who worked in medium-sized companies, and 79% of project managers who worked in small companies and microenterprises.

Data processing was done in MS Excel and Statistica software. Some statistical accounting was done in MATLAB software. AHP method was set as the main research method for quantifying competencies and skills of project managers.

The AHP method is based on the solution of an eigenvalue problem that is in more detail described in the next sentences. The results of the pair-wise comparisons are arranged in a matrix. That's the main principle of AHP comparison. Saaty's matrix is squared, which consists of estimated elements $s_{ij} \oplus \frac{w_i}{w_j}$ which are ratios of weights *i*-th and *j*-th criterion [47]. Elements on the main diagonal are equal to 1. In ideal case, for every *i*, *j*, *k* = 1, …, *n* should apply

$$s_{ij} = \frac{w_i}{w_j} = \frac{w_i}{w_k} . \frac{w_k}{w_j} = s_{ik} . s_{kj} \quad (1)$$

and then this matrix is called consistent and reciprocal. It is multiplied (from the right side) by normalized right Eigenvector (n is Eigenvalue) gives the ratio scale (weighting), the Eigenvalue determines the consistency ratio:

$$\begin{pmatrix} 1 & \frac{w_1}{w_2} & \dots & \frac{w_1}{w_n} \\ \frac{w_2}{w_1} & 1 & \dots & \frac{w_2}{w_n} \\ \dots & \dots & \dots & \dots \\ \frac{w_n}{w_1} & \frac{w_n}{w_2} & \dots & 1 \end{pmatrix} \cdot \begin{pmatrix} w_1 \\ w_2 \\ \dots \\ w_n \end{pmatrix} = \begin{pmatrix} w_1 \\ w_2 \\ \dots \\ w_n \end{pmatrix} \cdot n$$

In practice, it is very rare to obtain fully consistent matrix. Then, the maximal eigenvalue of matrix is taken and the index of consistency $\mathrm{CI} = \frac{\lambda_{\max} - n}{n-1}$. As an appropriate value for approximation of weights, Saaty suggested a normalized geometrical mean of row's numbers in Saaty's matrix:

$$w_i = \frac{\sqrt[n]{\prod_{j=1}^{n} s_{ij}}}{\sum_{k=1}^{n} \sqrt[n]{\prod_{j=1}^{n} s_{kj}}}. \tag{2}$$

4 Results of Research

Based on the results of the AHP method, it was found out which competencies and skills are most important in the development of the use of BIM technology in the management of construction projects. Project managers' competencies and skills were divided according to a range of literature review into three groups. Managerial competencies, digital competencies and BIM capabilities were acquired according to established criteria, the matrix of preferential criteria is in Table 4. There are project managers' competencies and skills and level of development by BIM technology in construction project management. These results point to the importance of BIM technology use on the development of some digital and managerial competencies, especially on BIM skills too. According to the results, development of BIM skills achieved a 0.499 value, which clearly demonstrates the connection between software works and development skills for project managers. The highest values were achieved by "designing" and "drawing." Very important competencies are rendering, share information, and understanding complexity of construction project.

Table 4 AHP method and development competencies and skill by BIM technology

Method	Group of competencies	Competencies and skills
AHP method	Technical competencies (0.324)	Ability to organize work to subordinates (0.057) Ability to make decisions (0.053) Ability to formulate goals (0.065) Ability to use project management software (0.052) Ability to design project (0.097)
	Behavioral and managerial competencies (0.177)	Ability to communicate (0.043) Ability to motivate team members (0.018) Help in solving problems (0.031) Ability to resolve conflicts (0.028) Ability to human resource planning (0.029) Ability to motivate (0.017) Ability to take responsibility (0.011)
	BIM skills (0.499)	Designing (0.148) Drawing (0.141) Rendering (0.059) Share information (0.059) Understanding complexity of construction project (0.047) Updating project data (0.045)

The next important group of competencies is closed to BIM skills; it's technical competencies. It includes the ability to organize work to subordinates, make decisions, formulate goals, use project management software, and design project. In connection with BIM skills, it's especially design project.

Behavioral and managerial competencies group achieved 0.177 value. In comparison with other groups, it's the lowest value, but some selected competencies are important, e.g., ability to communicate and solve problems in the team.

5 Conclusion

Competencies and skills of project managers are often the basics of success of construction project. Generally, human resources are a very important factor or key factor of success in construction industry. Development of managerial and digital competencies is very important, and it's generally necessary to continue in this process for every construction business. The digital age brings new challenges and technology, which can get a lot of processes done easier. One of this is developing the competencies of project managers and increasing their skills. That's valuable for every participant of the construction project and can be very important working on this. Based on AHP method, this research brings results of importance between competencies developed by BIM technology. BIM skills are very significant developed by use of BIM technology. Its main group of skills is developed by BIM technology for construction managers, namely, designing, drawing, and sharing project data. The role of project managers is strategic for the success of a construction project. Increasing and developing digital and managerial competencies should be the first for the satisfaction of everybody who is interested in the project. This research showed selected competencies, which are significantly developed by BIM technology, and it can be valuable for decisions in construction project management.

Acknowledgments The paper presents partial research results of project APVV-17-0549 “Research of knowledge-based and virtual technologies for intelligent designing and realization of building projects with emphasis on economic efficiency and sustainability.”

The paper presents partial research results of project KEGA no. 059TUKE-4/2019 “M-learning tool for intelligent modeling of building site parameters in a mixed reality environment.”

References

1. Maliene, V., Malys, N.: High-quality housing, a key issue in delivering sustainable communities. Build. Environ. **44**, 426–430 (2009)
2. Tabassi, A.A., Ramli, M., Bakar, A.H.A.: Effects of training and motivation practices on teamwork improvement and task efficiency: the case of construction firms. Int. J. Proj. Manag. **30**, 213–224 (2012)
3. Maqsood, A.: Applying project histories and project learning through knowledge management in an Australian construction company. Learn. Org. **13**(1), 80–95 (2006)
4. Bowen, P., Pearl, R., Akintoye, A.: Professional ethics in the South African construction industry. Build. Res. Inf. **35**, 189–205 (2007)
5. Mandičák, T., Mesároš, P., Tkáč, M.: Impact of management decisions based on managerial competencies and skills developed through BIM technology on performance of construction enterprises. Pollack Periodica. **13**(3), 131–140 (2018)
6. Avolio, B.J., Bass, B.M., Jung, D.I.: Re-examining the components of transformational and transactional leadership using the multifactor leadership questionnaire. J. Occup. Org. Psychol. **72**, 441–462 (1999)
7. Bass, B.M., Avolio, B.J.: Full Range Leadership Development: Manual for the Multifactor Leadership Questionnaire. Mindgarden, Palo Alto, CA (1999)
8. Daft, R.L., Pirola-Merlo, A.: The Leadership Experience, Asia Pacific Edition, 1st edn. Cengage Learning Australia, South Melbourne, VIC (2009)
9. Northouse, P.G.: Leadership: Theory and Practice, 4th edn. SAGE Publications, Thousand Oaks, CA (2007)
10. Tabassi, A.A., Roufechaei, K.M., Ramli, M., Bakar, A.H.A., Ismail, R., Pakir, A.H.K.: Leadership competences of sustainable construction project managers. J. Clean. Prod. **124**, 339–349 (2016)
11. Sajgalikova, H., Bajzikova, L., Polakova, M., Wojcak, E.: Revisiting new managerial competences: is there enough motivation to apply them? Procedia. Soc. Behav. Sci. **58**, 944–952 (2012)
12. Balula, A.: The use of DigComp in teaching and learning strategies - a roadmap towards inclusion. ACM Int. Conf. Proc. Ser. **13**, 275–282 (2016)
13. Sicilia, M.-A., Różewski, P., Royo, C., García-Barriocanal, E., Kieruzel, M., Uras, F., Sánchez-Alonso, S., Lipczyński, T., Hamill, C.: Digital skills training in higher education: insights about the perceptions of different stakeholders. In: ACM International Conference Proceeding Series, pp. 781–787. ACM, New York, NY (2018)
14. Dziekoński, K.: Project managers’ competencies model for construction industry in Poland. Proc. Eng. **182**, 174–181 (2017)

15. Atkinson, R., Crawford, L., Ward, S.: Fundamental uncertainties in projects and the scope of project management. Int. J. Proj. Manag. **24**(8), 687–698 (2006)
16. Gallstedt, M.: Working conditions in projects: perceptions of stress and motivation among project team members and project managers. Int. J. Proj. Manag. **21**(6), 449–455 (2003)
17. Ahadzie, D.K.: A model for predicting the performance of project managers in mass house building projects in Ghana. Dissertation, University of Wolverhampton (2007)
18. Hwang, B., Ng, W.J.: Project management knowledge and skills for green construction: overcoming challenges. Int. J. Proj. Manag. **31**, 272–284 (2013)
19. PMI: A Guide to Project Management Body of Knowledge (PMBOK Guide). Project Management Institute, Newton Square, PA (2013)
20. IPMA: ICB - IPMA Competence Baseline Version 3.0. International Project Management Association, Amsterdam (2001)
21. Fisher, E.: What practitioners consider to be the skills and behaviours of an effective people project manager. Int. J. Proj. Manag. **29**, 994–1002 (2011)
22. Yao, T.P., Luo, S.M.: Reform of engineering consulting education in 21st century. China Econ. **7**, 98–99 (2005)
23. Wang, Y.: Coordination issues in Chinese large building projects. J. Manag. Eng. **16**, 54–61 (2000)
24. Dulewicz, V., Higgs, M.: Assessing leadership styles and organisational context. J. Manag. Psychol. **20**, 105–123 (2005)
25. Liikamaa, K.: Developing a project manager's competencies: a collective view of the most important competencies. Proc. Manufact. **3**, 681–687 (2015)
26. Giudici, M., Filimonau, V.: Exploring the linkages between managerial leadership, communication and teamwork in successful event delivery. Tour. Manag. Perspect. **32**, 100558 (2019)
27. Shi, Y., et al.: Evaluation of the parameter sensitivity of a coupled land surface hydrologic model. J. Hydrometeorol. **15**, 279–299 (2014). https://doi.org/10.1175/JHM-D-12-0177.1
28. El-Baz, H., El-Sayegh, S.M.: Developing engineering management core competencies. In: Fifth LACCEI International Latin American and Caribbean Conference for Engineering and Technology "Developing Entrepreneurial Engineers for the Sustainable Growth of Latin America and the Caribbean: Education, Innovation, Technology and Practice" (2007)
29. Kerzner, H.: Project Management, 9th edn. John Wiley & Sons, Hoboken, NJ (2005)
30. EU Science Hib, The European Commission's science and knowledge service: (2019). https://ec.europa.eu/jrc/en
31. Lichtarski, J.M.: Specyfika doboru pracowników do struktur zadaniowych w teorii i praktyce [Specifics of employees selection to task structures in theory and practice]. Zarządzanie Zasobami Ludzkimi. **6**, 25–40 (2007)
32. American Society of Civil Engineers: Civil Engineering Body of Knowledge for the 21st Century. ASCE Report, 2nd edn. ASCE, Reston, VA (2008)
33. The Chartered Institute of Building: Code of Practice for Project Management for Construction and Development, 3rd edn. Wiley Blackwell, London (2014)
34. Uhm, M., Lee, G., Jeon, B.: An analysis of BIM jobs and competencies based on the use of terms in the industry. Auto. Constr. **81**, 67–98 (2017)
35. Davies, K., Wilkinson, S., McMeel, D.: A review of specialist role definitions in BIM guides and standards. J. Inf. Technol. Constr. **22**(10), 185–203 (2017)
36. Sijtsema, P.M., Gluch, P., Sezer, A.A.: Professional development of the BIM actor role. Auto. Constr. **97**, 44–55 (2019)
37. Rahman, R.A., Alsafouri, S., Tang, P., Ayer, S.K.: Comparing building information modeling skills of project managers and BIM managers based on social media analysis. Proc. Eng. **145**, 812–819 (2016)
38. Designing Buildings Wiki. Construction Project (2019). Available from https://www.designingbuildings.co.uk/wiki/Construction_project. Accessed 12 Oct 2019
39. Succar, B., Sher, W., Williams, A.: An integrated approach to BIM competency assessment, acquisition and application. Auto. Constr. **35**, 174–189 (2013)

40. Zuppa, D., Issa, R.R., Suermann, P.C.: BIM's impact on the success measures of construction projects. Comput. Civ. Eng. **24**, 503–512 (2009). https://doi.org/10.1061/41052(346)50
41. Barison, M., Santos, E.: The competencies of BIM specialists: a comparative analysis of the literature review and job ad descriptions. In: International Workshop on Computing in Civil Engineering ASCE, Miami, FL, USA, pp. 594–602 (2011). https://doi.org/10.1061/41182(416)73
42. Abdulkader, S.: Common BIM Roles and Their Responsibilities. Qatar BIM User Day (2013)
43. Joseph, J.: BIM Titles and Job Descriptions: How Do They Fit in Your Organizational Structure? Autodesk University (2011)
44. Knapčíková, L., et al.: Advanced materials based on the recycled polyvinyl butyral (PVB). In: MMS Conference 2017, pp. 1–9. EAI, Ghent (2018)
45. Knapčíková, L., et al.: Material recycling of some automobile plastics waste. Przemysl Chemiczny. **95**(9), 1716–1720 (2016)
46. Behún, M., Gavurova, B., Tkacova, A., Kotaskova, A.: The impact of the manufacturing industry on the economic cycle of European union countries. J. Competitiv. **10**(1), 23–39. ISSN 1804-171X (2018)
47. Saaty, H., et al.: Group Decision Making: Drawing Out and Reconciling Differences. RWS Publications. ISBN 978-1-888603-08-8, Pittsburgh, PA (2008)

Evaluation of Surface Roughness and Geometrical Characteristics of Flange Component for Driving of Robotic Arm Made by CAM Programming

Peter Michalik, Peter Tirpak, Jozef Macej, Michal Petrus, Martin Ambrozy, and Samuel Cehelsky

1 Introduction

Nowadays, robotics is an irreplaceable part of human life. With its development, the scope of application is constantly increasing. Robots replace human activity in industry, healthcare, transport, science and many other industries. The reasons for mass deployment of robots are mainly of an economic nature. An unforgettable advantage is the efficiency of work. With the use of robotics, one can replace the stereotypical activities of a human and increase the quality and precision of work. George Devol is considered to be the father of robotics. According to the Devol patent, in 1959, they created the first UNIMATE robotic arm. The robot weighed more

P. Michalik (✉)
1st TU Kosice, FVT with a seat in Presov, Bayerova, Slovakia
e-mail: peter.michalik@tuke.sk

P. Tirpak
2nd TU Kosice, FVT with a seat in Presov, Bayerova, Slovakia
e-mail: peter.tirpak@tuke.sk

J. Macej
3rd TU Kosice, FVT with a seat in Presov, Bayerova, Slovakia
e-mail: jozef.macej@tuke.sk

M. Petrus
4th TU Kosice, FVT with a seat in Presov, Bayerova, Slovakia
e-mail: michal.petrus@tuke.sk

M. Ambrozy
5th TU Kosice, FBERG Kosice, Letná 1, Slovakia
e-mail: martin.ambrozy@tuke.sk

S. Cehelsky
6th TU Kosice, FVT with a seat in Presov, Bayerova, Slovakia
e-mail: samuel.cehelsky@tuke.sk

L. Knapcikova et al. (eds.), *New Approaches in Management of Smart Manufacturing Systems*, EAI/Springer Innovations in Communication and Computing, https://doi.org/10.1007/978-3-030-40176-4_11

Fig. 1 Kuka robot KR1000 titan [1]

than 2 tons and was controlled by a magnetic tape program. The greatest development occurred in the 1970s. In 1973, Kuka developed the first robot to have SCANs with six electronically controlled axes. The KR 1000 Titan robots are designed with six-axis joint-arm cinematic systems. The structural components of the robot are made of iron castings (Fig. 1).

Richard Hohn developed the first microcomputer-controlled robot in 19874. The robot was called T3 (The Tomorrow Tool). In the same year, Kawasaki used Unimate robots to weld the skeleton of motorcycles. The first microprocessor-controlled robot was created. Towards the end of the 1990s, new technologies were introduced to enable load identification for performance improvement and collision detection. These technologies have led to increased occupational safety when handling robots. In 2002, the first case of direct collaboration between a worker and a robot appeared. The development of industrial robots focuses mainly on technology improvement, reducing the weight of robots and increasing efficiency and speed of work. The KUKA robot LBR 4 became a breakthrough in 2006. The body of the robot was built of aluminium, which reduced the weight to 16 kg. Currently, according to the number of installed robots of each company include the following. FANUC ranks first on the basis of currently available data. During its existence, Fanuc has managed to install more than 450,000 pieces of robots. The company thus successfully moved from the original fourth position to the first place. Stäubli is one of the largest robot manufacturers in the world. The use of their robots is wide and operates in many sectors, from food to pharmacy to heavy industry.

1. FANUC—450,000
2. Yaskawa—300,000
3. ABB—300,000
4. Kawasaki—110,000
5. Nachi—100,000
6. Denso—95,000
7. Kuka—80,000
8. Mitsubishi—70,000
9. Epson—55,000
10. Stäubli—45,000 [2]

For the production of robot components, CNC machine tools [3] are used whose architecture of the machining space design allows the machining of hose conveyor components and similar transport systems used in the mining and stone industry [4, 5]. Also, the production of measuring stand components needed for tracking and evaluating conveyor belts are used by Molnár et al. [6]. CAM programming for the production of robot components is very often used by manual programming, which allows fast programming of simple and very similar parts [7]. The same applies to cutting tools [8, 9] used for the production of individual robot components [10].

2 Flange Component

The flange component for driving the robotic arm in Fig. 2 is comprised of inner and outer cylindrical faces and front faces. Furthermore, there are fixing holes.

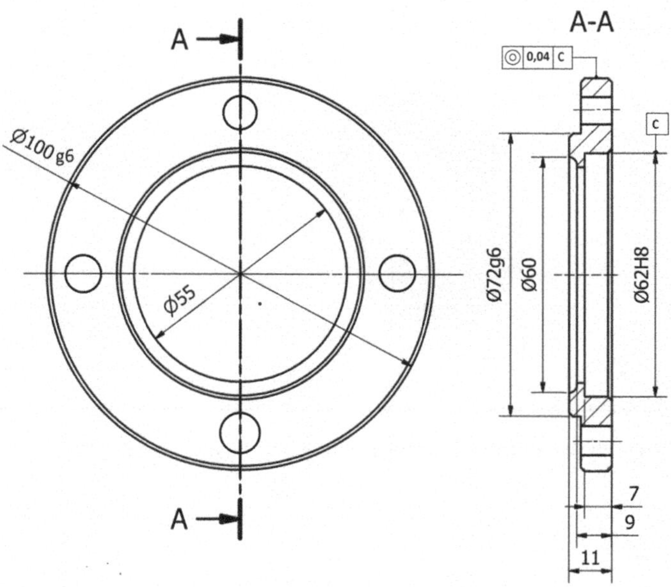

Fig. 2 Flange component manufacturing diagram

The material of the flange part is steel C45. It is used to cover and support the robot arm drive frame. D62 H8 and D72 g6 diameters also serve as sealing surfaces. When producing a diameter of 100g6 mm, the tolerance of 0.07–0.03 mm and the deviation of roundness of 0.06 mm must be observed. When producing a diameter of 62H8 mm, it is necessary to adhere to a tolerance of +0.04 to −0.025 mm and a deviation of 0.07 mm roundness and 0.04 mm coaxiality. The required surface roughness (Ra) is 3.2 μm.

3 CAM Program Creation Procedure

Software was used to create a 3D model of the flange part and then generate the production drawing, Autodesk Inventor Professional 2019. The 3D model was also used in HSM PRO 2019 to simulate and generate an NC program for the FANUC control system [11–13]. In the first step, the coordinate system was placed in the center of the part rotation and on the machined face in Fig. 3.

For the machining of the external face and cylindrical surfaces (Fig. 4), the right side removal tool has been selected. The feed rate was adjusted to 0.15 mm.ot-1 and speed value 700 rpm.

The next step was a simulation of the machining of the face from a diameter of 100 mm to a diameter of 72 mm, a thickness of 10 mm (Fig. 5), and a diameter of 72 mm, a thickness of 12 mm.

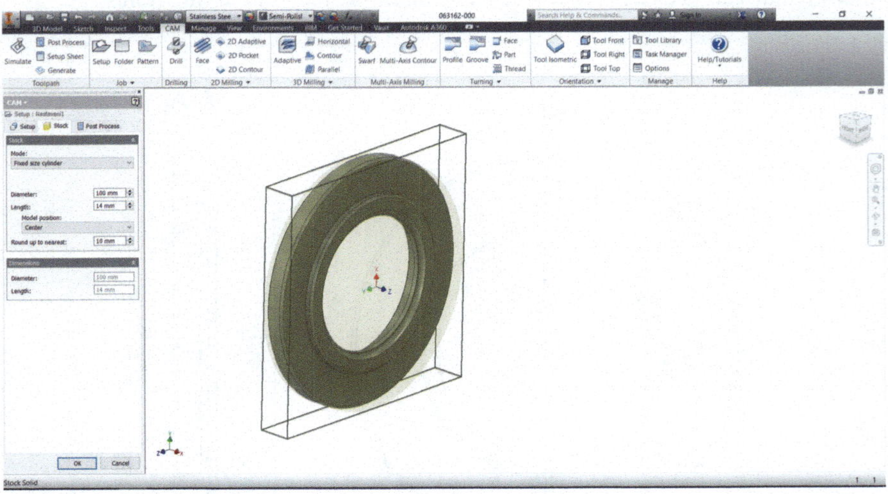

Fig. 3 Position of the coordinate system on the machined face

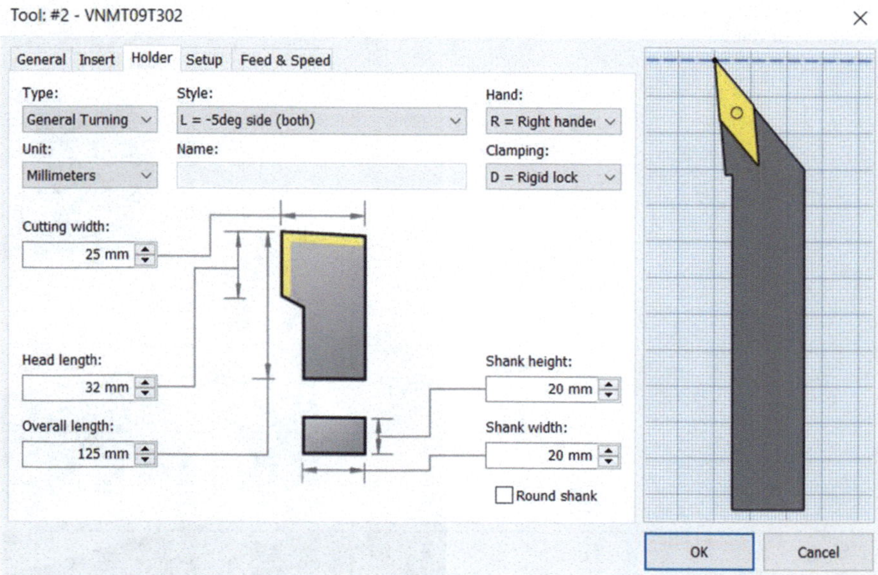

Fig. 4 Tool selection and cutting conditions selection

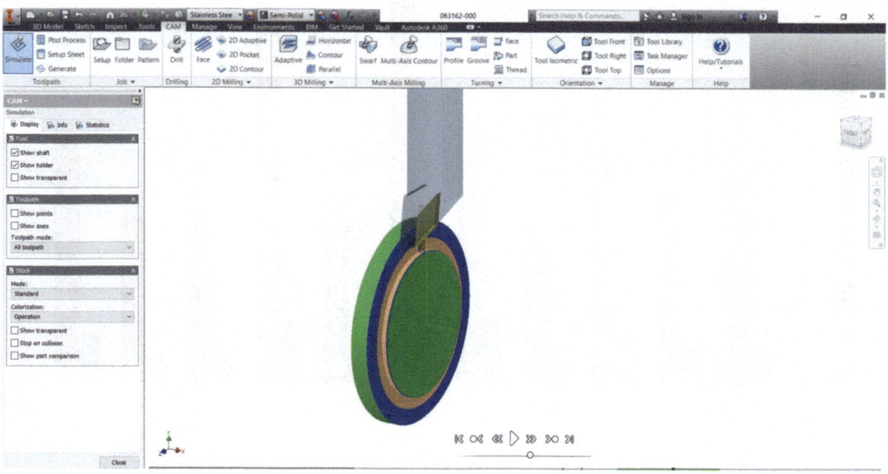

Fig. 5 Simulation of flange part face alignment

4 Measurement and Evaluation of Surface Roughness and Geometric Characteristics

Surface roughness measurements on the manufactured diameters of 100g6 mm and 62H804 mm were performed on a Mitutoyo SJ 400. The finished robotic flange was fixed in a hand vise with prismatic jaws (Figs. 6, 7 and 8).

Fig. 6 Measuring surface roughness of flange

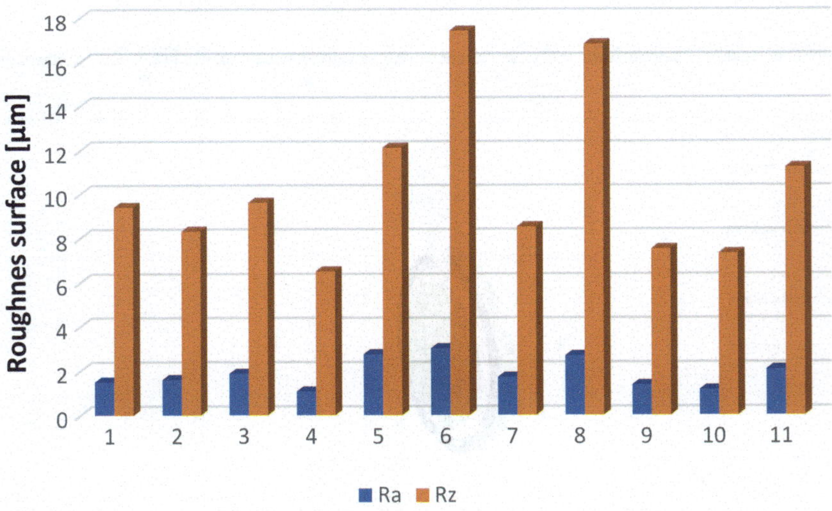

Fig. 7 Measuring value of surface roughness Ra and Rz for diameter D100g6 mm

The measurement of geometric characteristics, with cylindrical diameter of 100 and 62 mm, and alignment of these dimensions were performed on a Roundtest 120 RA (Fig. 9).

The measured values of cylindrical deviation are shown on cylindrical diameters 100g6 Fig. 10. The measured values of cylindrical deviation are shown on

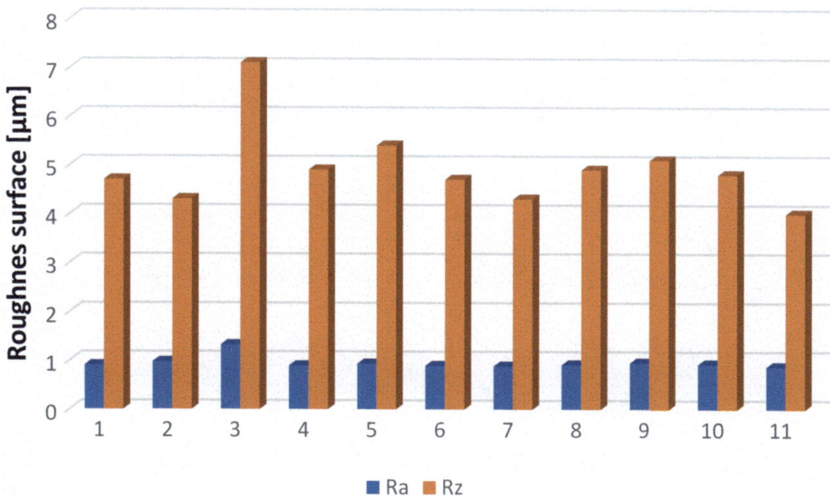

Fig. 8 Measuring value of surface roughness Ra and Rz for diameter D62H8 mm

Fig. 9 Measuring roundness and concentricity of flange

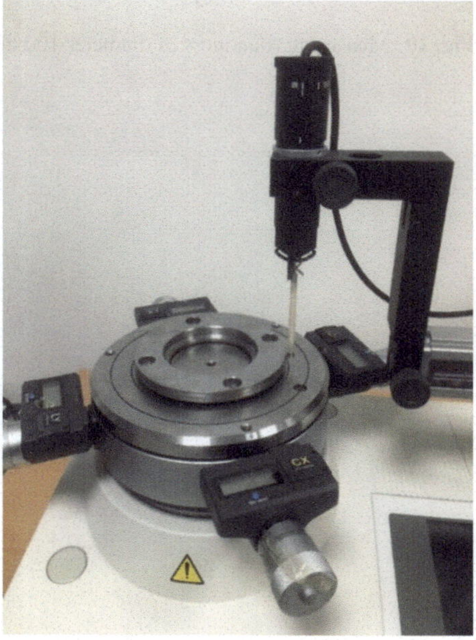

cylindrical diameters 62H8 Fig 11. The measured values of alignment are shown on cylindrical diameters 100g6 Fig. 10, mm and 62H8 Fig. 12. When working on a Leadwell T5 CNC lathe (Fig. 13), the operator must clamp the workpiece and tools correctly. The cutting conditions must be adapted to the size of the clamping surfaces of the blank and its rigidity.

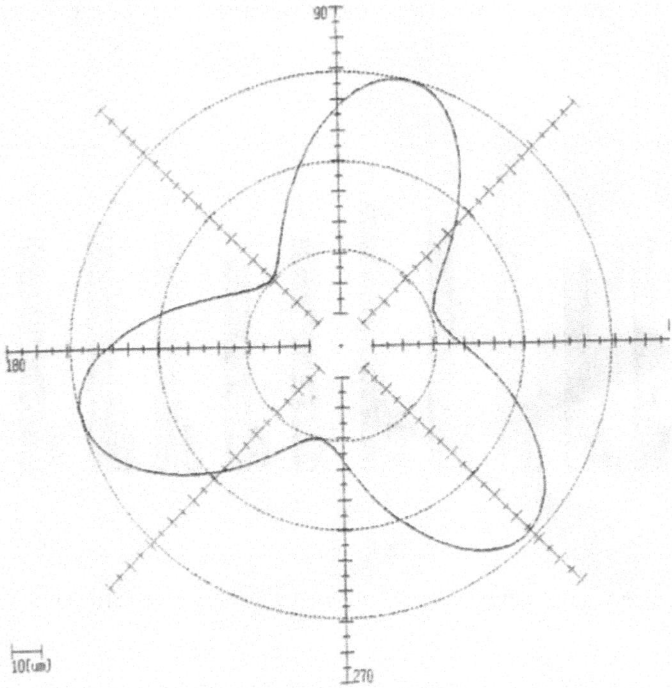

Fig. 10 Measuring roundness of diameter 100 mm

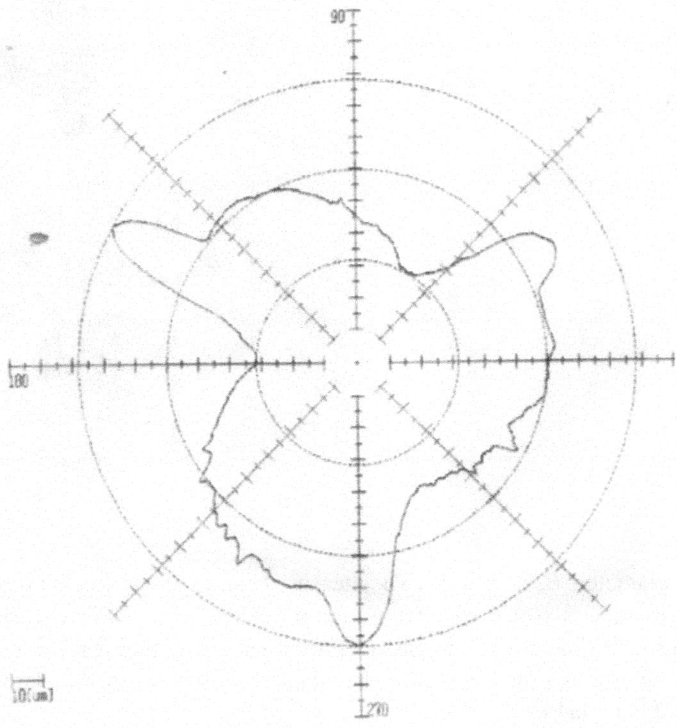

Fig. 11 Measuring roundness of diameter 62 mm

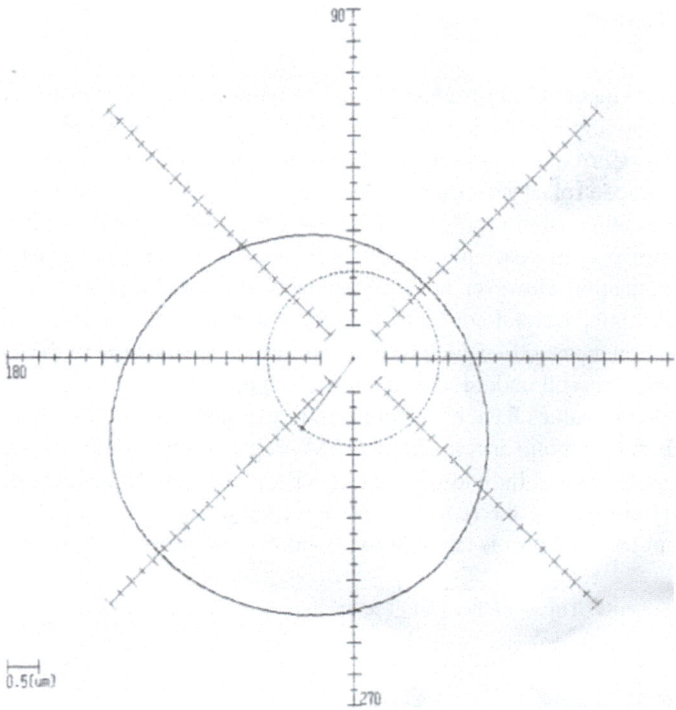

Fig. 12 Measuring concentricity of diameters 100 mm with 62 mm

Fig. 13 Production of flange face on CNC lathe

5 Conclusion

With the deployment of industrial robots, safety requirements have evolved. The classic conventional robots were deployed first, which may eventually replace new collaborative robots. Currently, robots are deployed in various areas of life, service, public sector, remote-controlled construction machines, autonomous submarine vessels at great depths, eating-assist robots, etc. Currently, there are various suppliers of CAD systems. We can say that each of them offers a basic CAD package and an extension of the CAM application. However, not every application is suitable for any part shape, for example, machining technology. It is given by the so-called internal core of the processor—application parasolid—which is designed by the manufacturer of the CAM application and which is still undergoing an innovative process.

The following values have been achieved in the manufacture of a flange component for driving a robotic arm made by CAM programming. Mean arithmetic value for diameter 100 mm of individual quadrants Ra = 1.92 μm. Roundness deviation is 22.3 μm. By comparing the required and achieved deviations, it can be concluded that the manufactured part is suitable for assembly and prescribed function.

Acknowledgements This work is a part of research project VEGA 1/0045/18.

References

1. https://www.turbosquid.com/3d-models/kuka-robot-titan-3ds/1034991
2. https://factoryautomation.cz/top-10-vyrobcu-robotu-na-svete-kolik-instalovali-robotu/
3. Stejskal, T., et al.: Specific principles of work area stiffness measurement applied to a modern three-axis milling machine. Int. J. Adv. Manuf. Technol. **102**, 2541–2554 (2019)
4. Michalik, P., et al.: Assessment of the production reducer for clamping the drilling tools. In: Advances in Manufacturing Engineering and Materials, pp. 557–566. Springer, Cham (2019). ISBN 978-3-319- 99352-2. https://link.springer.com/book/10.1007/978-3-319-99353-9?page=2#toc...
5. Michalik, P., et al.: Intelligently programming of holes machining. Manuf. Eng. **9**(4), 63–65 (2010)
6. Molnár, V., et al.: Monitoring of dependences and ratios of normal contact forces on hexagonal idler housings of the pipe conveyor. Measurement. **64**, 168–176 (2015)
7. Michalik, P., et al.: Numerically Controlled Machine Tools and Their Programming. TU Kosice, Presov (2016)
8. Balog, M., et al.: Plánovanie v Strojárskej Výrobe. Tribun EU, Brnno. ISBN 978-80-263-1078-5 (2016)
9. Dębski, H., Lonkwic, P., Różyło, P., et al.: Numerical and experimental analysis of the progressive gear body with the use of finite-element method [Badania numeryczne i doświadczalne konstrukcji chwytacza progresywnego z wykorzystaniem metody elementów skończonych]. Eksploatacja i Niezawodnosc – Maint. Reliab. **17**(4), 544–550 (2015)
10. Strohmandl, J.: Use of simulation to reduction of faulty products. UPB Sci. Bull. D Mech. Eng. **76**(3), 223–230 (2014)
11. Michalik, P., et al.: Comparison measurement of the distance between axes of holes with the Roundtest RA-120 and Thome Präzision-rapid. Appl. Mech. Mater. **616**, 284–291 (2014)
12. Michalik, P., et al.: Programming CNC machines using computer-aided manufacturing software. Adv. Sci. Lett. **19**(2), 369–373 (2013)
13. Michalik, P., et al.: CAM software products for creation of programs for CNC machining. Lect. Notes Electr. Eng. **141**(1), 421–442 (2011)

A Systematic Approach to Occupational Safety and Health Focusing on Prevention of Damaging Aspects and Risk Categories in Mining Company

Michaela Očenášová, Andrea Seňová, Henrieta Pavolová, Martin Rovňák, and Mária Shejbalová Muchová

1 Current Status of the Occupational Health and Safety Risk Assessment in Slovakia

The global economic growth results in continuous developing of new technologies, introduction of which often requires adopting changes in types and numbers of job positions regardless of the branch of industry. Moreover, new technologies produce also unanticipated health hazards, and often documented in the working environment are major occupational injuries or diseases arising out of employment. Hence, it is more than necessary that the employers would at all times pay close attention to establishing conditions ensuring occupational health and safety. An important aspect of supporting occupational health and safety is continuous assessment of the working environment condition and subsequent preventive actions that are based on both foreseeing the existing hazard factors and their eliminating. That safe worksites could be formed introduced must be a pool of legislative, social, economical, organizational, technical, educational, and health-promoting measures [1, 2].

The growing coal production evokes over the last 10 years requirements of abatement emissions and research of alternative energy sources. The International Energy Association expects an increase in the use of coal to 55% by 2030. This is in

M. Očenášová (✉) · A. Seňová · H. Pavolová · M. S. Muchová
Faculty of Mining, Ecology, Process Control and Geotechnologies,
Institute of Earth Resources, Technical University of Košice, Kosice, Slovakia
e-mail: michaela.ocenasova@tuke.sk; andrea.senova@tuke.sk;
henrieta.pavolova@tuke.sk; maria.shejbalova.muchova@tuke.sk

M. Rovňák
Faculty of Management, Institute of Environmental Management,
University of Prešov, Prešov, Slovakia
e-mail: martin.rovnak@unipo.sk

L. Knapcikova et al. (eds.), *New Approaches in Management of Smart Manufacturing Systems*, EAI/Springer Innovations in Communication and Computing, https://doi.org/10.1007/978-3-030-40176-4_12

consequence to increase in the energy intensity of national industries and an increase in oil and gas prices [3].

The support for coal mining in Slovakia influences the social situation in the associated regions not only for the 4464 employees in coal mining but also for the producers of electricity across the downstream sector, an estimated 15,000 employees [4].

Occupational health and safety policy is an integral part of every organization and company. It is necessary to ensure compliance with legal and technical regulations aimed at reducing the risk of accidents at the workplace. It is also necessary to take specific measures in this area which are specific and dependent on the activity of a particular company.

Relating to the health and safety aspects in the Slovak Republic (SR) is a set of the European Union (EU) directives. From the directives resulting obligations are further applied within the Slovak Republic legislation. Most important for the SR occupational health and safety issue are the following acts: Act No. 311/2011 Coll. (hereinafter referred to as the "Labour Code") and the Slovak National Council (NR SR) Act Nos. 124/2006 Coll., 335/2007 Coll., and 309/2007 Coll.

The Act on Protection, Support and Development of Public Health defines *risk* as "the probability of occurrence of a harmful effect on a human caused by exposure to a dangerous factor" [5].

Risk can be in more detail discussed and known as [6, 7]:

- Danger (probability) of suffering a loss, at certain level of being aware of the environment condition
- Possibility of occurring of an event that would prevent or endanger attainment of goals by an individual or an organization
- Danger (probability) of arising of negative deviations from set levels of goals of an individual or of an organization

Health and safety at work (HASAW) is a state of working conditions where the effects of various harmful and dangerous substances in the working environment on the human organism and its health are minimized or eliminated. Occupational safety also means a set of conditions that minimize or reduce incidents that can leave a negative impact on the health of employees [8].

Within the European Union, the Framework Directive 89/391/EEC of 12 June 1989 on the introduction of measures to encourage improvements in the safety and health of workers at work presents an important basis for assessing the risks. On the EU level, not rigidly stipulated are rules on how the risk assessment is to be performed, and hence the Framework Directive has been transposed into the national legislation. Practical implementation varies from a worksite to another and is always adapted to their specific features [6, 9].

At present, systematic OSH management methods are used in Slovakia, which are primarily based on the identification of hazards at the workplace, and in the second step, the risk rate is assessed. Consequently, it is necessary to adopt measures and criteria for safe work in the company. The whole procedure must be based on legal and technical standards of the Slovak Republic. It is also necessary to systematically manage these risks in the company.

The following three types of OSH management policy are used in Slovakia [8]:

A safe enterprise is a tool aimed at establishing an effective OSH management system. When the conditions are all established, the company is certified to be a safe enterprise in all field, such as fire safety of technical equipment, health conditions at the workplace, safety of the environment, etc. It consists of two parts: assessment of the OSH management system (verification of the functionality of the OSH management system) and assessment of the achieved OSH level (evaluation of the actual OSH status level).

OHSAS 18001 is a globally recognized standard that enables management to identify weaknesses in an organization. It is a simple tool for achieving OSH certification.

Own enterprise management system is mainly used by small- and medium-sized enterprises.

The basic principle applied in OSH is the man-machine-medium system, and the basic element of this principle is man. The human factor is influenced by its physical and mental predispositions. Machine represents any work equipment used in the enterprise. Medium are all factors affecting a person during working hours.

The basic objective of OSH policy is occupational safety, which is characterized as a state of the workplace, which provides a high level of assurance that the risk of danger to worker's lives and health is minimized. The objective of OSH is to prevent unforeseen external influences, which could lead to the destruction and demage of property or to health hazards at work [10].

Important principles of health and safety at work include:

- Apply the same safety priority to the protection of the environment, quality, and economic aspects.
- Ensure a high level of OSH and its continuous improvement.
- Raise awareness of OSH employees.
- Ensure that other stakeholders are aware of the risks as well as of the confirmation.
- Provide information to customers how to dispose of goods.
- Cooperate with other stakeholders to prevent accidents and improve OSH [11].

The most important tasks in health and safety at work (HASAW) management include [12]:

- Coping with the safety rules when working remains an important ethical and humane task.
- The gradual expansion of HASAW from manufacturing industries into all other spheres.
- HASAW needs strong advocates in all aspects of social life.
- Health and safety at work must be part of the corporate culture within the IMS of the company.

- Rather than merely focusing on the technical aspect of the matter, HASAW legislation should be able to offer systematic solutions.
- The need to use the communication and information technology.
- The legislation that can be effectively applied even in challenging and non-standard situations.
- Integration of national structures and practices into the common European structural framework.
- HASAW system accounts for one of the fundamental pillars of each business.

To fulfill all of the above-listed points, we need flexible management who would be able to shift away from old practices and adapt to new HASAW conditions. The managers are required to keep their employees updated and communicate with them, and they are responsible for their education and supervision. Nevertheless, they are still expected to be open to the opinions of their co-workers.

The term "safety culture" expresses the sum of those human activities that aim to create working conditions, where HASAW stands for the common effort of employees and employers at all levels of management to create conditions for safe work. To ensure such conditions, it is necessary that the following should be true and borne in mind [12]:

- One must not only act in agreement with laws but also take an extra step.
- One should realize that the risk will not disappear.
- The extent of acceptable risk is not constant, but it derives from the current technical and scientific level of knowledge in the given sector.
- Proactive risk management must also include a set of measures applicable in case an undesirable event should take place.
- All people concerned must have sufficient knowledge of the potential risks and appropriate procedures that they need to apply when the circumstances are such.
- The risk-maker must have sufficient knowledge so that they can take appropriate action whenever an undesirable event urges such an intervention.
- Education and innovation are unmissable constituents of the health and safety.

The task of the management of each company in the field of health and safety is to create the so-called safe enterprise. Such an enterprise is expected to meet the health and safety requirements as well as to fulfill all attributes of prevention of risky or dangerous events, and at the same time, they should ensure the gradual minimalization of individual risks through the application of safety innovations. A safe enterprise should also ensure that the health of workers as well as environmental protection is of great importance in the decision-making process, while the economic aspects come only second. The next attribute of a safe enterprise is the effective encouraging of employees to maintain the highest possible level of safety in the workplace, especially by using personal protective equipment (Table 1).

Table 1 Development of OSH management systems

Title	Name of the document	Country/ organization	Year
ISO 45001	Occupational health and safety management systems: Requirements	ISO	2016
BS OHSAS 18001	Occupational health and safety management systems: Requirements	Great Britain	2007
ILO OSH 2001	Manual for OSH management system	MOP	2001
OHSAS 18001, 18002	Occupational health and safety management systems. Related Standard (18002): Guidance on the implementation of OHSAS 18001	International institutions	1999, 2000
BS 8800	Guidelines on occupational health and safety management systems	Great Britain	1996
ISA 2000	Requirements for safety and health management systems	SGS Yardley ICS	1996
AIHA OHSMS 96/3/26	Occupational Health and Safety Management System	USA	1996
Australia standard DR 96311	OSH management system—general guidelines on system principles and techniques	Australia, New Zealand	1996
NPR 5001	Instruction on occupational health and safety management system	Netherlands	1997
MLN 1999	Manual for OSH management system	Japan	1999
Code of Practice	Management of health and safety at work	Great Britain	1992, 1999
ISRS	International Safety Assessment and Certification System	AGA/DNV Norway	1995
VCA	Occupational Safety System—Checklist for subcontractors	Benelux	1997
ASCA	Occupational Safety Inspection Program in Enterprises	Germany	1996
BTP 0011/98	Internal security management system in the company	Czech Republic	1998
PN-N-18000	Polish standard for occupational health and safety management system	Poland	1999
MALWA	Safety and risk management program	Poland	1996
Internal control AFS 1996	Obligatory regulation to safety, health and environmental management system	Sweden	1996
Internal control	Obligatory regulation to safety, health and environmental management system	Norway	1991
TUTTAVA	Program for evaluation and motivation for safe work and health protection	Finland	1994
MIL-STD-882C	System requirements for the security program	USA	1993
Fed. reg.59:3904-3916	Manual for OSH management system	USA	1989

Source: Manažérstvo rizík 2 [12]

ISO 45001:2018 (issued on 12 March 2018) that is currently coming to force is aimed at helping organizations to manage HASAW-related opportunities and risks as well as at improving HASAW performance to protect workers of all organizational levels. Having substituted the original OHSAS 18001:2007 norm, the abovementioned regulation is still being aligned with other management system standards, ISO 9001:2015 and ISO 14001:2015. Companies that are currently certified by the regulation OHSAS 18001:2007 will be expected to transfer and implement the newly set requirements of ISO 45001:2018 within a 3-year transition period.

1.1 New Approach to the Occupational Safety Issue

A new approach to the occupational safety requires a methodical style targeted to prevent occurrence of negative aspects—prevention of consequences of accidents. Prioritized is participative management focusing on the human factor and on safety and health culture. Hence, within a company, the occupational safety and health (OSH) is becoming an issue not only for the management but for the employee as well. The Slovak Republic intentions in the field are based on the OSH European strategy concept for the years 2008–2012, which contains the following challenges [9]:

- *Holistic approach* with respect to individual classes of work and inclusive of psycho-social aspects and ease at work
- *Promoting the culture* of risk prevention and building partnerships in the OSH, improving the social dialogue, economic stimulation, and social responsibilities
- *Advocating social policy* as an aspect of competitiveness

New approaches of the EU to protecting the employee have become the ground of the Slovak Republic state policy on occupational safety and health. Corporate occupational safety policy connects to the national policy, objectives of which are run down below:

- *Strengthening of the legal responsibility* relating to the OSH issues
- *Expanding positive attitudes* of the top-ranking political and state representatives, employers, employees, and general public to the OSH
- *Advocating systematic solutions, planning, complexity*, and introducing good practices in the OSH management
- *Ensuring consistent implementation of regulations* in everyday practice and compliance with the OSH principles
- *Supporting economic involvement of employers* so that safe work would become "profitable"
- *Advocating holistic approach to* resolving the OSH issues, as well as securing satisfactory working conditions
- *Ensuring effective labor inspections* and continuous supervision over applying the OSH regulations [9]

1.2 *Hazardous Work and Damaging Factors of the Working Environment*

Hazardous work is the work with an increased risk of occurrence of occupational illness, professional poisoning, and other health damage related to work. When defining the risk, it is important to identify risk factors first as these may adversely influence health of the employee; included into the group can be [6]:

- Chemical factors (dust, chemical agents, allergens, toxic substances, carcinogens, and mutagens)
- Biological factors (microorganisms and human endoparasites)
- Physical factors
- Physical stress and psychical stress

To evaluate the influence of a risk factor on health, performed must be both qualitative and quantitative measurements, and results of these shall be compared against the permissible limit values. When presence of a risk factor in the working environment is confirmed, the employer becomes obliged to elaborate and implement operating rules for manipulations with the risk factor, when integral parts of these shall be [5]:

- Risk assessment
- Evaluation of possibilities of eliminating the damaging factor effects
- Measures for diminishing the exposure
- The way of ensuring appropriate medical supervision for the concerned employees by way of occupational health service

The risk assessment results in inclusion of individual professions into risk factor categories. Developed for individual working environments, there were four categories of work activities. If the risk assessment result points to the third or fourth category, the employer is obliged to introduce measures that are to cut down the hazardous factor effects to the lowest attainable level, as well as to submit a proposal to include the work task among risk works as classified by the pertinent regional public health authority [5].

The risk work task-performing employee is, generally, affected by damage to his or her health—by occupational disease. In case of employees performing work activities falling into categories 3 and 4, the employer is obliged to arrange for their regular medical preventive examinations (once in 2 years for category 3 and once a year for category 4), the objective of which is to early establish onset of the disease and to temporarily disclose the employee from the risk work up to his/her full recovery and reversal of his/her health condition to normal. The employer, employees of whom are performing risk work tasks, is under the obligation to prepare and submit to the regional public health authority by 31 December of each year information on the results of evaluation of health risks and on the measures implemented for their reduction or elimination at workplaces where the employees are performing such risk tasks [6].

2 Risk Factors at the Workplace in a Mining Company

The area of mining and quarrying is one of the primary sectors of the economy, in which the rocks are separated in different ways, and subsequently these materials are loaded and transported to places where they are further processed. Working in mining is one of the physically and mentally demanding works, and undoubtedly all mining activities also entail the dangers and risks associated with this work. In the past, these workplaces have experienced a high number of injuries. Therefore, there is a need to continually analyze and predict possible hazards and risks at these workplaces.

Negative impacts in these workplaces include especially high dust load, noise, excessive vibration, heat-climatic and humidity conditions of workplaces, size and dullness of workload, lack of lighting, intensity of work process, and eventually impact of long-term stress in underground conditions, if we talk about deep mining.

The health status of mining workers, which is related to their physical and mental resistance, also plays an important role in taking into account the risk in the area of mining activities. It is precisely in the area of mining that constant monitoring of the health condition is necessary, prior to taking up employment, in the form of an initial checkup as well as of the entire health prevention during the employment period. It is precisely the prevention and creation of an optimal working environment that should be a priority in these companies [10].

It is essential that every worker working in these companies knows these hazards and is able to avoid these dangers also through OSH policy.

Responsibilities of mining workers in the workplace are to [10]:

- Verify the safety status of the workplace before and during work, discontinue the operation in case of danger, and notify the foreman of this fact
- Use protective equipment and follow the prescribed regulations
- Comply with and follow safety instructions such as safety labels, warning signs, etc.
- Move away from the workplace only with the agreement of supervisor or if necessary
- In case of accident or breakdown, report these serious facts to his supervisor

The basic obligations of an undertaking carrying out mining activities are laid down by the Mining Act and also by the Decree of the Slovak Mining Office on Occupational Health and Safety and Operational Safety in Mining and Underground Mining Activities.

Dustiness and noise of these workplaces could be included among the greatest risks in this area as hazardous work.

2.1 Dustiness as a Risk Factor at the Workplace in a Mining Company

Dust and dust particles affect the body in every environment. They can enter the human body by inhalation, orally and percutaneously. With increased amounts, deposited dusts in the lungs can cause various diseases, especially if their mass and particulate basis is composed of quartz and minerals of similar structure. The problem with these particulate pollutants is that they act in the lungs for a long time as they are insoluble or only slightly soluble [13].

Samples for measuring the concentration of airborne dust are carried out at the workplace by passing a certain amount of air through the filters. Regulation and flowmeter are used as the vacuum source. The samples taken are evaluated in the laboratory gravimetrically. Samples are taken from internal and external devices using conditioned probes, e.g., from closed transfer shredders, crushers, separators, and the like.

The issue of dustiness in the workplace generally has two basic groups [14]:

1. The issue of dust exposure due to the risk of fire and thus combustible dust
2. The issue of dust exposure due to hygiene and work categorization

The dust quantification is carried out with the aid of dust measurement and thus an analytical evaluation of the dust. According to the type of sampling, we distinguish four types of measurement [14]:

1. Measurement of airborne dust concentration inside technologies
2. Measurement of sedimentation dust at and around the technology
3. Special dust measurements in conveying piping isokinetic sampling and measurement of separator efficiency
4. Hygiene measurements of airborne dust at the workplace for the specified occupational risk category

The following types of sampling are distinguished:

– Personal sampling in the worker's breathing zone (whole shift sampling), total dust, for dust that does not contain respiratory fractions; and respiratory dust, dust containing respiratory particles.
– Stationary sampling at reference points to determine total and respiratory dust. Samples are made with a small light dust meter. The minimum number of shifts during a shift is 3 [14].

In relation to the risk of dustiness, it is necessary to address dust prevention in mining areas and workplaces and to improve these technologies to help reduce dustiness. Technical means of dust reduction:

(a) Reduce the amount of dust particles that arise during technological processes.
(b) Eliminate dust at the source of dust particles.
(c) Reduce the amount and concentration of dust in air by efficient means such as ventilation, scrubbing, filtration, etc. [10].

2.2 *Noise as a Risk Factor at the Workplace in a Mining Company*

Noise is an undesired sound intensity (noisiness) which is quantified in decibels (dB). The decibel scale is a logarithmic one, which means that increase in the level of noise by 3 dB presents in fact twofold increase in the noise intensity. Nonetheless, it is not only intensity that is a gauge of noise harmfulness; another and more important factor is the duration of exposure to noise. For that reason, used for measuring are time-weighted average levels of noise [15].

Noise at work presents a global issue identified in a host of branches of industry. Exposure to excessive noise may result in damages to hearing. Loss of hearing due to noise may result either from on-off exposure to an acoustic pulse in excess of 140 decibels (dB(C)) or from exposure to extreme intensity noise over 85 decibels (dB(A)) for several hours of each workday for a prolonged period of time. According to the estimates, exposed to noise in the EU countries are some 60 million workers—30% of the labor force [16, 17].

According to the level and nature of the work and work environment factors that can potentially be of impact upon the health of the employees, and according to evaluation of health risks and of changes in the health status are work tasks included into four categories [5]:

1. **First category**—includes work tasks not involving risk of damage to health of an employee due to work tasks and/or to the working environment.
2. **Second category**—includes work tasks at which, considering the risks, no damage to health can be presumed (un-exceeded limits governed by special regulations).
3. **Third category**—includes work tasks as which:

   (a) Exposure of the employee is not reduced by the work task factor and by the working environment to the level of limits stipulated by special regulations. To decrease the risk, introduced must be organizational measures inclusive of using personal protective work instruments (PPWI).
   (b) Work tasks at which exposure of the employee is reduced by technical measures to the level of prescribed limits but at which mutual combination of the work factors can result in damages to health of the employee.
   (c) Works tasks with no prescribed limits, though at which exposure to work factors can induce damage to health of the employee.
   (d) Work tasks performed in a controlled zone.

4. **Fourth category**—includes exceptionally, and for a limited period of time not in excess of a period of 1 year, the following:

   (a) Works at which it is impossible to reduce, by introduced technical and/or organizational measure, exposure of the employee to the work factor and the working environment factor to the level of set limits; at which exposure to the work factor and the working environment exceeds limits; and at which experienced are changes in the health status of the employee related to effects of factors and introduced have to be technical measures and other specific protective measures inclusive of using personal protective work instruments.

(b) Works that by the rate of exposure to individual work factors and working environment should fall in the third category but where mutual combination of work factors and of working environment increases the health damage risk.

Deciding on framing a risk or hazard into a category is a Slovak Republic regional public health authority based either on a proposal of the employer or at its own discretion.

The most commonly used risk assessment methods include [14]:

- **The point method of risk assessment** is the simplest of the methods of risk assessment, which is also the basis of several methods for semi-quantitative risk assessment. The assessor aims to prepare an expert estimate, i.e., to determine the likelihood of an accident. It can be described by relative word values. This is the frequency of occurrence of threats, usually using five degrees.
- **Data sheets (risk cadaster)** are a form of registration of causal dependence for a particular group of risks; they must contain at least information about the type of danger and its consequences.
- **The interactive catalog sheet for processing** is primarily the list of hazards listed in Regulation of the Government of the Slovak Republic No. 395/2006 Coll. on the conditions of providing personal protective equipment. The first part consists of the general information about the system under assessment; the legislative framework, which the processor can supplement with its own directives; and the technological procedures, which include a clause concerning safety and health at work, using hyperlinks to link to useful websites, an overview of the safety markings and signals, a sample form for providing PPE, and a template for specifying risk situations. The second part contains the identification of specific hazards in which the physical, chemical, and biological risk factors are summarized. Each identified hazard is accompanied by a commentary describing the hazard and the source of the hazard.

3 The Risk Category Noise Evaluation Process for the Worker Profession in Processing Industrial Minerals

The extent of work tasks of employees of the laborers in processing non-ferrous ores usually involves the below outlined activities:

- Inspections of the worksite and machinery at the beginning and end of the shift
- Monitoring the operation of belt conveyors

- Operating the belts, feeding tables, belt conveyors, and pumping units
- Cleaning and removal of foreign objects off the belt conveyers
- Securing commencing, operating, and shutting down of machinery and technologies and supervising the magnesite ore charging
- Monitoring proper run of devices when in operation
- Cleaning of floors and workplaces
- Evaluating appropriateness of the working environment conditions
- Managing the teamwork
- Recording technical data on progress and results of work tasks
- Securing dispensing of meals
- Keeping the working environment in proper order

Included among the risk factors having an impact upon the non-ferrous ores processing workmen can be as *noise* so *the microclimate*. Further, difficult working conditions characterizing factors are height, physical stress, night work, and lifting and carrying loads [16].

3.1 Level, Type, and Duration of Exposure to Noise

In question is the processing of raw magnesite ore in heavy suspensions. Operations are controlled from the control room and are continuous (24/7).

Elaborated for the purpose of illustrating exposure of the workman at processing non-ferrous raw materials to noise in a selected company has been a table showing how was the work-time during a shift scheduled for the profession, the purpose of which was to determine how long an employee is performing his or her professional operations when working in a noisy environment.

Based on observing operations of workmen at processing industrial minerals in a selected processing plant, it has been determined that a workman is for some 6 h and 30 min a day performing "excessively noisy" work tasks, as documented in Table 2.

3.2 Measuring Exposure to Noise

During measuring exposure to noise was the operator holding the position of workman at processing industrial minerals performing supervision of operations, pumps, washing units, vibration and resonance screens, and shifting of materials and of separators. As shown in Table 2, in the case of the profession, exposure to noise continued for 6.5 h. Established during the measuring were equivalent noise level $L_{\mathrm{Aeq,T}}$ of 100 dB and peak level $L_{\mathrm{CPk,T}}$ C of acoustic pressure of 144.6 dB. Calculated by measuring the determined values was standardized level of exposure to noise for an 8 h work-shift, $L_{\mathrm{AEX,8h}}$, of the workman processing industrial minerals of 99.1 dB,

Table 2 Overview of exposure to noise in a plant for the profession of workman at treatment of industrial minerals

Description of performed activities	Duration of individual activities		
	Time		Time
	From	To	Total
Arrival to the workplace and scheduling of tasks	7:00	7:15	0:15
Inspection of the machinery	**7:15**	**7:45**	**0:30**
Work cycle	**7:45**	**11:00**	**3:15**
Lunch	11:00	11:30	0:30
Work cycle	**11:30**	**14:15**	**2:45**
Cleaning works; entering records in books	14:15	14:30	0:15
Preparations for leaving the workplace; leaving	14:30	15:00	0:30
Noisy operations—6 h 30 min	8 h in aggregate		
Non-noisy operations—1 h 30 min			

Source: original (personal) processing

which was increased by measurement error of 3.6 dB(U). Simultaneously, assessed during measurements was efficiency of hearing protection safety products—so-called standard exposure to noise behind hearing protectors $L_{\mathrm{AEX,8h,R}}$. Measurements were performed with using hearing protectors Classic E.A.R. EN 352-2, SNR - 28 dB, with equivalent level C ($L_{\mathrm{Aeq,\,T}}$) of 99.1 dB for mining industry.

Upon deducting the effect of hearing protectors: 99.1 + 3.6 − 28 = 74.7 dB

Determined further, during measurements performed, was the peak level C behind protectors $L_{\mathrm{CPk,T,R}}$ for 144.6 + 3.6 − 28 = 120.2 dB.

3.3 Comparison of Measurement Results

3.3.1 Comparison of Correspondence and/or Non-correspondence of Results with the Standardized Level of Exposure to Noise Limit Level $L_{\mathrm{AEX,8h,L}}$

- *Protectors on* → 102.7 − 28 = 74.7 < 87—limit value of exposure to noise *was not exceeded.*
- *Without protectors* → 102.7 > 87—limit value of exposure to noise *was exceeded by* 15.7 dB.

Determination of the hearing protection degree is shown in Table 3.

Whereas $L_{\mathrm{AEX,8h,R}}$ 74.7 dB falls within the 72–78 dB interval, the degree of hearing protection proves to be *acceptable.*

Comparison of correspondence and/or non-correspondence of results with the exposure to noise operational values:

Table 3 Determination of the hearing protection degree

Standardized level of exposure to noise $L_{AEX,8h,R}$	Degree of protection
$L_{AEX,8h,R} > 87$ dB	Inadequate
(87–82 dB)	Acceptable
(83–77 dB)	Good
(78–72 dB)	Acceptable
$L_{AEX,8h,R} < 72$ dB	Too high (excessive)

Source: Guideline No. 30/2010 of the SR Ministry of Health of 07.09.2010 [18]

- Upper operational value: 85 dB
- Lower operational value: 80 dB

For the worker operating in processing industrial minerals, the standardized exposure to noise level increased by measurement error (U), reached the value of 102.7 dB > 85 dB, *and hence was exceeded by* 17.7 dB.

3.3.2 Comparison of Results with Peak Levels of C Acoustic Pressure

- Comparison with limit value of peak level of C acoustic pressure—140 dB:
  - For the workman at processing industrial minerals, the $L_{CPk,T+U}$ amounts to:

    $148.2 > 140$
    $148.2 - 28 = 120.2\ (L_{CPk,T,R+U}) < 140$
- Comparison with upper performance value of the peak level of C acoustic pressure ($L_{CPk,h}$)—137 dB:
  - For the workman at processing industrial minerals, the $L_{CPk,T+U}$ 148.2 > 137 dB, which means that the level was *exceeded* by 11.2 dB.

3.4 Inclusion of the Profession into the Risk Work Category

The governing *noise quantities* at workplaces are standardized level of exposure to noise and peak level of C acoustic pressure. For the purposes of the occupational health and safety stipulated for the noise, exposure risks are:

1. ***Exposure to noise limit values***
2. ***Exposure to noise operational values***$L_{AEX,8h,L} = 87$ dB a $L_{CPk} = 140$ dB:

(a) *Upper* operational values of exposure $L_{AEX,8h,a} = 85$ dB a $L_{CPk} = 137$ dB
(b) *Lower* operational values of exposure $L_{AEX,8h,a} = 80$ dB a $L_{CPk} = 135$ dB

While considered when exercising the limit value is also attenuation produced by hearing protectors, the contrary, i.e., disregarding the attenuation, is the case when exercising operational values. The noise limit value may not be exceeded for the worker regardless if she/he uses hearing protectors or not.

Performed within the profession is an activity (control room) necessitating mental concentration or hearing involving (verbal) communication—in question is an activity that may fall under individual categories of work.

Established for the protection of health of employees against non-specific, especially disturbing or abusing, effects of noise are operational values of standardized exposure to noise levels for individual groups of works, as categorized in Table 4.

Special groups of employees, per wording of the SR National Council Act No. 124/2006 *Coll.* on occupational safety and health, as amended, are:

- Pregnant women
- Mothers up to the end of the ninth month after childbirth
- Nursing mothers
- Juveniles
- Physically challenged employees

According to Decree No. 448/2007 *Coll.* regarding the details of factors of work and the work environment in relation to categorization of works from the viewpoint of risks to health and regarding the requisites of the proposal for classification of works into categories, included into risk category 4 for the noise risk factor are:

Table 4 Operational values of standardized exposure to noise L_{AEX}, 8 h levels A for work categories

Work category	Activity	Noise at workplace $L_{AEX,8h,R}$ (dB)
I	Activity requiring continuous concentration or unhindered communication; creative activity	40
II	Activity, in which communication stands for an important integral part of the work being performed; activities characteristic with enormous demand on precision, speed, and/or attention or alertness	50
III	Routinely performed activities, in which communication is a part of the work being performed; activities performed based on partial auditory information	65
IV	Activities executed while using noisy machinery or instruments, performed in noisy environment that does not meet the requirement to be included into categories I, II, or III	80

Source: SR Government Regulation No. 115/2006 *Coll.* [19]

(a) Works at which exceeded are upper operational values of exposure to noise and where the standardized level of exposure to noise $L_{AEX,8h}$ is exceeded by 10 dB or more or where the peak level C of acoustic pressure L_{CPk} is exceeded by 3 dB or more
(b) Works at which the standardized level of exposure to noise and peak level C of acoustic pressure correspond with the criteria of category 3 but when at a time observed in the employee are changes in hearing capacity due to effect of noise

Whereas in the profession of workman operating in processing industrial minerals exceeded was the peak level C of acoustic pressure L_{CPk}, by 11.2 dB, the profession will be included into ***category 4***. Exceeded by 17.7 dB was at a time the standardized level of exposure to noise $L_{AEX,8h}$.

In the selected plant, the employees performing in the profession of workman operating in processing of industrial minerals will be pursuant to provisions of Section 30, Sub. 5, Lit. a) of Act No. 355/2007 *Coll.* for category 4 once a year participating on preventive medical examination in relation to their work tasks.

4 Measures for Elimination or Reduction of Exposure to Noise

Indirect influences upon health and safety of employees present any influences on health and safety resulting from mutual effects of noise and warning acoustic signals, respectively, or other noises that must be monitored to reduce the risk of accidents.

4.1 Proposed Preventive and Safety Measures

Used at the control rooms of the evaluated workplaces are acoustic signals that are aired by the control unit at whatever deviation from the preset range of respective parameters and also whenever an exceptional event occurs. Installed on the outside of the plant are acoustic alarms, while double-paned windows that are facing the plant make the control room soundproof. Appointed employees are supposed to regularly check general conditions at individual workplaces using a portable noise-level meter that is capable of recording also peak values (max measurable short-term or momentary values of noise at a workplace). If, based on the risk assessment, the employer detects a condition hazardous to the health of employees, he shall provide an appropriate medical supervision, integral part of which are preventive medical checkups.

Entitled to examination of hearing are employees exposed to noise where the upper operational values are exceeded and also the employee exposed to noise

Table 5 Measures for elimination or reduction of exposure to noise

Measures for elimination or reduction of exposure to noise according to Government Regulation 115/2006 *Coll.*—so-called general principles of prevention	Performance of measures by the employer
The work methods designed to alleviate exposure to noise and selection of appropriate devices with minimum noise pollution	Replace resonance screen by less noisy ACH screens
Structural and spatial arrangement of the workplace and workstations	Sources of noise to be situated in the external environment and to be remotely manipulated from the control room/station
Appropriate information for and practical training of employees focused on proper use of working devices	Regular maintenance—scheduled temporary stoppages of operations
Technical measures, e.g., noise curtains, covers, noise-absorbing linings, dampers, or insulation	Control room—double-pained windows—soundproofing
Proper maintenance of processing devices, workstations, and on-site systems	Importance of regular lubricating—to reduce noise coming from among friction surfaces
Work organization focused on noise reduction: 1. Limited duration and level of noise the workers are exposed to 2. Appropriate schedule of work with adequate breaks for rest	Equipping the control room with PCs. Breaks—0.5 h for lunch

Source: original (personal) processing

where the lower operational values are exceeded provided the assessment and measurement have proven the existence of possible hazard to their health.

The employer shall provide to the employee suitable and appropriate personal protective safety instruments for hearing protection, so-called hearing protectors, in the case the risk of hearing damage of the employee cannot be eliminated by implementing other measures. Hearing protectors are selected so that the risk to hearing capacity would be eliminated or so that the possible risk would be reduced to the least possible extent (Tables 5 and 6).

4.2 *Proposed Manner of Informing the Employees About Risks Associated with Performing Hazardous Work Tasks*

In accordance with the SR Government Regulation No. 115/2006 *Coll.*, the employer shall make sure that employees exposed to noise that equals to or exceeds the lower operational values of exposure as well as delegates of employees for the OSH were reasonably informed about and practically trained in possible hazards arising out of exposure to noise.

Table 6 Measures resulting from exceeding the upper operational values

Measures necessitated by exceeding the upper operational values	Measure to be adopted by the employer
The employer shall forthwith adopt measures ensuring elimination or alleviation of exposure to noise	Through providing PPWI (personal protective work instruments)
Identified shall be workplaces or worksites at which employees are exposed to or at which it can be anticipated that they will be exposed to noise that exceeds the upper operational values of noise	Identify the workplaces that are hazardous as to occurrence of excessive noise
Such worksites or workplaces *shall be clearly delineated, marked by* safety health hazard signs, while used shall be warning signs, and, wherever technically practicable, will be declared *restricted areas*	Caution and warning signs
If exposure to noise equals to or exceeds the upper operational values of exposure, the employees *must use hearing protectors* The employer shall ensure and supervise usage of these protective instruments and will be held responsible for effectiveness of measures relating to providing PPWI (personal protective work instruments)	– Plugs with mean attenuation SNR 28 dB – Daily checked by the shift supervisor and the foreman – Once or twice a week checked by qualified OSH technician

Source: original (personal) processing

The practical training and information pertain especially to [7]:

- Natures of such risks
- Measures adopted in relation to elimination or mitigation of exposure to noise to the lowest possible level, inclusive of conditions under which such measures are being implemented
- Exposure to noise limit and operational values
- Results of assessing and measuring of noise along with substantiation of their importance and possible hazards
- Proper use of hearing protectors
- Reason for and manner of determining and publicizing symptoms of hearing damages
- Circumstance under which the employees are entitled to appropriate preventive medical examinations
- Safe work procedures from the point of decreasing exposure to noise to the lowest level possible

5 Conclusions

Although having been eliminated, there is a risk and danger of injury in connection to every work activity. The scope of work in mining is regarded as especially risky, which is mainly caused by the working conditions and the environment where miners perform their daily tasks. It is necessary to realize that protection and prevention of risk impacts must be addressed and taken into account at the early stage of the mining project, even before the workers start their mining for raw materials. Health and safety at work is an area that takes advantage of knowledge from several scientific disciplines. At present, it is no longer merely viewed as a legal obligation, but it often accounts for another form of competitive advantage among individual companies. It is also necessary to realize that the HASAW policy is not only an area to be dealt with solely by the employer, but the employees, whose role is to comply with these regulations, are also prompted to give it their full attention.

Health and safety at work should account for one of the fundamental pillars of each company. The critically important task for each risk management is to perform a job of high quality. When assessing individual risks, the managers ought to provide the measures that create a favorable environment in which individual risks are minimized to an acceptable level and thus prevent many work-related injuries. Each employee is responsible for making themselves familiar with the safety standards of individual operations and possible risks. They are expected to learn about procedures that prevent the risk occurrence or at least they minimalize the danger. Precautionary measures that are applicable by each employee can prevent many undesired events. Therefore, it is crucial that each worker is appropriately trained and pays maximum attention to the job they perform. We have focused on the noise risk factor in the field of mining and quarrying because employees working in mining companies often underestimate this risk. Employees frequently fail to acknowledge impacts of the factor upon their health, which is in contrast with factors such as dust or extreme physical load that are perceived as so-called tangible or clearly visible ones. Hence, from the part of employers, it is necessary to pay sufficient attention to the risk factor discussed.

Based on the steps taken, we arrived to the results that within the profession of workman at treatment of industrial minerals, exceeded was the peak level C of acoustic pressure L_{CPk} by 11.2 dB, when at a time exceeded was the standardized level of exposure to noise $L_{\mathrm{AEX,8h}}$ by 17.7 dB, which means that the profession should be included into risk category 4. According to provisions of Act No. 355/2007 Coll., based on the proclaimed risk category 4, the employer must adopt measures for elimination of these risks at the workplace without any delay and within a period of 1 year adopt any possible measures to degrade the category.

Contribution is partial result of projects solving VEGA MŠVVaŠ SR 1/0515/18 “The decision-making model of process of evaluating raw material policy of regions.”

References

1. Cehlár, M., Teplická, K., Seňová, A.: Risk management as instrument for financing projects in mining industry. In: SGEM 2011: 11th International Multidisciplinary Scientific Geoconference, vol. 1, pp. 20–25, 913–920. STEF92 Technology Ltd., Albena, Bulgaria (2011)
2. Jurkasová, Z., Cehlár, M., Khouri, S.: Tools for organizational changes managing in companies with high qualified employees. In: Production Management and Engineering Sciences, pp. 409–412. CRC Press/Balkema, Leiden (2016)

3. Laciak, M., Skvarekova, E., Durdan, M., Kostur, K., Wittenberger, G.: The study of underground coal gasification in laboratory conditions. (Studium podzemneho splynovania uhlia v laboratornych podmienkach). Chemicke Listy. **106**(5), 384–391. ISSN 0009-2770 (2012)
4. Skvarekova, E., Kozakova, L.: Brown coal and lignite issues from the perspective of sustainable development. Mineral Res Manage. **28**(2), 31–42 (2012)
5. Act No. 355/2007 Coll. on Protection, Support and Development of Public Health and on Amendments and Supplements to Certain Acts
6. Seňová, A., Čulková, K.: Risk management in area of security and protection of health during the work 2012. In: New Technologies—Trends, Innovations and Research, pp. 347–376. InTech, Rijeka (2012)
7. Hnilica, J., Fotr, J.: Aplikovaná Analýza Rizika ve Finančním Managementu a Investičním Rozhodování. (Applied Risk Analysis in Financial Management and Investment Decision-Making), p. 124. Grada Publishing, Praha (2009)
8. Turňová, Z., Harangozó, J., Tureková, I., Rusko, M.: BOZP v malých a stredných podnikoch. In: Problemy Profesjologii, p. 100. Zielona Góra, Polska, Polskie Towarzystwo Profesjologiczne, Instytut Inżynierii Bezpieczeństwa i Nauk o Pracy, Uniwersytet Zielonogórski (2012)
9. Šukalová, V.: Riadenie OSH—moderný prístup. (OSH management—a modern approach). In: Práce a Mzdy č. 1/2007. Poradca s. r. o., Žilina (2006). http://www.poradca.sk/SubPages/OtvorDokument/Clanok.aspx?idclanok=59733&zor=1 2010/11/14
10. Harangozó, J., Menčík, M.: Bezpečnosť a ochrana zdravia pri práci z pohľadu banskej činnosti. In: Integral Safety of Environs 2016. Proceedings of the International Scientific Conference. Žilina (2016)
11. Tureková, I.: Comparison OSH management system in Slovakia. In: Management in Environment 2014. Proceedings of the 14th International Conference, Bojnice. Strix, Žilina (2014)
12. Pačajová, H.: Manažérstvo Rizík 2, p. 131. Technická Univerzita v Košiciach, Košice (2016)
13. Bobro, M.: Funkcie minerálnych disperzoidov v pracovnom prostredí slovenských baní. Acta Montanistica Slovaca. **7**, 74–78 (2002)
14. Turňová, Z., Balog, K.: Complex safety evaluation of welding workplace. In: Proceedings from International Science Conference. Integral Safety 2011, p. 123. Vydavateľstvo AlumniPress, Trnava (2011)
15. OSHA European Agency: Historical background on occupational safety and health legislation. http://osha.europa.eu/sk/legislation/directives. Accessed 2 Feb 2011
16. Seňová, A., Teplická, K., Pavolová, H.: The process of risk assessment method RBMI in the mining Enterprise. In: SGEM 2015, pp. 705–712. STEF92 Technology, Sofia (2015)
17. European Risk Observatory: Outlook 1 New and Emerging Risks in Occupational Safety and Health, p. 24. The Publications Office of the European Union, Luxemburg (2009). http://osha.europa.eu/sk/publications/outlook/te8108475enc_osh_outlook 2013/28/01
18. Guideline No. 30/2010 of the SR Ministry of Health of 07.09.2010 that adjusts the procedure used at assessing the exposure to noise level in employees using hearing protectors
19. SR Government regulation No. 115/2006 Coll. on minimal health and safety requirements for protecting employees against the exposure to health related risks and hazards

Ensuring the Reliability of Separation Equipment Based on Parameter Identification of the Operation Process

Ivan Pavlenko, Oleksandr Liaposhchenko, Vsevolod Sklabinskyi, Vitalii Ivanov, and Oleksandr Gusak

1 Introduction

The processes of formation and separation of heterogeneous dispersed systems (e.g., emulsions, suspensions, and aerosols) are associated with the hydromechanical, heat, and mass transfer processes [1] in industrial machines and apparatuses. Consequently, the separation equipment is an integral part of industrial enterprises. In this case, applying modern approaches to ensuring the reliability and energy efficiency of the main process is an integral part of designing the corresponding technological equipment [2]. Investigation of hydrodynamic separation processes complicated by the combined heat and mass transfer processes is impossible without solving the fundamental problems of hydromechanics, as well as heat and mass transfer. Due to this fact, the formulated theme is topical, which has scientific and practical importance in the fields of energy efficiency, resource conservation, and environmental safety of technical and industrial systems.

It should be noted that due to the strategy of optimization of static and dynamic simulations of the separation process, the following main problems should be solved for designing the corresponding technological equipment [3]:

- Development of physical models of separation processes with the consequent analysis of the influence of technological by the performance criterion as a target function.
- The choice of types of technological equipment and its parameters for obtaining the required efficiency of the separation process.

I. Pavlenko · O. Liaposhchenko · V. Sklabinskyi · V. Ivanov (✉) · O. Gusak
Sumy State University, Sumy, Ukraine
e-mail: i.pavlenko@omdm.sumdu.edu.ua; o.liaposhchenko@pohnp.sumdu.edu.ua; v.sklabinskyi@pohnp.sumdu.edu.ua; ivanov@tmvi.sumdu.edu.ua; o.gusak@pgm.sumdu.edu.ua

L. Knapcikova et al. (eds.), *New Approaches in Management of Smart Manufacturing Systems*, EAI/Springer Innovations in Communication and Computing, https://doi.org/10.1007/978-3-030-40176-4_13

– Optimization of the separation process within the different types of simulations by the complex criterion of efficiency.

Industrial samples of highly efficient two- and three-stage separators previously developed at the Scientific and Research Laboratory of Granulation and Mass Transfer Equipment of Sumy State University have been successfully passed all the pilot-scale and acceptance tests. As a result, these devices have been implemented at the industrial enterprises PJSC "Ukrnafta" and PJSC "Ukrgazvydobutok" (Ukraine), "LUKOIL Group" (Russia), and "Regal Petroleum Ltd." (UK). Moreover, the results of the acceptance tests proved that in terms of the main technical characteristics, the proposed inertial-filtering separators occupy an intermediate position between gravitational-inertial separators of PJSC "Gazprom" (Russia) and filters-coalescers of "Pall Corporation" (USA).

According to the results of preliminary tests of experimental samples, the main characteristics of the modular separation equipment and combined multifunctional separators were determined. Particularly, for ensuring the efficient separation of particles with the average size of 5 μm, the average velocity is in a range of 0.18–0.72 m/s, and the hydraulic resistance is in a range of 0.15–2.0 kPa for units and 15–30 kPa—for apparatuses. In this case, the efficiency of a single stage of the modular separation equipment is in a range of 70–90%, and this indicator for the combined multifunctional separators rises up to 99.5–99.9%.

Previous studies have shown that combining the inertial-filtering method of separation with the vibrational one allows enhancing the separation efficiency and extending a range of using the corresponding technological equipment. However, theoretical investigations of mechanisms of vibrational-inertial and vibrational-inertial-filtering separation are insufficiently studied and require developing advanced techniques for designing the technological equipment.

2 Literature Review

The high importance of the problem is highlighted by recent research projects and relevant publications in the mentioned scientific area. Particularly, the research article [4] is aimed at mathematical modeling of the hydromechanical and mass transfer process in the field of deposition of aerosols from gas streams in chemical, oil, and gas industries.

The research [5] presents the application of the Eulerian-Lagrangian methodology for studying the steam-water separation. The article [6] presents spatial CFD analysis for studying the mal-distribution steam flow in using the module "Fluent" in the ANSYS platform.

The gravitational separation of oil-water mixtures using super-hydrophobic fabrics with extremely high flux was investigated in the research work [7]. The effects of ultrasonic flotation on the separation selectivity and kinetics of the process were analyzed in the article [8]. The continuous oil-water separation by means of the magnetic drive was proposed in the paper [9].

Experimental research of gas-liquid turbulent swirling flow and designing the corresponding separation equipment were realized in the paper [10]. The research work [11] is aimed at experimental and analytical solving the coupled aerodynamic and thermal problems for modeling the air flow through perforated elements. The article [12] presents an experimental assessment of water hammer-induced column separation in a pipe flow. Detailed experimental analysis of adsorption and desorption mechanisms was made in the paper [13].

In the article [14], the recent scientific approaches in water treatment were analyzed. The approach for energy-efficient separation of oil-water mixtures by a silica nanoparticle coating was proposed in the paper [15]. A study on an interfacial surface in the modified absorption tower was presented in the research work [16]. The article [17] is devoted to ensuring the reliability of the pneumatic classification process for granular materials in a rhomb-shaped apparatus.

Finally, the special means of separation were presented in a row of researches. Particularly, cryogenic air separation was proposed in the article [18]. Gravitational solid-liquid separation of water-based fluids using the gamma-ray technique was proposed in the article [19]. The general approach for ensuring the vibration reliability of the technological equipment using artificial intelligence systems was proposed in the research work [20]. The paper [21] is aimed at the intensification of interfacial adsorption by the ultrasonic method. Method for the separation of water-oil emulsions using ultrasonic standing waves was proposed in the research work [22].

Additionally, the following scientific problems have been previously solved [23]:

- A scientific approach for the physical and mathematical modeling of hydrodynamics and heat and mass transfer processes during the separation of heterogeneous multicomponent systems has been developed.
- Understanding of the fundamental methods and mechanisms of the gas-dynamic, inertial-filtering, and vibrational-inertial separation.

Due to the abovementioned and in order to improve the approaches and methods for designing the highly efficient separation equipment, the purpose of the proposed research is to improve the approach for ensuring the reliability of technological equipment of energy-efficient modular separation devices by means of identifying the parameters of the refined mathematical models describing the main and secondary hydromechanical processes. In this case, the object of study is the separation processes for heterogeneous systems and the corresponding separation equipment. The working hypothesis is the assumption that energy costs can be significantly reduced and the separation efficiency can be increased by the implementation of perspective directions in the organization of flow modes and interaction of two-phase turbulent flows. In this case, the avoidance of the negative impact of secondary processes on the separation process should be ensured with the development and research of the corresponding experimental samples of the modular technological equipment [24].

The research novelty of the research is highlighted by the brand-new class of separation process and the corresponding designs of the proposed modular separation equipment realizing the following processes, methods, and approaches:

- Vibrational-inertial and vibrational-inertial-filtering separation
- Dynamic control of the value and direction of the flow rate
- Control of gas-liquid flow pulsations
- Semi-permeable double blinds with filter-coalescers
- Combination of processes of the vibrational-inertial-filtering separation with the condensation

3 Research Methodology

Theoretical study and numerical simulation of the movement of the main and dispersed phases as well as of the flexible and moving separation elements are based on the fundamental physical laws of classical mechanics [25], fluid and gas mechanics [26], fundamentals of aeromechanics [27, 28], and theory of aeroelasticity [29]. According to the results of theoretical and experimental studies and the chosen strategy of the research, scientifically based practical recommendations were developed. These recommendations are valuable for realizing computer-integrated design technologies and automated calculations of separation equipment. These means, which are associated with the evaluation of optimal design parameters and the layouts of modular separation units as parts of combined multifunctional devices, allow one to increase the separation efficiency and to reduce the hydraulic losses in the separation equipment.

The motion of a dropping liquid in a plane channel (Fig. 1) is described by the following system of differential equations:

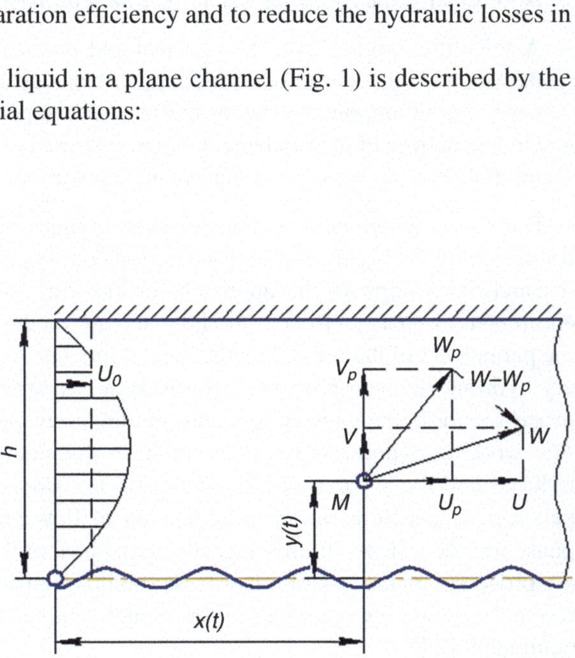

Fig. 1 The design scheme of the operating channel for the separation of a gas-liquid mixture

$$\begin{cases} u = u_0 + \dfrac{6\kappa_1 a}{\lambda h}\left[\sin\omega_0 t - \dfrac{1}{2}\sin\left(\omega_0 - \lambda u_0\right)t - \dfrac{1}{2}\sin\left(\omega_0 + \lambda u_0\right)t\right]; \\ v = \dfrac{\kappa_2 a}{2}\left[\cos\left(\omega_0 - \lambda u_0\right)t - \cos\left(\omega_0 + \lambda u_0\right)t\right], \end{cases} \quad (1)$$

where a is the amplitude of the vibration velocity, m s^{-1}; ω_0, the vibrational frequency, s^{-1}; and $\lambda = 2\pi/L$, the wavelength parameter, m^{-1}, and κ_1, κ_2 are the dimensionless parameters.

The velocity field is described by its components by the following equations obtained previously:

For modeling the unsteady motion of the gas-dispersed flow, Oseen's equations [30] are used. In two-dimensional formulation after considering the assumptions and simplifications [31], these equations are simplified [32]:

$$\frac{du_p}{dt} = -\beta\left(u - u_p\right); \frac{dv_p}{dt} = -\beta\left(v - v_p\right) \pm g, \quad (2)$$

where u_p, v_p are the horizontal and vertical components of the velocity field of a dropping liquid; $\beta = 18\mu/(\rho_p \cdot d_p^2)$ is the inertia factor [s^{-1}], which includes the dynamic viscosity μ [Pa s] of the gas-liquid mixture, density ρ_p [kg m^{-3}], and diameter d_p [m] of particles; and g is the gravitational acceleration, m s^{-2} (the sign "+" or "–" depends on the relative position of the stationary and vibrating walls).

4 Results

Considering Eq. (1) allows one to obtain the general solution of the system Eq. (2):

$$u_p = \Sigma u_p^{\langle i \rangle}; v_p = \Sigma v_p^{\langle j \rangle} \quad (3)$$

This solution is a superposition of the partial solutions $u_p^{\langle i \rangle}$ ($i = 0$–4) and $v_p^{\langle j \rangle}$ ($j = 0$, 3–5). Consequently, the motion of a dropping fluid in gas flow with the vibrating wall is a response to a polyharmonic impact with the following different frequencies: ω_0 and a couple of $\omega_0 \cdot (1 \pm \text{Ho})$.

The further analysis of the proposed model can be realized after considering the homochronicity number [33]:

$$\text{Ho} = \frac{\lambda u_0}{\omega_0} \quad (4)$$

Particularly, it should be proved that in the case of relatively small wavelengths ($\lambda \gg \omega_0/u_0$), when the homochronicity number is significant (Ho = ≫1), a relative motion of particles in the inertial coordinate system occurs in a sinusoidal trajectory. In other words, the quasi-steady motion occurs. Otherwise, close to the vibrating wall, when the homochronicity number is negligible (Ho $\ll$ 1), there is a relative equilibrium of liquid particles. Additionally, in the transient case (Ho $\approx$ 1), the relative trajectories of dispersed droplets are Lissajous curves:

$$x_{\mathrm{p}}^{r}(t) = -\frac{A_2}{\omega_0}\cos(\omega_0 t - \varphi_2) + \frac{A_4}{2\omega_0}\cos(2\omega_0 t - \varphi_4);$$
$$y_{\mathrm{p}}^{r}(\mathrm{t}) = -\frac{B_4}{\omega_0 + \lambda u_0}\sin(2\omega_0 t - \varphi_4), \quad (5)$$

where amplitudes A_2, A_4, and B_4 and phase shifts φ_2 and φ_4 are determined by the following expressions:

$$A_2 = \frac{\alpha}{\sqrt{1+\left(\frac{\omega_0}{\beta}\right)^2}}; A_4 = \frac{\alpha/2}{\sqrt{1+4\left(\frac{\omega_0}{\beta}\right)^2}}; B_4 = \frac{\alpha'}{\sqrt{1+4\left(\frac{\omega_0}{\beta}\right)^2}};$$
$$\varphi_2 = \mathrm{arctg}\frac{\omega_0}{\beta}; \varphi_4 = \mathrm{arctg}\frac{2\omega_0}{\beta}, \quad (6)$$

which include the dimensionless parameters:

$$\alpha = \frac{6\kappa_1 a}{\lambda h}; \alpha' = \frac{\kappa_2 a}{2}. \quad (7)$$

Ensuring the reliability of separation equipment can be realized using the parameter identification of the operation process based on the proposed mathematical model with the consequent applying a probabilistic approach.

Particularly, the time T_s [s] for capturing dispersed particles in a channel with the vibrating wall as well as the effective length ΔL_s [m] of gutters for capturing dispersed liquid is determined by the following dependencies, correspondently (Fig. 2):

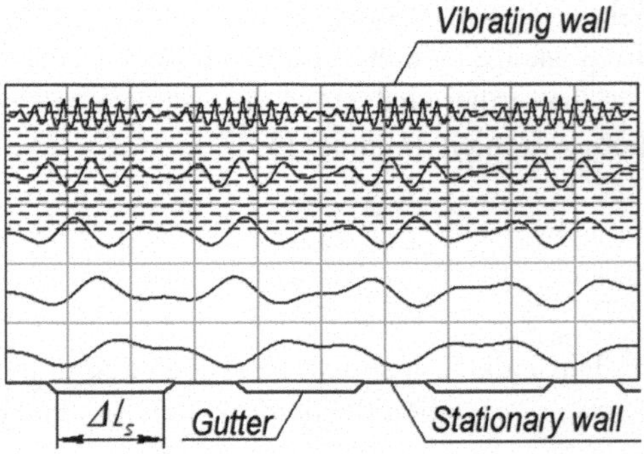

Fig. 2 The design scheme of the operating channel for the separation of a gas-liquid mixture

$$T_s = \frac{18\mu h}{\rho_p g d_p^2}; \Delta L_s = \frac{3}{5}\frac{u_0 h}{g\tau}. \tag{8}$$

Consequently, this methodology allows one to evaluate the geometrical parameters and effective area for locating of gutters on the stationary wall in the separation channel.

5 Conclusions

According to the carrying out critical overview of recent achievements in the field of designing the separation equipment, it should be concluded that the proposed combined multifunctional apparatuses equipped with modular separation devices have relatively high separation efficiency in a wide range of changes in flow rate, pressure, temperature, and phase ratio. In terms of the main technical characteristics, this equipment is not inferior to the analogs produced by "Koch-Glitsch," "Monsanto Enviro-Chem Systems Inc.," and "ACS Industries, Inc." (USA), "Sulzer AG" (Switzerland), and "Shell Global Solution" (Netherlands).

However, designing a brand-new technological equipment should be based on studying the hydromechanical processes occurring in separation channels, particularly, development of the comprehensive approach for studying the impact of the main and secondary hydromechanical as well as heat and mass transfer processes on the separation efficiency considering the effects of turbulized flow and combined condensation in a channel with an oscillating flow. In this case, the development of technological equipment should be complemented by ensuring the reliable mathematical and numerical modeling of the vibrational-inertial-filtering separation considering the secondary processes, which make a significant contribution to the separation efficiency.

In this paper, the main parameters of the modular separation equipment and combined multifunctional separators were analyzed. Particularly, the average velocity of 0.18–0.72 m/s and the hydraulic resistance of 0.15–2.0 kPa allow one to ensure the efficiency of modular separation equipment more than 90% for particles 5 μm.

The refined mathematical model of the movement of a dispersed phase in a gas-liquid flow was developed for identifying the parameters of the main and secondary hydromechanical processes for the consequent ensuring the reliability of the separation processes and the corresponding separation equipment for multicomponent heterogeneous systems. As a result, it was discovered that the motion of a dropping fluid in gas flow with the vibrating wall has a superposition of the main component with the frequency ω_0 of the vibrating wall, as well as both sub- and superharmonic components. In this case, the shift of the superimposed harmonics is symmetrical and is proportional to the homochronicity number. Additionally, it was obtained that the trajectories of dropping particles also depend on the mentioned dimensionless criterion. Particularly, in the case of a quasi-steady motion of the gas-liquid flow, the relative trajectories of particles in the inertial coordinate system occur in a sinusoidal trajectory. Vice versa, close to the vibrating wall, there is a relative equilibrium of liquid particles. Finally, in the case of the transient mode, the relative trajectories of liquid particles are Lissajous curves.

The proposed model additionally allows one to obtain analytical dependencies for determining the effective length of gutters for capturing dispersed liquid and the time of occurrence of the separation process by the average size of dispersed particles and the main physical characteristics of the gas-liquid flow.

The scientific results and practical recommendations were partially implemented at the machine-building enterprises of chemical, oil, and gas industries PJSC "Sumy NPO," JSC "Nasosenergomash," PJSC "Ukrhimproekt," and PJSC "Ukrgazvydobutok" (Ukraine), "HMS Group" and "LUKOIL Group" (Russia), and "Regal Petroleum Ltd." (UK).

Acknowledgments The results of the presented research were obtained at the Faculty of Technical Systems and Energy Efficient Technologies of Sumy State University, Sumy, Ukraine, within the research project "Development and Implementation of Energy Efficient Modular Separation Devices for Oil and Gas Purification Equipment" (State reg. No. 0117U003931) ordered by the Ministry of Education and Science of Ukraine (Scientific Advisor—DSc., Principal Researcher, Professor Oleksandr Liaposhchenko).

References

1. Yu, J., Zhang, Y., Chen, Y., Chen, X.: Simulation of gas-liquid separation characteristics of waxy natural gas in axial flow guide vane cyclone separator. Petrochem. Equip. **48**(4), 1–8 (2019). https://doi.org/10.3969/j.issn.1000-7466.2019.04.001
2. Khmelev, V.N., Shalunov, A.V., Nesterov, V.A., Golykh, R.N., Dorovskikh, R.S.: Increase of separation efficiency in the inertial gas-purifying equipment by high-intensity ultrasonic vibrations. In: International Conference of Young Specialists on Micro/Nanotechnologies and Electron Devices, EDM, pp. 233–239 (2014). https://doi.org/10.1109/EDM.2014.6882519
3. Liaposhchenko, O.O., Sklabinskyi, V.I., Zavialov, V.L., Pavlenko, I.V., Nastenko, O.V., Demianenko, M.M.: Appliance of inertial gas-dynamic separation of gas dispersion flaws in the curvilinear convergent-divergent channels for compressor equipment reliability improvement. IOP Conf. Ser. Mater. Sci. Eng. **233**, 012025 (2017). https://doi.org/10.1088/1757-899X/233/1/012025
4. Plyatsuk, L.D., Ablieieva, I.Y., Vaskin, R.A., Yeskendirov, M., Hurets, L.L.: Mathematical modeling of gas-cleaning equipment with a highly developed phase contact surface. J. Eng. Sci. **5**(2), F19–F24 (2018). https://doi.org/10.21272/jes.2018.5(2).f4
5. Fang, D., Li, L., Li, J., Wang, M., Yu, H., Zhang, J., Qiu, S., Tian, W., Su, G.H.: Full-scale numerical study on the thermal-hydraulic characteristics of steam-water separation system in an advanced PWR UTSG. Part 2: Droplets separation process. Prog. Nucl. Energy. **118**, 103139 (2020). https://doi.org/10.1016/j.pnucene.2019.103139
6. Fang, D., Wang, M., Duan, Y., Li, J., Qiu, G., Tian, W., Zuo, C., Su, G.H., Qiu, S.: Full-scale numerical study on the flow characteristics and mal-distribution phenomena in SG steam-water separation system of an advanced PWR. Prog. Nucl. Energy. **118**, 103075 (2020). https://doi.org/10.1016/j.pnucene.2019.103075

7. Pan, Y., Liu, L., Zhang, Z., Huang, S., Hao, Z., Zhao, X.: Surfaces with controllable superwettability and applications for smart oil-water separation. Chem. Eng. J. **378**, 122178 (2019). https://doi.org/10.1016/j.cej.2019.122178
8. Mao, Y., Bu, X., Peng, Y., Tian, F., Xie, G.: Effects of simultaneous ultrasonic treatment on the separation selectivity and flotation kinetics of high-ash lignite. Fuel. **259**, 116270 (2020). https://doi.org/10.1016/j.fuel.2019.116270
9. He, S., Zhan, Y., Zhao, S., Lin, L., Hu, J., Zhang, G., Zhou, M.: Design of stable superhydrophobic/super-oleophilic 3D carbon fiber felt decorated with Fe_3O_4 nanoparticles: facial strategy, magnetic drive and continuous oil/water separation in harsh environments. Appl. Surf. Sci. **494**, 1072–1082 (2019). https://doi.org/10.1016/j.apsusc.2019.07.258
10. Tianxing, Z., Khezzar, L., AlShehhi, M., Xia, Y., Hardalupas, Y.: Experimental investigation of air–water turbulent swirling flow of relevance to phase separation equipment. Int. J. Multiphase Flow. **121**, 103110 (2019). https://doi.org/10.1016/j.ijmultiphaseflow.2019.103110
11. Khovanskyi, S., Pavlenko, I., Pitel, J., Mizakova, J., Ochowiak, M., Grechka, I.: Solving the coupled aerodynamic and thermal problem for modeling the air distribution devices with perforated plates. Energies. **12**(18), 3488 (2019). https://doi.org/10.3390/en12183488
12. Jansson, M., Andersson, M., Pettersson, M., Karlsson, M.: Experimental assessment of water hammer-induced column separation in oil-hydraulic pipe flow. J. Fluids Eng. Trans. ASME. **141**(10), 101107 (2019). https://doi.org/10.1115/1.4043854
13. Gupta, N.K., Choudhary, B.C., Gupta, A., Achary, S.N., Sengupta, A.: Graphene-based adsorbents for the separation of F-metals from waste solutions: a review. J. Mol. Liq. **289**, 111121 (2019). https://doi.org/10.1016/j.molliq.2019.111121
14. Plyatsuk, L.D., Roy, I.O., Chernysh, Y.Y., Kozii, I.S., Hurets, L.L., Musabekov, A.A.: Clarification of the recent scientific approaches in magnetic water treatment. J. Eng. Sci. **6**(1), F12–F18 (2019). https://doi.org/10.21272/jes.2019.6(1).f3
15. Wang, F., Pi, J., Li, J., Song, F., Feng, R., Wang, X., Wang, Y.: Highly-efficient separation of oil and water enabled by a silica nanoparticle coating with pH-triggered tunable surface wettability. J. Colloid Interface Sci. **557**, 65–75 (2019). https://doi.org/10.1016/j.jcis.2019.08.114
16. Ochowiak, M., Wlodarczak, S., Pavlenko, I., Janecki, D., Krupinska, A., Markowska, M.: Study on interfacial surface in modified spray tower. Processes. **7**(8), 532 (2019). https://doi.org/10.3390/pr7080532
17. Lytvynenko, A., Yukhymenko, M., Pavlenko, I., Pitel, J., Mizakova, J., Lytvynenko, O., Ostroha, R., Bocko, J.: Ensuring the reliability of pneumatic classification process for granular material in a rhomb-shaped apparatus. Appl. Sci. (Switzerland). **9**(8), 1604 (2019). https://doi.org/10.3390/app9081604
18. Portillo, E., Alonso-Farinas, B., Vega, F., Cano, M., Navarrete, B.: Alternatives for oxygen-selective membrane systems and their integration into the oxy-fuel combustion process: a review. Sep. Purif. Technol. **229**, 115708 (2019). https://doi.org/10.1016/j.seppur.2019.115708
19. Fagundes, F.M., Santos, N.B.C., Martins, A.L., Damasceno, J.J.R., Arouca, F.O.: Gravitational solid-liquid separation of water-based drilling fluids weighted with hematite through the gamma-ray attenuation technique. J. Pet. Sci. Eng. **180**, 406–412 (2019). https://doi.org/10.1016/j.petrol.2019.05.054
20. Pavlenko, I., Ivanov, V., Kuric, I., Gusak, O., Liaposhchenko, O.: Ensuring vibration reliability of turbopump units using artificial neural networks. In: Trojanowska, J., Ciszak, O., Machado, J., Pavlenko, I. (eds.) Advances in Manufacturing II, Vol. 1 - Solutions for Industry 4.0, Lecture Notes in Mechanical Engineering, pp. 165–175. Springer, Cham (2019). https://doi.org/10.1007/978-3-030-18715-6_14
21. Yang, L., Li, X., Li, W., Yan, X., Zhang, H.: Intensification of interfacial adsorption of dodecylamine onto quartz by ultrasonic method. Sep. Purif. Technol. **227**, 115701 (2019). https://doi.org/10.1016/j.seppur.2019.115701
22. Luo, X., Gong, H., Cao, J., Yin, H., Yan, Y., He, L.: Enhanced separation of water-in-oil emulsions using ultrasonic standing waves. Chem. Eng. Sci. **203**, 285–292 (2019). https://doi.org/10.1016/j.ces.2019.04.002

23. Liaposchenko, O., Pavlenko, I., Nastenko, O.: The model of crossed movement and gas-liquid flow interaction with captured liquid film in the inertial-filtering separation channels. Sep. Purif. Technol. **173**, 240–243 (2017). https://doi.org/10.1016/j.seppur.2016.08.042
24. Liaposhchenko, O., Pavlenko, I., Ivanov, V., Demianenko, M., Starynskyi, O., Kuric, I., Khukhryanskiy, O.: Improvement of parameters for the multi-functional oil-gas separator of "Heater-Treater" type. In: 2019 IEEE 6th International Conference on Industrial Engineering and Applications (ICIEA), pp. 66–71 (2019). https://doi.org/10.1109/IEA.2019.8715203
25. Panchenko, A., Voloshina, A., Boltyansky, O., Milaeva, I., Grechka, I., Khovanskyy, S., Svynarenko, M., Glibko, O., Maksimova, M., Paranyak, N.: Designing the flow-through parts of distribution systems for the PRG series planetary hydraulic motors. Eastern-Eur. J. Enterprise Technol. **3**(1(93)), 67–77 (2018). https://doi.org/10.15587/1729-4061.2018.132504
26. Syomin, D., Rogovyi, A.: Features of a working process and characteristics of irrotational centrifugal pumps. Proc. Eng. **39**, 231–237 (2012). https://doi.org/10.1016/j.proeng.2012.07.029
27. Pavlenko, I., Liaposhchenko, O., Sklabinskyi, V., Ivanov, V., Gusak, O.: Hydrodynamic features of gas-liquid flow movement in a separation device plane channel with an oscillating wall. Problemele Energeticii Regionale. **3**(38), 62–70 (2018). https://doi.org/10.5281/zenodo.2222360
28. Wang, Y., Guo, L.: Theoretical and experimental research on interfacial shear stress and interfacial friction factor of gas-liquid two-phase wavy stratified flow in horizontal pipe. Heat Mass Transf. **55**(8), 2117–2135 (2019). https://doi.org/10.1007/s00231-019-02560-x
29. Lee, H.M., Kwon, O.J.: Performance improvement of horizontal axis wind turbines by aerodynamic shape optimization including aeroelastic deformation. Renew. Energy. **147**, 2128–2140 (2020). https://doi.org/10.1016/j.renene.2019.09.125
30. Parmar, M., Haselbacher, A., Balachandar, S.: Generalized Basset–Boussinesq–Oseen equation for unsteady forces on a sphere in a compressible flow. Phys. Rev. Lett. **106**, 084501 (2011). https://doi.org/10.1103/PhysRevLett.106.084501
31. Daitche, A.: On the role of the history force for inertial particles in turbulence. J. Fluid Mech. **782**, 567–593 (2015). https://doi.org/10.1017/jfm.2015.551
32. Kelbaliev, G.I., Ibragimov, Z.I., Kasimova, R.K.: Deposition of aerosol particles in vertical channels from an isotropic turbulent flow. J. Eng. Phys. Thermophys. **83**(5), 908–916 (2010). https://doi.org/10.1007/s10891-010-0413-4
33. Zamuraev, V.P., Kalinina, A.P.: Numerical and analytical simulation of the structure of a supersonic gas flow in a variable-section channel with power supply. J. Eng. Phys. Thermophys. **88**(1), 214–223 (2015). https://doi.org/10.1007/s10891-015-1184-8

Influence of Time Structure on Production Process Efficiency

Henrieta Pavolová, Andrea Seňová, Tomáš Bakalár, Michaela Očenášová, and Zuzana Šimková

1 General Characteristics of Production Process and Time Structure

The present brings ever-increasing competitiveness of companies and increasing the quantity and quality of products while reducing the time horizon in the production process. The time structure plays an important role not only in the production process but also in the time receiving the order and delivery of goods to the customer. It is therefore necessary to take into account the time factor at each stage of the decision-making process as well as the planning of the company's objectives. Whether it is an operational, tactical, or strategic goal, time management needs to be taken into account. In every company, it is necessary to constantly improve the time structure, especially in the manufacturing company, where reducing the time consuming production process leads to cost reduction and thus contributes to increasing the competitiveness of the company.

The main component of the manufacturing company is the production that consists of the performance of each manufacturing plant; therefore, it is a key area in order to succeed in the ever-growing competitive environment. Given this, it is in the interest of manufacturing plants to monitor the organization and management of production processes in order to identify the negative aspects that affect their efficiency and ensure their subsequent elimination. One of the essential aspects that directly affect the economy efficiency of manufacturing plants is also the time structure of their production process, which basically reflects the flow of products through different workplaces, depending on the time factor, and addresses

H. Pavolová · A. Seňová · T. Bakalár · M. Očenášová (✉) · Z. Šimková
Faculty of Mining, Ecology, Process Control and Geotechnologies, Institute of Earth Resources, Technical University of Košice, University of Košice, Kosice, Slovakia
e-mail: henrieta.pavolova@tuke.sk; andrea.senova@tuke.sk; tomas.bakalar@tuke.sk; michaela.ocenasova@tuke.sk; zuzana.simkova@tuke.sk

L. Knapcikova et al. (eds.), *New Approaches in Management of Smart Manufacturing Systems*, EAI/Springer Innovations in Communication and Computing, https://doi.org/10.1007/978-3-030-40176-4_14

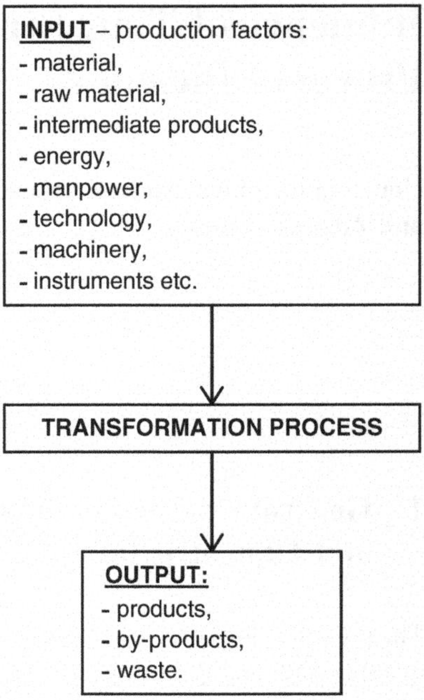

Fig. 1 Production process diagram. (Source: Processed by authors)

compliance of deadlines in line with the plans, the capacity utilization, and determines the time necessary for performance, control, and handling [1]. The time structure of production is given, irrespective of the type of production, by aggregating the time distribution of the production of products within a particular manufacturing plant. In general, it can be stated that the time structure of production is determined by a particular structure of the continuous production time of a particular product, given by a summary of the times needed for the work, downtimes, and the transformation process [2, 3].

The production process creates the basic means of satisfying the current needs of the anthropogenic society by creating goods of a material nature, i.e., products. The production process can be considered as the result of purposeful human activity, where the purposeful combination of production factors transforms input elements (inputs, i.e., raw materials, other materials, semi-finished products, energy, and information) into outputs (i.e., products) as generally described in Fig. 1. The production process is of a creative nature, and it is necessary to react flexibly to changing conditions under the influence of random factors.

The production process consists of three phases:

– The pre-production phase (pre-production) consists of the preparation and purchase of materials and raw materials, the following two ones are dependent on this phase.
– The making phase (pre-assembly) consists of different stages of product assembly.

- The finishing phase (assembly) includes all types of assembly and testing of finished products.

It is in these phases of the production process that it is necessary to determine the time structure of individual operations, which is what the technical-economic standards of time serve us for. The basic goal of labor standardization is to determine the optimum time consumption for a performance of work at individual workplaces.

The main indicators of the production process, where the basic aspect is time, are [4]:

- Labor productivity
- Labor intensity
- Continuous production time

Factors influencing the time structure of the production process include [4]:

- Aspect of coherence of production process and operations
- Human participation in the production process (qualification of workers)
- Degree of complexity and length of technological operations
- Batch size and method of sorting production equipment
- Length of transport routes and speed of handling equipment
- Time of control operations and time of interruption

The manufacturing process consists of individual smaller working processes, which are divided into working operations. The time structure of the production process depends on whether these operations are performed simultaneously or sequentially.

The basic principles of time management include [4]:

- Ensuring direct transport routes and eliminating unnecessary transport and handling
- Minimization of transfer of input materials and other materials
- Rhythmic and continuous flow of materials and raw materials
- Overall streamlining of handling and automation in the production process
- Ensuring proportionality of production process capacities
- Ensuring ergonomic, hygienic, and safety requirements

The choice of a suitable analysis of time consumption depends on the importance of the individual factors in the production process. The most common limiting and decisive factor of the production process is human and therefore the most often used analysis of time consumption from the perspective of the worker.

The material flow through the workplace shall be such as to ensure smooth production of the final product, which consists of individual components. Therefore, this step is regulated by the time ties. A particular, pre-planned flow of unfinished parts through the plant creates a temporal structure of the production process [5]. The time structure of the production process takes into account, in addition to the length of time of the product flow through the manufacturing process, also the time of human work and the time used by machinery and equipment. The time structure of the production process is expressed by continuous time of production. Continuous time of production is a time indicator reflecting duration of the production process [6], which is formed by cycles indicated below [7]:

- Pre-production cycle, i.e., the time of preparatory work to the startup of the product manufacture (the structural preparation and development work on a new product; technological preparation of production, including the preparation of tools and materials; the economic and planning preparation of new manufacturing; and so on).
- The production cycle, i.e., the time of basic manufacturing operations, time of transport, control and other operations in the manufacturing process, and durations of breaks (times of interruptions and delays) occurring during the product manufacturing.

A particular problem in the context of production timing is in idle and waiting times, i.e., times of necessary breaks in the production process, which are given by the mode of work and a complicated combination of partial production operations [8].

Analysis of the temporal structure of the production process. When analyzing the time structure of the production process, we consider the time factor as a most significant factor of the work. This parameter directly affects commercial and economic results of each manufacturing plant; it is based on direct measurement of partial working times, which are commonly used for the determination of wage costs or development of work schedules [9]. One of the analytical methods of the time structure of production processes is the method of continuous study of the time consumption, which we used in tracking the time structure in the production of aluminum castings. This particular method can detect, record, and evaluate the real-time consumption of a particular employee or manager, as well as production facility. This method is based on measurements of time consumptions forming the image of [9, 10]:

- Operation, which is used to assess the workplace or production cycle organization
- Working day, which is used to assess the organization of the workplace, working hours used losses, and the like

Methodology and methods of assessment of the continuous study of production time consumption. To obtain results of the real-time consumption in the production of a production batch, we used the following indicators [6, 11]:

Occupancy rate (U_1):

$$U_1 = \frac{T1 + T2}{T} \times 100 \quad (1)$$

A share of breaks conditionally necessary (U_2):

$$U_2 = \frac{T3}{T} \times 100 \qquad (2)$$

A share of needless time consumption caused by the worker (U_3):

$$U_3 = \frac{T'2 - T2 + \mathrm{TD}}{T} \times 100 \qquad (3)$$

A share of needless time consumption due to technical and organizational losses (U_4):

$$U_4 = \frac{\mathrm{TE}}{T} \times 100 \qquad (4)$$

A percentage of the possible labor efficiency increase by eliminating unnecessary time consumption caused by workers:

$$U_5 = \frac{T'2 - T2 + \mathrm{TD}}{T - (T'2 - T2 + \mathrm{TD} + \mathrm{TE})} \times 100 \qquad (5)$$

A percentage of the possible labor efficiency increase by eliminating unnecessary time consumption caused by technical and organizational losses:

$$U_6 = \frac{\mathrm{TE}}{T - (T'2 - T2 + \mathrm{TD} + \mathrm{TE})} \times 100 \qquad (6)$$

Percentage of the total possible increase in labor efficiency:

$$U_7 = U_5 + U_6 \qquad (7)$$

where:

*T*1—time of work
*T*2—the time of generally required breaks
*T*3—the time of conditionally required breaks
TD—personal loss of time
TE—technical and organizational loss of time
T—time of the shift

In view of the foregoing, the results of the above observations and quantifiers can be used to [6]:

- Quantify various activities expressed by the time consumption
- Analyze the structure of the work time consumption
- Analyze the lost times by causes

– Develop the performance curves throughout the shift, especially if also monitoring the amount of the production output (performance)

2 Analysis of the Temporal Structure of Aluminum Casting Production

The aluminum casting production process (Fig. 2) consists of the following partial manufacturing operations:

1. Manufacturing operation = preparation of the melt, which is melted from Al alloy bricks to a liquid state and subsequently transported by trolleys to pressure presses (responsible melter).
2. Manufacturing operation = casting of Al mold according to predefined requirements of the customer, when the founder (responsible for the manufacturing operation) is obliged to clean the mold cavities and also the separating plane with compressed air, possibly stronger contamination, and then blow it with compressed air. In interaction with the activities performed, the semi-automatic casting cycle of the Al casting is started, which is removed from the mold by means of casting throats and deposited on the manipulation table. During the semi-automatic cycle of the next operation, it also performs a visual inspection of the just released Al casting and puts it in the so-called gitterbox (cage metal pallet), where possible unsatisfactory Al castings are placed in the trolley for return material.
3. Manufacturing operation = grinding with a belt grinder, when it grinds four sides along the circumference of the Al casting, whereby the individual Al castings entered in the dispatch are separated by cardboard.
4. Manufacturing operation = deburring (sawing or finishing) carried out by hand grinders, where employees reveal the edges on both sides, scrape the ejection marks, and of course contouring. The cracked surfaces must be sanded with a handheld vibrating grinder while separating the Al castings specified in the guide.
5. Manufacturing operation = tumbling, the aim of which is to obtain a marble surface by degreasing with shampoo, water, and abrasive stones. The individual castings are deposited on the conveyor of the tumbling device (tumbling drum)

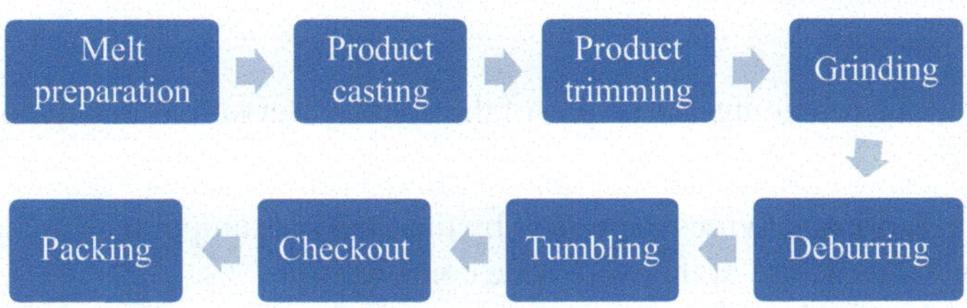

Fig. 2 Production process of aluminum castings. (Source: Processed by authors)

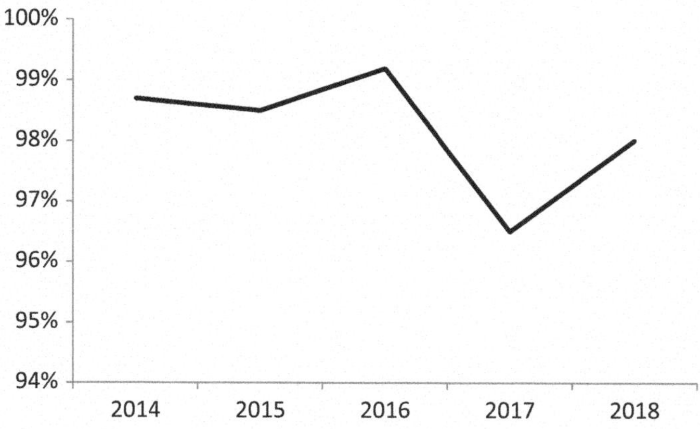

Fig. 3 Development of production process efficiency. (Source: Processed by authors)

down the bottom, where the tumbling process is followed by drying in the dryer on the continuous belt. The number of pieces shall be entered in the dispatch note.

6. Manufacturing operation = final inspection, which is performed visually, of the cleanliness and compactness of the surface of the final Al refraction. If deficiencies are found, the employee concerned will make minor surface repairs by grinding. In the case of non-repairable pieces of Al castings, these are discarded. Quality-compliant products are placed in cartons in which the individual layérs are corrugated.
7. Manufacturing operation = packaging, when part of the carton is put on, it is strapped to the Euro pallet two times and one time around the perimeter. Carobes are placed under the tape, and the shipping label is glued. The numbers of the dispatch notes contained in the carton shall be entered in the dispatch document.

During the analysis period, the production process efficiency of aluminum castings showed on average 98.2% per year, with the highest value in 2016 at the level of 99.2% and the lowest in 2017, specifically 96.5% (Fig. 3). Taking into account the fact that efficiency is generally referred to as the principle of minimum, which expresses how it was possible in the reporting period to produce aluminum castings with a minimum consumption of production factors, we could conclude that on average 0.98 € of production inputs were consumed per 1 € aluminum castings. This fact also pointed to the need for a more detailed analysis of the time structure of the entire production process.

For the purpose of construction of the production cycle, we compiled partial time consumptions of partial production operations through the personal observation as given Table 1.

In view of the above-detailed analysis, we concluded that the most critical point of the production process of aluminum castings was in the operations of mechanical machining of finished products that gave rise to the most losses of the batch-production consumption time. Therefore, we have decided to dedicate more to the

Table 1 Time consumptions in partial operations in production of a production batch of aluminum castings

Workplace	Working operation	Consumption of time (min/production cycle)	Time of downtime (min/production cycle)	Percentage of downtime out of the production cycle time
Melting plant	Transportation (as part of melting)	1:37	0	
	Raw material charged into the furnace	0:54	0	
	Melting	9:00	0	
	Transport to further processing	1:59	0	
For the workplace		**11:53**	**0**	**0**
Casthouse	Cleaning, lubrication of cavities	0:19	0	
	Casting	1:05	0	
	Transport to the warehouse (melter as part of melting)	1:10	0	
For the workplace		**1:24**	**0**	**0**
Hydraulic press	Manipulation with parts	105	0	
	Transport	2	2	1.87
For the workplace		**107**	**2**	**1.87**
Grinding plant	Delivery of parts	1:29	1:29	3.67
	Grinding	37:30	0	
	Transport	1:28	1:28	3.63
For the workplace		**40:27**	**2:57**	**7.30**
Mechanical machining	Delivery of parts	10	10	2.23
	Takeover of grinding rolls	10	10	2.23
	Machining of castings	379	0	
	Furnishing of the area to store the production batch to	10	10	2.23
	Placement of castings	20	20	4.45
	Removal	20	20	4.45
For the workplace		**449**	**70**	**15.59**
Tumbling	Tumbling and drying	25:00	0	
	Transfer to storage	1:28	0	
For the workplace		**26:28**	**0**	**0**

Source: Processed by author

Table 2 Observation sheet of the work shift image

Number	Consecutive time	Unit time	Time symbol	Title of time consumption
1	5:45			Start of the shift
2	6:00	0:15	TE	Identifying difficulties the night shift
3	6:10	0:10	TE	Bringing in a pallet with castings
4	6:20	0:10	TE	Takeover of grinding rolls
5	8:20	2:00	TB1	Continuous work—product machining
6	8:30	0:10	TD	Products fell down due to improper storage—stacking
7	8:40	0:10	TE	Furnishing an empty transport box for the products machined
8	8:50	0:10	TB1	Placing products into the transport box
9	10:03	1:13	TB1	Continuous work—product machining
10	10:06	0:03	TE	Defects found on products
11	10:20	0:14	TE	Consultation with the foreman
12	10:25	0:05	TE	Defective product taken away
13	11:35	1:10	TB1	Work assumed
14	11.40	0:05	TB1	Placing products into the transport box
15	11:45	0:05	*T2*	Leaving the workplace—toilet
16	11:50	0:05	TB1	Work assumed
17	12:20	0:30	*T2*	Lunch break
18	13:20	1:00	TB1	Completion of the production batch
19	13:25	0:05	TB1	Placing products into the transport box
20	13:45	0:20	TE	Removal of the transport box with products machined
21	13:45			End of the shift—the end of the observation

Source: Processed by author

image analysis of the staff performance from the beginning to the end of the production shift and keep time records of all mechanical operations with the time data rounded to whole minutes (Table 2).

The final stage of observation was to evaluate the image of the work shift. At this stage, we compute the unit time from the consecutive time. Each unit time is evaluated according to the activity carried out or inactivity. The same activities are summarized into an actual balance of time consumption. The actual balance reflects the amount of time in minutes and percentage per each category of the time of the work shift observed. A developed balance of the real-time consumption is given in Table 3.

Development of the actual balance of the time resulted in the values needed to calculate the utilization of an employee, time of breaks, losses of time, and increase in efficiency. Partial calculations leading to the increased efficiency of production, which were quantified upon formulas (1)–(7), are indicated in Table 4. Results of the analysis made us to conclude that the efficiency could be increased by 25.33% in the mechanical machining of aluminum castings while, due to the elimination of

Table 3 Balance of the actual time consumption

Type of the time	Time symbol	Minutes	Percentage of the shift time
Time of unit labor	TA1	0	0
Time of batch labor	TB1	348	72.5
Time of shift labor	TC1	0	0
Working time	***T*1**	**348**	**72.5**
Time of generally required breaks	*T*2	35	7.29
Time of conditionally required breaks	*T*3	0	0
Personal losses of time	TD	10	2.08
Technical and organizational losses of time	TE	87	18.13
Time of the shift	***T***	**480**	**100**

Source: Processed by authors, *Working time* real worked time without breaks and downtime, *Time of the shift*- total shift time of worker

Table 4 Quantification of the real-time consumption indicators

Time consumption indicator	Symbol	[%]
Level of occupancy	U_1	79.79
Share of conditionally required breaks	U_2	0
Share of needless consumption of time caused by employee	U_3	2.08
Share of needless consumption of time due to technical and organizational losses	U_4	18.13
Percentage of possible increase in labor efficiency by eliminating the unnecessary time consumption caused by employee	U_5	2.61
Percentage of possible increase in labor efficiency by eliminating the unnecessary time consumption due to technical and organizational losses	U_6	22.72
The overall percentage of possible increase in labor efficiency	U_7	**25.33**

Source: Processed by authors

unnecessary time consumption due to technical and organizational reasons, it would be possible to increase efficiency of product process by 22.72% and the needless consumption of time caused by employee by 2.61%. The most significant factor determining the actual efficiency was the level of utilization of a particular employee that was 79.79% and the unnecessary consumption of time, which was negligible, accounting only for 2.08% of the total consumption of time (Fig. 4).

The analyses carried so far resulted in measures increasing the efficiency of the production process, designed to eliminate the consumption of time in the mechanical stage of actual production of aluminum castings in the context of addressing any shortcomings of technical and organizational losses, e.g., arrival and/or removal of castings as well as arrival and/or removal of the transport box, including the transfer of finished aluminum castings to the warehouse or takeover of necessary tools and the like. In this way, consumption of time would be reduced at the workplace by 70 min per employee; if having four employees a shift, it would be 280 min per shift. Given the standard production output of 24 pieces per hour, then, if eliminating the consumption of that time, it could be possible to increase production by 112 pieces per hour, i.e., by about 2240 pieces per month.

Fig. 4 Share of the actual use of the shift time in the production process of aluminum castings. (Source: Processed by authors)

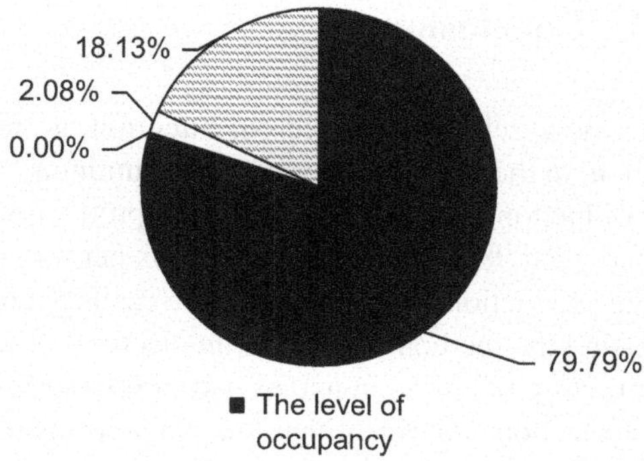

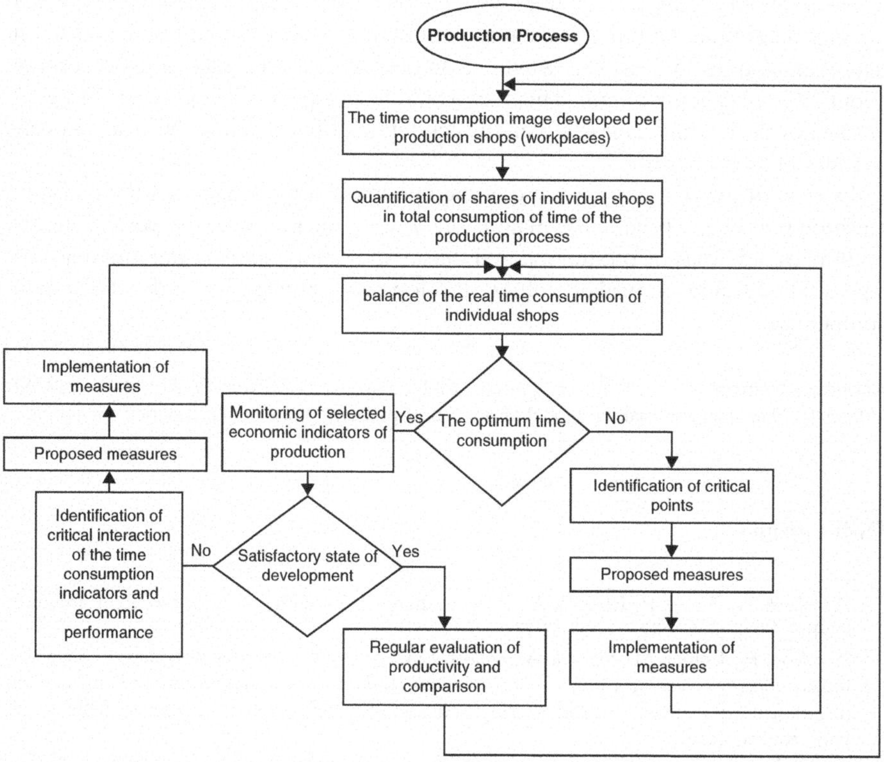

Fig. 5 Model proposed for the management of the time consumption in the production process. (Source: Processed by Authors)

By summarizing the results, we concluded that the production of aluminum castings should be constantly monitored and regularly assessed through the indicators of consumption of time indicated above, as presented in the model proposed for managing the partial production operations forming the time images of partial time consumption in the production of aluminum castings (Fig. 5).

3 Conclusion

In the article, we focused our attention on the issue of time consumption management in the production process of aluminum castings and its impact on efficiency of production process. Efficiency of production process of aluminum castings was averaged 98.2% during the analysis period per year, which pointed to the fact that the production 1 € from aluminum castings consumed 0.982 € of production factors. Using the method of the continuous time consumption, which consisted in the construction of a time image of partial operations in the comprehensive process of that production, we identified the most critical location of production, which was mechanical machining of aluminum castings. Upon the balance of the actual time consumption at work in this workshop, we quantified individual indicators to point to a working occupancy of an employee, the time of breaks, and time losses, which directly determine the possible increase in efficiency, which could be increased in that workshop by 25.33%; this would mean ultimately an increase in production by about 2240 pieces per month. This fact would have a positive impact on the development of the overall economy, i.e., the economic efficiency, and thus the competitiveness of the enterprise.

In view of the foregoing, we are of the opinion that the balance of the real-time consumption in any production plant is in close interaction with the possibilities of optimizing all areas of production and thus also gaining some competitive advantage and defining a further future development strategy in order to remain competitive.

Acknowledgement Contribution is partial result of projects solving VEGA MŠVVaŠ SR 1/0515/18 "The decision-making model of process of evaluating raw material policy of regions."

References

1. Pribulová, A., Futas, P., Rosová, A., et al.: Influence of foundry dust on moulding mixtures quality. Metalurgija. **52**, 51–54 (2013)
2. Pavolová, H., Cehlár, M., Seňová, A., Dugas, J.: Evaluation of the time structure of the manufacturing process using the method of continuous study of the time consumption in the context of productivity increase. In: Engineering Sciences and Production Management 2015, p. 68. Petit, Košice (2015)
3. Straka, M., Trebuna, P., Rosová, A., et al.: Simulation of the process for production of plastics films as a way to increase the competitiveness of the company. Przemysl Chemiczny. **95**, 37–41 (2016)
4. Mihok, J., Janeková, J., Vidová, J.: Podnikové Hospodárstvo - Ekonomika Podniku, p. 212. Strojnícka Fakulta Technickej Univerzity, Košice (2005)
5. Dubovická, L.: Vybrané Aspekty Produkčného Manažmentu, p. 102. Ekonóm (Economist), Bratislava (2010)
6. Ďudák, J.: Časová Štruktúra Výrobného Procesu: Teoretické Základy. SPU, Nitra (2012)
7. Leščišin, M., Stern, J., Dupaľ, A.: Manažment Výroby, p. 328. Sprint víra, Bratislava (2008)

8. Kukpkovič, M., et al.: Podnikové Hospodárstvo: Komplexný Pohľad na Podnik, p. 452. Sprint víra, Bratislava (2003)
9. Kavan, M.: Výrobní a Provozní Management, p. 424. Grada Publishing, Praha (2002)
10. Pavolová, H.: Manažérstvo Výroby, p. 90. TU, Košice (2014)
11. Novák, J., Šlampová, P.: Racionalizace Výroby, p. 75. Technická Univerzita Ostrava, Ostrava (2007)

Business Process Modeling in Industry 4.0 Using Transformation Accelerator Tool

Dragan Peraković, Marko Periša, Ivan Cvitić, and Petra Zorić

1 Introduction

With the development of new innovative information and communication services (IICS), it is possible to contribute more efficiently to all processes of system operation in the fields of networking and providing all relevant information to the stakeholders of the business process. Today's concept used in manufacturing processes and the development of logistics processes based on modern information and communication (IC) technologies is called Industry 4.0. It focuses on the process of transition from traditional business to digital business [1]. To establish an ecosystem and to conduct all business processes and models, it is important to choose the optimal information and communication service (ICS), where it is necessary to define and link infrastructure elements, people, and other stakeholders. Unlike previous research, this research is focused on the identification of important elements of the ecosystem with the aim of selecting and forming optimal elements. For this purpose, it is necessary to satisfy all the requirements (technical and technological, organizational, and legislative) in the function of selecting the appropriate elements of the ecosystem. This research is focused on the following elements: users, device (sensors and user device), communication network and technology, environment, and content. Necessity for integration and connection of the all the abovementioned ecosystem elements into a complex with the establishment of new value chains in the delivery of services is needed.

The aim of this research is to identify the relevant parameters required for modeling ecosystem elements within the Industry 4.0 concept. The identification of relevant parameters provides a starting point in the field of modeling ecosystem

D. Peraković · M. Periša · I. Cvitić · P. Zorić (✉)
Faculty of Transport and Traffic Sciences, University of Zagreb, Zagreb, Croatia
e-mail: dperakovic@fpz.unizg.hr; mperisa@fpz.unizg.hr; icvitic@fpz.unizg.hr; pzoric@fpz.unizg.hr

L. Knapcikova et al. (eds.), *New Approaches in Management of Smart Manufacturing Systems*, EAI/Springer Innovations in Communication and Computing, https://doi.org/10.1007/978-3-030-40176-4_15

elements for the purpose of creating a unique integrated system. In the process of designing a unique integrated system, it is important to create new business models with the aim of more efficient business within the concept of Industry 4.0. A scientific literature review provides a comparative analysis of the business transition from traditional business to digital business through the presentation of current business models.

2 Literature Review

The term Industry 4.0 first appeared in 2011 at the technology fair in Hanover, Germany, where it was presented as an idea of strengthening the competitiveness of the German industry. The initiative became an integral part of the High-Tech Strategy 2020 for Germany, which aimed to promote Germany as a global leader in the field of technological innovations [2]. In the field of modeling and development of innovative services, numerous researches are focused exclusively on partial results, which refer to individual elements of the ecosystem.

Therefore, ecosystem users need to be segmented appropriately (user requirements for business processes and end users) and meet certain requirements needed for delivering content [3–5]. Sensors and user devices play a key role in creating an integrated ecosystem based on the Internet of Things (IoT) concept where Device-to-Device (D2D) communication is important. Researching the possibilities of the integration and interoperability of D2D communication, the authors have put the focus on network layer functionality such as addressing, routing, mobility, security, and resource optimization, resulting in the identification of the elements of the network layer limitations for the purpose of D2D communication in IoT concept [6]. The sensors used in the IoT concept can be classified into three basic categories: motion sensors, ambient sensors, and position sensors [7]. Devices in Industry 4.0 represent cybernetic-physical systems, i.e., integrated systems with information and communication components. These are autonomous systems that can make their own decisions based on machine learning algorithms and real-time data collection, analysis of results, and historical data on system behavior [8]. Communication networks and technologies that can connect these elements are divided into three categories. The first category includes widely used short- or medium-range radio technology. This category includes technology and networks: Automatic Identification and Data Capture (AIDC) technology, IEEE 802.15.14 (e.g., ZigBee), IEEE 802.11 (Wi-Fi), LoRa, and others [9]. Radio-frequency identification (RFID) technology, which belongs to a group AIDC technology, represents the most common technology within the concept of Industry 4.0. Numerous authors conducted various experimental measurements where the regression models of the external factors influencing the operation of RFID technology were created [10, 11]. The second category represents technology and networks based on mobile communication systems (e.g., 2G/GSM, 3G/UMTS, 4G/LTE, 5G) that can provide greater coverage and far greater bandwidth (>100 Mbps) but also consume significant energy in the device itself. The third category represents networks with wide area coverage (IoT networks) and with low power emitting, such as LPWAN (Low Power Wide Area Network), NB-IoT, Sigfox, and others. Cloud Computing as data storage and data processing technology and the IoT network are the basic support networks and technologies of the Industry 4.0. Typical resources such as employees, devices, materials, and manufacturing activities in the Industry 4.0 turn into smart manufacturing facilities so they can feel and react to each other in an intelligent environment [12]. A major impact on the application of innovative services through Industry 4.0 concept relates on business models and processes as well as on the future company's business development plans [13–15]. The field of artificial intelligence also has a significant contribution in supporting the work of a highly automated system which allows them greater autonomy in decision-making and defining digital services [16]. The research on sustainable systems as a support in designing services consider three key segments: the system's sustainability and complexity, the dynamics of the communication network development, and the integration of these segments through the ecosystem [17]. The concept of digital innovation and Big Data in the development of IC services has a great significance in the application of contemporary IC concepts such as IoT; hence, the conceptual methodological framework has been developed for this purpose [18].

For the purpose of modeling innovative services, it is necessary to include a certain degree of generic models to assist in the understanding of the work process. Successful business performance in an environment of smart solutions requires the integration of all relevant parameters to satisfy the end user and the business success of the company. For this purpose, it is possible to use elements of various framework TM forums such as the Business Process Framework (eTOM), Information Framework (SID), and Application Framework (TAM) [19]. These frameworks, with their functionalities and specific characteristics, can be the basis for designing decision-making models, all with the aim of more effectively implementing the business strategy and maintaining organizational management.

Visualization of business processes and scenarios takes place using the 3D tool (Transformation Accelerator) available through TM forum web pages and corresponding licenses [20].

3 Identification of the Relevant Parameters for Modeling Ecosystem Elements in Industry 4.0 Environment

By conducting an analysis of the results of the scientific research from the topic of research, the aim is to determine the relevant parameters necessary for modeling the elements of the ecosystem in the Industry 4.0 environment. The scope of the Industry 4.0 concept application is observed from the scenario of the smart factory and its production processes.

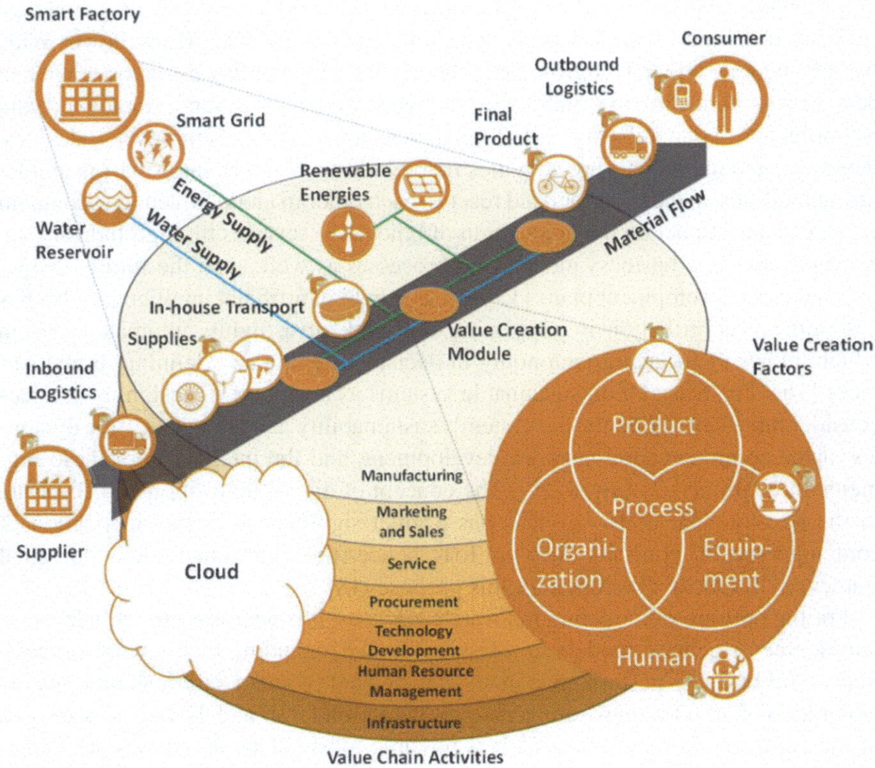

Fig. 1 Microperspective of Industry 4.0 [23]

When modeling ecosystem elements, defining a scenario (smart factory) creates a starting point for selecting relevant ecosystem parameters. For this purpose, it is necessary to define the stakeholders and their roles that have an impact on the business model of the company [21, 22]. The value chain elements on the example of a smart factory can be defined through the vertical and horizontal elements, whereby one element becomes dependent on the other (Fig. 1).

Key elements include infrastructure, human resources management, technology development, procurement, service, marketing and sales, and manufacturing [23].

The research in this paper focuses on the following elements of ecosystems: users, device (sensors and user device), communication network and technologies, environment, and content. Users are also categorized as stakeholders within the value chain of the Industry 4.0. User experience (UX) model that has a role to impact on users through marketing, real-time data analysis and information, and design services is very important for the users [24]. UX also refers to the user device and the application solution that serves the user to obtain all relevant information about the product. For the identification of factors affecting the life cycle of smart mobile devices and sensors, and the correlation between maintenance and

Table 1 Relevant parameters for modeling ecosystem elements in Industry 4.0 [40]

Parameter	Description of the parameter
User	Segmentation of users, depending on the role within the system: customer, worker
Devices	Classification on the user device and sensors in the smart factory environment
Technology	Communication network and technologies
Environment	Application scenario for which ecosystem elements are modeling
Content	The type of content and its delivery
Economic sustainability	Development strategy and business models
Security	User security, reliability of service
Ecological sustainability	Impact on the environment
Life cycle	Lifetime of the elements
Integration	Connecting all elements, process management, security aspects
Forecast	Using artificial intelligence, Big Data, data mining, and machine learning

application solutions, it is important to understand the concept of the application lifecycle [5, 25]. Sensors are used to enable data collection from the ambient environment, and their role in creating a smart factory is extremely important [26].

The communication network has the role of connecting all elements of the ecosystem. The scientific research in this field points to IoT networks, LoRaWAN, Lora, NB-IoT, and Sigfox, as reference networks for delivering defined services in the field of Industry 4.0 [27–32]. Also, research on the impact of the 5G network and the ability of blockchain technology to develop industry 4.0 has been presented [33–35]. Interoperability in the concept of a smart factory, Cloud Computing, and IoT networks is important from the point of view of access to data regardless of location and coverage network [36].

For the modeling of ecosystem elements, the user safety area in the smart factory environment is also important, as well as the reliability of the service. Services based on concepts of IoT and Cloud Computing improve the automation processes in the fields of logistics and Industry 4.0. Absence or unavailability of such services may result in negative user experience in using certain types of services and may also represent a potentially high security risk and endangered user safety.

The level of availability of the service in such cases is influenced by various factors identified and classified by the authors in the analyzed research through guidelines for achieving the required level of availability of information and communication services [37]. Detection of DDoS (Distributed Denial of Service) traffic is of great importance for the availability of services and other information and communication resources [38]. The research presented in the analyzed work shows the application of artificial neural networks in developing models of detection and classification of three types of DDoS attacks and legitimate network traffic [39].

By analyzing the abovementioned researches, Table 1 shows the relevant parameters for modeling the ecosystem elements in industry 4.0 environment on the example of a smart factory.

All elements of the ecosystem and their connectivity to the integrated information and communication system are important for all processes within a smart factory and therefore for a business model. The impact of industry 4.0 development on business models has an increasing importance, mostly due to automation and business process optimization [41–43].

4 TM Forum and Frameworx

TM Forum was created from a group of telecommunication professionals in 1995. The goal of that is to define a set of standard processes aiming to obtain a common language for all those involved in the business development process in the telecommunication industry. The Frameworx, formerly known as New Generation Operation System and Software (NGOSS), provides a set of best practices to support companies to develop their internal and external business processes [44]. Frameworx is created and evolved by TM Forum members who participate in the Forum's Collaboration Community.

The Frameworx:

- Business Process Framework (eTOM), a comprehensive, industry-agreed, multilayered view of the key business processes required to run an efficient, effective, and agile digital enterprise.
- The Information Framework (SID), which operates on all information flowing through the organization and between service providers and their business partners.
- Application Framework (TAM), which supports successful business transformation and provides a common language and means of identification for buyers and suppliers in all areas of application development.
- Open Digital Framework provides the blueprint for service providers to deliver intelligent operations fit for the 5G era. The framework enables a step-change in business agility, operating cost, and customer centricity by revolutionizing the operating model of service providers and their technology partners. Created through continuous global expert collaboration, this living framework combines the tools, knowledge, and essential standards needed to transform legacy ways of working, processes, and underlying IT systems in a single framework, to enable zero-touch interoperability and ecosystem partnering.

This set of standards comes as a response to the growing complexity of new services. It allows users to analyze their operations through the application of market benchmarks. Frameworx presents a set of guidelines and specifications for the software creation industry, by analyzing historical data and current projects, thus responding to the growing technological evolution and consumer demands [45].

eTOM represents a set of categorized business process specifications of a typical Internet and Computer Service Providers (ICSP). The eTOM focus is the interconnection between business processes: customers, services, resources, partners, and suppliers. Analyzing all phases of business processes makes eTOM a key tool to improve ongoing development by enabling:

- The creation of a common language for use across departments, systems, external partners, and suppliers, reducing cost and risk of system implementation, integration, and procurement.
- The adoption of a standard structure, terminology, and classification scheme for business processes to simplify internal operations and maximize opportunities to partner within and across industries.
- The disciplined and consistent application of business process development enterprise-wide, allowing for cross-organizational reuse.
- The understanding, design, development, and management of IT applications in terms of business process requirements so applications will better meet business needs.
- The consistent creation of high-quality end-to-end process flows, eliminating gaps and duplications in process flows.
- Identification of opportunities for cost and performance improvement through re-use of existing processes and systems.

The eTOM presents a business process structure, but it is not considered a business model of the service provider, as it does not address the organization strategic objectives [45].

The Information Framework (SID) provides a reference model and common vocabulary for all the information required to implement eTOM processes.

The Information Framework allows to [44]:

- Reduce integration costs by adopting standards-based information models and use them in applications and interfaces.
- Save project hours, through a mature structure and 1500 entities developed and evaluated by subject matter experts.
- Reduce time-to-market, using easy-to-understand integration interfaces based on the Information Framework, eliminating the need to convert data between systems.
- Avoid unnecessary discussions with project teams, partners, and suppliers, resulting in loss of implementation time, by adopting a rich and extensible industry-accepted information model.
- Comply with Information Framework, thus saving time and money during vendor evaluation and procurement processes.

The TAM provides a bridge between eTOM and SID and can be used for the following purposes:

- Optimize procurement using definitions and common language to specify and evaluate solutions.

- Document and rationalize the application inventory during the project transformation phase.
- Integrate quickly and at lower costs by clearly defining and communicating the functions provided within each application.
- Reduce development costs by modeling and standardizing application requirements.
- Increase automation and efficiency with standard, deployable components.

This way of visualization provides a clear insight into all relevant processes and business models for a precisely defined service or information system. Using the abovementioned solutions in the development of innovative services for Industry 4.0 environment processes, it is possible to create more efficient process management system, especially in the business transition from traditional to digital business.

The telecommunication network is the basis for the efficient delivery of innovative services in the field of Industry 4.0. The digitalization of industry is causing a complete change in the paradigm of production and innovation. In the new wave of digitalization, innovation systems and technology management are facing new challenges. Digitalization in Industry 4.0 focuses on the use of digital technologies to change business models, generate new revenue, and create value. By implementing the TM Forum frameworks, existing processes can be prepared for upcoming next-generation telecommunications networks (5G, IoT) and innovative services.

5 Overview of the Parameters with the 3D Modeling Process Tool

Relevant parameters are through a simulation tool used for overview of the eTOM, SID, and TAM processes. The stakeholders defined in Fig. 2 are a part of the smart factory environment: worker, suppliers, users, product, organization, equipment, development engineers, I 4.0 platform, and CSP. I 4.0 platform is the integrator of

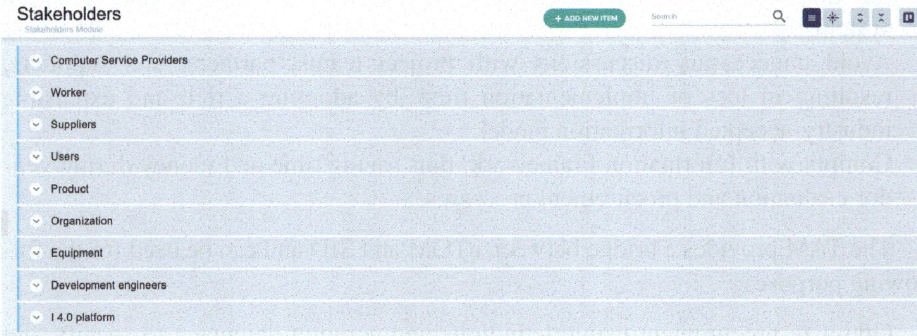

Fig. 2 An overview of the stakeholders

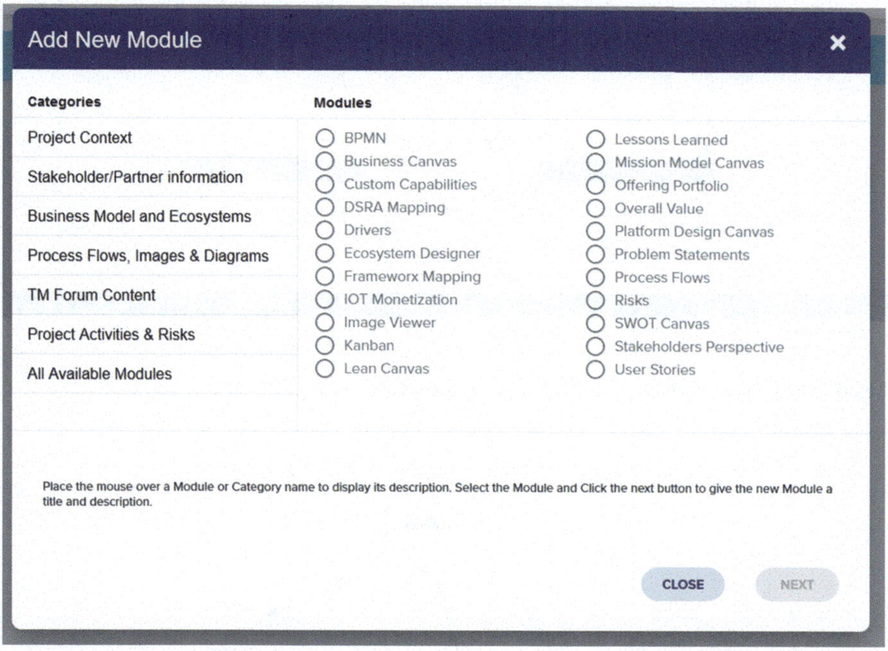

Fig. 3 Categories and modules for business process modeling

all the data needed for delivering the required services in the smart factory environment.

After defining stakeholders and their roles, it is necessary to select the category and modules required for modeling. In order to define the business model, it is necessary to define a usage scenario. Figure 3 shows all available modules in the modeling process of the business model.

The key values in defining a business model are shown in Fig. 4. It is important to note that figures show only part of the value as well as the roles of some of the defined stakeholders.

The values to be entered are divided by key partners, activities, resources, propositions, customer relationships, advertising channels, and structure of the costs and revenues. The tool also has the ability to use the Unified Modeling Language (UML) use case diagram to show all functionality within the ecosystem and its elements. Using the visual elements of the UML diagram, an ecosystem model was created, as shown in Fig. 5. The next step allows using the reference model Digital Services Reference Architecture (DSRA), which provides creation of:

- Illustrating how some example business scenarios implement the DSRA concepts.
- Mapping TM Forum APIs to Support Services.

Fig. 4 The process of defining key values of business scenario

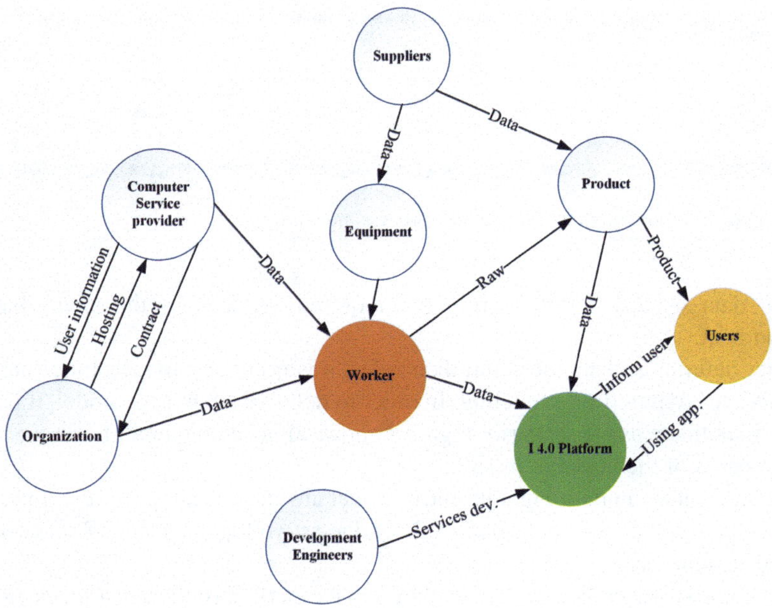

Fig. 5 An overview of the ecosystem model

- Introducing the Digital Service component and Digital Service Management API.
- Extending some of the Support Service Descriptions.

An overview of the all eTOM, SID, and TAM processes is generated at the end of the modeling, while the last phase defines a document describing all the tasks and roles of each process.

6 The Impact of Digital Transformation on Business Processes in Industry 4.0

All manufacturing organizations in the field of industry can be seen as a business process that transforms input values (material, labor, hours, use of plant and equipment) into products or services to meet customer requirements. Industry 4.0, with its features and functionality, introduces the capabilities that transform traditional manufacturing approaches toward a more efficient flow of all processes [46]. The use of the next-generation telecommunication networks introduces several changes. These are reflected in changes in the industrial ecosystem and new business processes, as well as enterprise structures that require new business models.

Business process management is especially important for production and business management because this approach applies methods to analyze, measure, model, automate, optimize, and improve production processes. In the Industry 4.0 ecosystem, business process management will focus on supporting the strategic goals of the organization, aligning resources within the company and between companies (partner organizations), as well as across the supply chain. This integrated business approach requires a change in approach to process management in order to improve the overall performance of the organization and increase the efficiency, effectiveness, and flexibility of the day-to-day operations of the company.

The four critical levels shown in Fig. 6 relate all relevant organizational activities of an enterprise within Industry 4.0. The success of the enterprise digitalization process can be measured through these levels, i.e., through the implementation of activities inside and outside the enterprise, all with the aim of improving the digital value chain of the enterprise itself.

The levels are described according to their activities within Industry 4.0:

The Customer Solutions ecosystem—a level that represents the business model and the measure of success of the end-user satisfaction. Depending on the characteristic values of the product or service, companies can offer the best to customers. Such an approach is defined by personalization according to the user's needs and provides an innovative approach in the design of end solutions (applications). It is also possible to involve external entities, thus creating added value for the enterprise itself.

The Operations ecosystem—level within the value chain of an enterprise is measured. Performance measurement includes physical activities and flows that support ecosystem customer solutions. This may include product development, planning, sourcing, manufacturing, warehousing, logistics, and services. Parts of this level of ecosystem are also enterprise partners: contract manufacturers, logistics partners, and academic institutions.

The Technology ecosystem—level that demonstrates the capabilities of modern IC technologies. It also shows the architecture of the system with all its elements, all with the aim of connecting all levels of the ecosystem into one. It incorporates

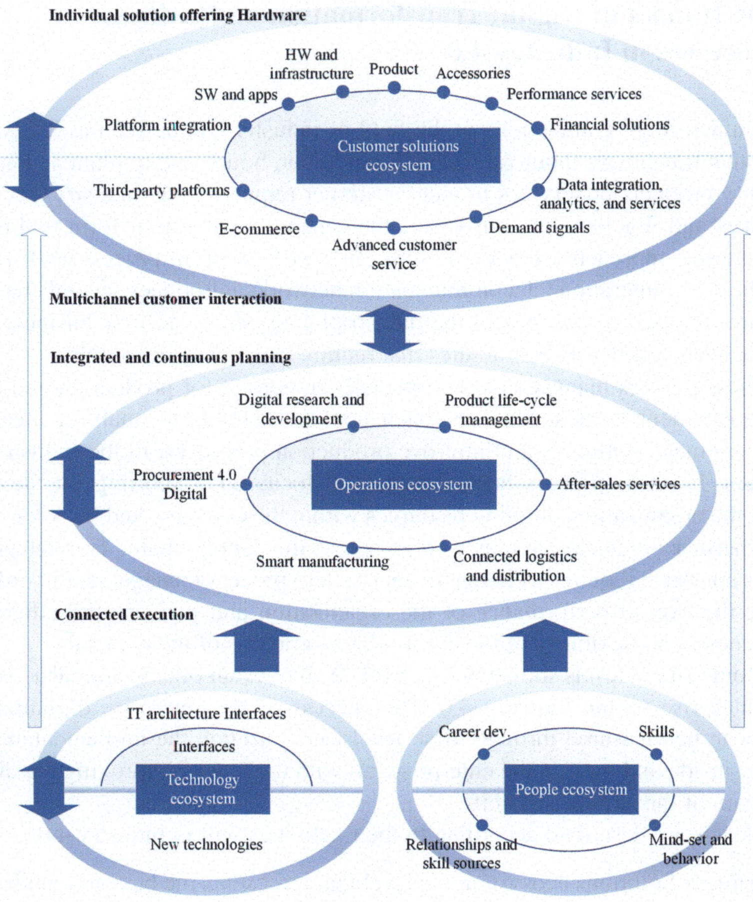

Fig. 6 Critical levels of the Industry 4.0 digital ecosystem [47]

technologies in Industry 4.0 such as artificial intelligence, 3D print, Industrial Internet of Things (IIoT) and sensors, augmented and virtual reality, and robots.

The People ecosystem—level of organization and cultural competence. This ecosystem encompasses skills, mood, thinking, and behavior as well as human resources careers important for the development and support of digital transformation within the enterprise.

A research conducted by the PWC consortium on the impact of enterprise digital transformation in Industry 4.0 defines the described four levels. These levels, through their interconnectedness, influence the business model and decision-making processes within the companies themselves.

By analog and digital culture, we can describe the situation within the company before and after the implementation of the digital transformation of the company (Table 2).

Table 2 Comparison of analog and digital culture in Industry 4.0 [47]

Analog culture		Digital culture
Push products into the market, strongly purchasing-/supply-driven	**Customers and demand**	Pull: Customer demand drives supply of products
• Strong hierarchy and slow decision-making • Process and task orientation and employees with strongly predefined work ("do your tasks") • Carry out, keep status quo, and accept barriers	**Organization**	• Flat hierarchy and fast decision-making • Results and productorientation and empowered employees ("find a way to achieve a goal") • Innovate, improve, and try to overcome barriers
• Good understanding of analog customers and learnings from the past • Experience and stability count • Homogeneous teams and work in departments • Working within tasks and silos • Career progression within predefined paths	**Way of working**	• Good understanding of digital customers and new trends • Potential, vision, curiosity, motivation, flexibility, and adaptability count • Mixed teams and work in integrated communities • Strong collaboration • Rapid, unpredictable career progression

Table 2 also represents differences in relation to customer demand, company organization, and workflows. Companies that are not able to follow the processes of digitalization will not make the necessary rise in their business.

The importance of implementing the digital transformation process in the field of Industry 4.0 has also been identified through four key ways of conducting it from traditional to developing a new business model [41]. According to the authors, the business model is described as "the logical or argumentative way an organization creates, delivers, and records value." The possibilities of digital transformation are presented in four ways, from the traditional business model to the new business model (Fig. 7).

The possibilities shown in Fig. 7 are described as:

- Internal and External Process Optimization (incremental innovation)—the first step in implementing transformation involves the introduction of co-modern IC technologies (Big Data, Cloud Computing, Collaborative Robots, Additive Manufacturing, Artificial Vision, or Augmented Reality). The aim is to optimize current business processes with contemporary IC technologies without major changes in the production process and business organization. This could be the first step for traditional manufacturing companies to embrace Industry 4.0 without dealing with high risks.
- Customer Interface Improvement—the next incremental innovation is focused on delivering added value (value proposition through product and service offerings, customer segment enhancement, sales and marketing channels, and customer relationship). The introduction of contemporary IC technologies (Big Data, Cloud Computing, Collaborative Robots, Additive Manufacturing,

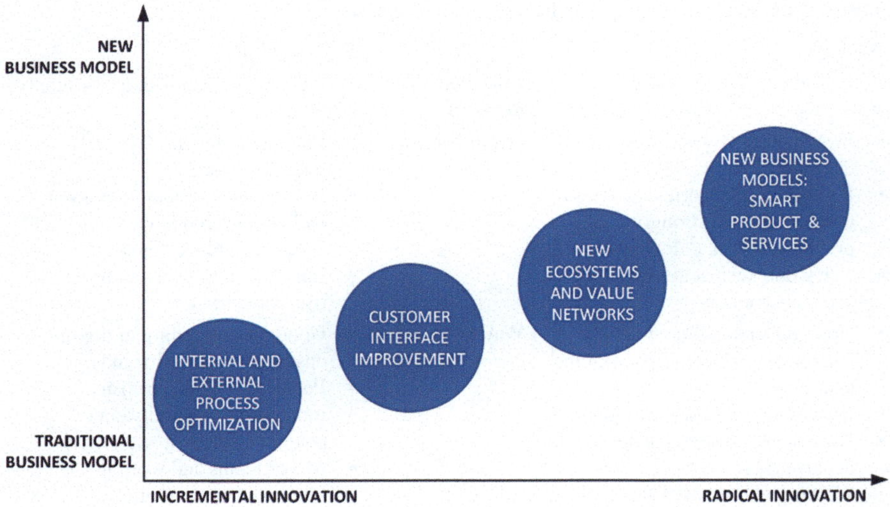

Fig. 7 Possibilities of conducting digital transformation [41]

Artificial Vision, or Augmented Reality) creates new customer-oriented interaction methods. The aim is to improve understanding of customer needs and create a customer experience. After the internal and external processes are optimized, this could be the next investment that will increase the value of traditional business.

- New Ecosystems and Value Networks—the introduction of modern IC technologies creates a transformation of the current business model. This is focused at key or characteristic activities of the company, the achievement of new required skills and resources from associates and system stakeholders. The aim is to define new values in the ecosystem, increase stakeholder knowledge, and introduce radical changes to many elements of the business model.
- New Business Models: Smart Products and Services—unlike the previous steps, a completely new business model based on contemporary IC technologies is proposed, which enables innovative products and services. There is a need to change almost all elements of the business model. This type of new business models can give businesses the opportunity to expand into new markets. It could also be applied in parallel with the actual business model. That way, the company could experiment with a new business model, while the old one is still generating revenue.

The abovementioned digital transformation possibilities influence the creation of new business models. The organization and governing body has solely the choice which one of the proposed new business models to choose.

7 Conclusion

This paper presents the identification of relevant parameters for the modeling of ecosystem elements in Industry 4.0 environment. By this, starting point for the modeling ecosystem elements in order to create a unified integrated system is enabled. The great opportunity to develop innovative services within Industry 4.0 has enabled the transition from traditional to digital business. The use of relevant tools for modeling the example of smart factory in operation, whose business is based on the Industry 4.0 concept, allows possibility to propose an ecosystem model that will be optimal for business processes that are performed. It is important to mention that the relevant parameters are key not only to designing innovative services but also to modeling all relevant elements of the ecosystem. Moreover, during the process of creating an effective ecosystem within the Industry 4.0 environment, it is necessary to take care of selecting the optimal communication network and technology as well as other elements. Also, business process represents an important factor in the process of implementing the concept of Industry 4.0.

The transition from the traditional way of business to the opportunities offered by digital transformation brings major changes in the definition of new business models. New features and characteristics of contemporary IC technologies bring many advantages in designing innovative services and solutions. Designing new business models and processes requires customization of business and services to meet customer requirements. In this process of alignment of all elements and values of a company with the needs of clients, it is possible to use the eTOM, TAN, and SID standards of the TM Forum.

The Transformation Accelerator tool has a large selection of functionality, all with the aim of optimizing business processes. For the organizational unit that bases its business on the elements of Industry 4.0, it can provide creating a business model according to clearly defined rules. It also enables the certification process of certain elements of an enterprise information system, such as processes large quantity of data. Using this tool, it is possible to create an effective ecosystem in the field of digital transformation within Industry 4.0.

References

1. Oztemel, E., Gursev, S.: Literature review of Industry 4.0 and related technologies. J. Intell. Manuf. (2018). https://doi.org/10.1007/s10845-018-1433-8
2. Klitou, D., Conrads, J., Rasmussen, M.: Germany: Industrie 4.0 Fact box for Germany's Industrie 4.0 policy initiative (2017)
3. Takeda, A., Hatakeyama, Y.: Conversion method for user experience design information and software requirement specification. In: Markus, A. (ed.) Design, User Experience, and Usability: Design Thinking and Methods, pp. 356–364. Springer, Cham (2016). https://doi.org/10.1007/978-3-319-40409-7_34
4. Guo, Y., Wu, J., Yang, K., Yu, L.: Research on requirement elicitation model of high-end equipment based on requirement classification under Internet and big data environment. In:

Advances in Computer Science Research, vol. 71, pp. 685–692 (2017). https://doi.org/10.2991/icmmita-16.2016.127
5. Häikiö, J., Koivumäki, T.: Exploring digital service innovation process through value creation. J. Innov. Manag. **4**(2), 96–124 (2016)
6. Bello, O., Zeadally, S., Badra, M.: Network layer inter-operation of Device-to-Device communication technologies in Internet of Things (IoT). Ad Hoc Netw. **57**, 52–62 (2017). https://doi.org/10.1016/j.adhoc.2016.06.010
7. Sikder, A.K., Petracca, G., Aksu, H., Jaeger, T., Uluagac, A.S.: A survey on sensor-based threats to Internet-of-Things (IoT) devices and applications. http://arxiv.org/abs/1802.02041 (2018)
8. Rojko, A.: Industry 4.0 concept: background and overview. Int. J. Interact. Mobile Technol. **11**(5), 77 (2017). https://doi.org/10.3991/ijim.v11i5.7072
9. Sethi, P., Sarangi, S.R.: Internet of Things: architectures, protocols, and applications. J. Electr. Comput. Eng. **2017**, 1–25 (2017). https://doi.org/10.1155/2017/9324035
10. Balog, M., Szilágyi, E., Dupláková, D., Minďaš, M.: Effect verification of external factor to readability of RFID transponder using least square method. Measurement. **94**, 233–238 (2016). https://doi.org/10.1016/j.measurement.2016.07.088
11. Kolarovszki, P.: Research of readability and identification of the items in the postal and logistics environment. Transp. Telecommun. J. **15**(3), 196 (2014). https://doi.org/10.2478/ttj-2014-0017
12. Ren, L., Zhang, L., Tao, F., Zhao, C., Chai, X., Zhao, X.: Cloud manufacturing: from concept to practice. Enterp. Inf. Syst. **9**(2), 186–209 (2015). https://doi.org/10.1080/17517575.2013.839055
13. Immonen, A., Ovaska, E., Kalaoja, J., Pakkala, D.: A service requirements engineering method for a digital service ecosystem. SOCA. **10**(2), 151–172 (2016). https://doi.org/10.1007/s11761-015-0175-0
14. Abeywickrama, D.B., Ovaska, E.: A survey of autonomic computing methods in digital service ecosystems. SOCA. **11**(1), 1–31 (2017). https://doi.org/10.1007/s11761-016-0203-8
15. Sklyar, A., Kowalkowski, C., Tronvoll, B., Sörhammar, D.: Organizing for digital servitization: a service ecosystem perspective. J. Bus. Res. (2019). https://doi.org/10.1016/j.jbusres.2019.02.012
16. Pakkala, D., Spohrer, J.: Digital service: technological agency in service systems. In: Proceedings of the 52nd Hawaii International Conference on System Sciences, vol. 6, pp. 1886–1895 (2019)
17. Barile, S., Lusch, R., Reynoso, J., Saviano, M., Spohrer, J.: Systems, networks, and ecosystems in service research. J. Serv. Manag. **27**(4), 652–674 (2016). https://doi.org/10.1108/JOSM-09-2015-0268
18. Chae, B.(.K.).: A general framework for studying the evolution of the digital innovation ecosystem: the case of big data. Int. J. Inf. Manag. **45**, 83–94 (2019). https://doi.org/10.1016/j.ijinfomgt.2018.10.023
19. Mochalov, V.P., Bratchenko, N.Y., Yakovlev, S.V., Gosteva, D.V.: Distributed management system for infocommunication networks based on TM Forum Framework. CEUR Workshop Proc. **2254**, 81–93 (2018)
20. Tr3Dent: Digital transformation accelerator. https://www.tr3dent.com/ (n.d.)
21. Flatscher, M., Riel, A.: Stakeholder integration for the successful product–process co-design for next-generation manufacturing technologies. CIRP Ann. Manuf. Technol. **65**(1), 181–184 (2016). https://doi.org/10.1016/j.cirp.2016.04.055
22. Thoben, K.-D., Wiesner, S., Wuest, T.: "Industrie 4.0" and smart manufacturing—a review of research issues and application examples. Int. J. Autom. Technol. **11**(1), 4–16 (2017). https://doi.org/10.20965/ijat.2017.p0004
23. Stock, T., Seliger, G.: Opportunities of sustainable manufacturing in Industry 4.0. Proc. CIRP. **40**, 536–541 (2016). https://doi.org/10.1016/j.procir.2016.01.129

24. Zheng, P., Wang, H., Sang, Z., Zhong, R.Y., Liu, Y., Liu, C.: Smart manufacturing systems for Industry 4.0: conceptual framework, scenarios, and future perspectives. Front. Mech. Eng. **13**(2), 137–150 (2018). https://doi.org/10.1007/s11465-018-0499-5
25. Raihanian Mashhadi, A., Behdad, S.: Ubiquitous life cycle assessment (U-LCA): a proposed concept for environmental and social impact assessment of Industry 4.0. Manuf. Lett. **15**, 93–96 (2018). https://doi.org/10.1016/j.mfglet.2017.12.012
26. Nihtianov, S., Luque, A.: Smart Sensors and MEMS Intelligent Sensing Devices and Microsystems for Industrial Applications, 2nd edn. Woodhead Publishing, Cambridge, UK (2018)
27. Mekki, K., Bajic, E., Chaxel, F., Meyer, F.: Overview of cellular LPWAN technologies for IoT deployment: Sigfox, LoRaWAN, and NB-IoT. In: 2018 IEEE International Conference on Pervasive Computing and Communications Workshops, PerCom Workshops 2018, March, pp. 197–202 (2018). https://doi.org/10.1109/PERCOMW.2018.8480255
28. Sinha, R.S., Wei, Y., Hwang, S.H.: A survey on LPWA technology: LoRa and NB-IoT. ICT Exp. **3**(1), 14–21 (2017). https://doi.org/10.1016/j.icte.2017.03.004
29. Aernouts, M., Berkvens, R., Van Vlaenderen, K., Weyn, M.: Sigfox and LoRaWAN datasets for fingerprint localization in large urban and rural areas. Data. **3**(2), 13 (2018). https://doi.org/10.3390/data3020013
30. Periša, M., Sente, R.E., Cvitić, I., Kolarovszki, P.: Application of innovative smart wearable device in Industry 4.0. In: Proceedings of the 3rd EAI International Conference on Management of Manufacturing Systems, pp. 1–10. EAI, Ghent, Belgium (2018). https://doi.org/10.4108/eai.6-11-2018.2279105
31. Peraković, D., Periša, M., Sente, R.E.: Information and communication technologies within Industry 4.0 concept. In: Ivanov, V., et al. (eds.) Advances in Design, Simulation and Manufacturing, pp. 127–134. Springer International Publishing, Cham (2018). https://doi.org/10.1007/978-3-319-93587-4_14
32. Peraković, D., Periša, M., Zorić, P.: Challenges and issues of ICT in Industry 4.0. In: Advances in Design, Simulation and Manufacturing II, pp. 259–269. Springer, Cham (2020). https://doi.org/10.1007/978-3-030-22365-6_26
33. Jovović, I., Husnjak, S., Forenbacher, I., Maček, S.: Innovative application of 5G and blockchain technology in Industry 4.0. EAI Endorsed Trans. Ind. Netw. Intell. Syst. **6**(18), 157122 (2019). https://doi.org/10.4108/eai.28-3-2019.157122
34. Cheng, J., Chen, W., Tao, F., Lin, C.L.: Industrial IoT in 5G environment towards smart manufacturing. J. Ind. Inf. Integr. **10**(March), 10–19 (2018). https://doi.org/10.1016/j.jii.2018.04.001
35. Boyes, H., Hallaq, B., Cunningham, J., Watson, T.: The industrial internet of things (IIoT): an analysis framework. Comput. Ind. **101**(June), 1–12 (2018). https://doi.org/10.1016/j.compind.2018.04.015
36. Mittal, S., Khan, M.A., Romero, D., Wuest, T.: Smart manufacturing: characteristics, technologies and enabling factors. Proc. Inst. Mech. Eng. B J. Eng. Manuf. **233**(5), 1342–1361 (2019). https://doi.org/10.1177/0954405417736547
37. Cvitić, I., Peraković, D., Kuljanić, T.M.: Availability factors in delivery of information and communication resources to traffic system users. In: Mikulski, J. (ed.) Smart Solutions in Today's Transport, pp. 28–41. Springer International Publishing, Cham (2017)
38. Cvitić, I., Peraković, D., Periša, M., Botica, M.: Novel approach for detection of IoT generated DDoS traffic. Wirel. Netw. **1**, 1–14 (2019). https://doi.org/10.1007/s11276-019-02043-1
39. Peraković, D., Periša, M., Cvitić, I., Husnjak, S.: Model for detection and classification of DDoS traffic based on artificial neural network. Telfor J. **9**(1), 26 (2017). https://doi.org/10.5937/telfor1701026P
40. Peraković, D., Periša, M., Zorić, P.: Identification of the relevant parameters for modeling the ecosystem elements in Industry 4.0. In: Knapcikova, L., Balog, M., Peraković, D., Periša, M. (eds.) 4th EAI International Conference on Management of Manufacturing Systems, p. 260. Springer International Publishing, Cham (2020). https://doi.org/10.1007/978-3-030-34272-2

41. Ibarra, D., Ganzarain, J., Igartua, J.I.: Business model innovation through Industry 4.0: a review. Proc. Manuf. **22**, 4–10 (2018). https://doi.org/10.1016/j.promfg.2018.03.002
42. Mittal, S., Khan, M.A., Romero, D., Wuest, T.: A critical review of smart manufacturing & Industry 4.0 maturity models: implications for small and medium-sized enterprises (SMEs). J. Manuf. Syst. **49**(November), 194–214 (2018). https://doi.org/10.1016/j.jmsy.2018.10.005
43. Sjödin, D.R., Parida, V., Leksell, M., Petrovic, A.: Smart factory implementation and process innovation: a preliminary maturity model for leveraging digitalization in manufacturing. Moving to smart factories presents specific challenges that can be addressed through a structured approach focused on people. Res. Technol. Manag. **61**(5), 22–31 (2018). https://doi.org/10.1080/08956308.2018.1471277
44. David, S.: Information Systems Investment Decision Making Using Return of Investment Application of CRUDi Framework with eTOM Business Process Framework in Portuguese Telecommunication Industry. Universidade Nova de Lisboa, Lisbon (2018)
45. TM Forum: TM Forum. https://www.tmforum.org/ (2019)
46. Da Xu, L., Xu, E.L., Li, L.: Industry 4.0: state of the art and future trends, 7543. https://doi.org/10.1080/00207543.2018.1444806 (2018)
47. Giessbauer, R., Lübben, E., Schrauf, S., Pillsbury, S.: Global Digital Operations Study 2018—How Industry Leaders Build Integrated Operations Ecosystems to Deliver End-to-End Customer Solutions. Strategy & White paper (2018)

Analysis of Customer Acquisition Cost and Innovation Performance Cost After BIM Technology Implementation

Jana Smetanková, Katarína Krajníková, Tomáš Mandičák, and Annamária Behúnová

1 Introduction

Efforts to increase efficiency and profit lead to the improvement of traditional processes, products, and services; therefore, research into innovative technologies is increasingly supported. Research and development of innovation contributes to process improvement and environmental protection. Research is a systematic inquiry to describe, explain, predict, and control the observed phenomenon. Research involves inductive and deductive methods [1].

The main aim of the research is exploration, description, explanation, and application the obtained results. Application of research is the creation of new innovative processes and technologies.

Innovation is the key to gaining a competitive advantage and is a major aspect of economic growth. Innovation is the provision of more efficient products, processes, services, technologies, or business models that are available to markets, governments, and community [2].

Innovation can be:

- Improving or replacing processes, products, and services and improving the quality of existing products and services. These processes result in increased efficiency and productivity.

J. Smetanková · K. Krajníková · T. Mandičák
Faculty of Civil Engineering, Technical University of Košice, Košice, Slovakia
e-mail: jana.smetankova@uke.sk; katarina.krajnikova@tuke.sk; tomas.mandicak@tuke.sk

A. Behúnová (✉)
Department of Industrial Engineering and Informatics,
Technical University of Kosice, Prešov, Slovakia
e-mail: annamaria.behunova@tuke.sk

L. Knapcikova et al. (eds.), *New Approaches in Management of Smart Manufacturing Systems*, EAI/Springer Innovations in Communication and Computing, https://doi.org/10.1007/978-3-030-40176-4_16

- Development of new and improved products and services—meeting the needs or requirements of customers or consumers.
- Adding "value" to existing products or services in order to differentiate an company from its competitors [3].

One of the innovative ways of building construction is building information modelling. Building information modelling is a process for creating and managing information on a construction project across the project life cycle. BIM is a way of working. BIM is information modelling and information management in a team environment; all team members should be working to the same standards as one another. BIM creates value from the combined efforts of people, process, and technology. BIM brings together all of the information about every component of a building, in one place. The future of the construction industry is digital, and BIM is the future of design and long-term facility management [4].

2 Total Cost

Costs are the inevitable expenses, incurred at business. Companies plan, monitor, and evaluate their costs. Many types of costs are easily traceable and quantifiable. One type of cost is one that is characterized by a proportional relationship between the amount of input costs and the amount of outputs. Another type of cost is costs that must be identified or allocated—relationship between input costs and output units need not be directly observable and quantified.

Costs arise not only in production activities but also in business applications such as financial accounting, cost accounting, budgeting, capital budgeting, and valuation. There are different ways to categorize costs, for example, in relation to production, the context in which they are used, etc. [5].

Total costs are all costs (fixed, variable, direct, indirect) incurred in realization and managing the project. Total costs are divided into:

- *Variable cost*—costs that change as production changes—material, wages of labor, energy used in production, etc.
- *Fixed cost*—costs that do not change with production change—rent, salaries of employees, advertising, promotional campaigns, etc.
- *Direct cost*—costs that can be attributed to the production of products and services—labor, material, fuel, power, or any other expense related to the production of a product is a direct cost.
- *Indirect cost*—costs that can't be directly attributed to the production of products and services—for example, depreciation, supervision, security, maintenance, and administrative that are incurred which cannot be assigned to a specific product or department hence are classified as an indirect cost.
- *Product and period costs*—these are similar to direct and indirect costs. Product costs are costs that the accounting system of a company associates directly with output and that are used to cost estimating to inventory. Period costs are charged

as expenses to the current period. Under direct costing, period costs are not viewed as costs of the products being manufactured, so they are not associated with valuing inventories.
- *Other types of costs* [5, 6].

3 Key Performance Indicators

Key performance indicator (KPI) is a type of performance measurement. Key performance indicators are tools to evaluate the "success" of an organization or a specific activity (projects, programs, products, and other initiatives). The success of a project is defined to a repeated or regular achievement of some levels of an operational objective. Sometimes success is defined as progress toward strategic objectives [7].

Key performance indicator (KPI) is a type of performance measurement that helps understand how an organization or department works. Individual KPIs:

- Are clearly defined and quantified
- Support communication throughout the organization and departments
- Support business activity and departments [8]

Project that reach success are with high profits and good reputations. Successful projects have characteristics that guarantee a certain level of success. These characteristics are:

- People—manage projects
- Setting objectives—observe the level of efficiency
- Integration of the company with customers

The people, who lead the project, determine the company culture, attitudes, and values, which can lead to the success or unsuccess of the company.

Successful organizations create their business objectives, plan activities, and create business plans. Individual companies set objectives with regard to market needs and strategic objectives of the company. A successful organization plans human resources, product, and marketing strategies and creates a learning background for its employees [9].

Successful project management requires a thorough analysis of work, sales, and financial results. It is important to monitor relevant key performance indicators. It is very important to categorize key performance indicators. KPIs define way of measurement. There are two measurement categories for KPIs:

- Quantitative indicators—represent objective facts and are measurable
- Qualitatively indicators—represent subjective facts (feelings, opinion, taste) and are presented as numeric or textual values that represent the interpretation of these elements [10]

Business metrics, also called KPIs (key performance indicators), display a measurable value that shows the progress of a company's business goals. KPIs show if the company has achieved its planned targets. There are many publications that define key performance indicators. Table 1 provides an overview of the key performance indicator definitions by various authors.

The contribution deals with concretely customer acquisition cost (CAC), innovation performance cost (IPC), and their impact on total cost of construction company.

3.1 Customer Acquisition Cost (CAC)

Customer acquisition costs are types of key performance indicator, concretely co-called customer metrics.

Customer acquisition cost is one of the most important parameters in trading. Customer acquisition cost reflects the effectiveness of marketing campaigns. When creating marketing campaigns, it's important to focus on three basic parameters:

- Customer acquisition cost
- Lifetime value
- Payback periods

The company's priority is effective project management. One of the tools to promote efficiency is to determine the precise procedures. Companies are able to maintain a high level of productivity while reducing costs. For example, customer service contributes to increasing efficiency. Customer service carries out repairs and conducts marketing campaigns. Customer service is an important tool to support the company's success. Communicating with customers can greatly help increase profits. The way customer service is performed has a direct impact on the company's reputation [9].

Customer acquisition cost (CAC) presents to the costs that are required to acquire a new customer. CAC is the sum cost of sale and marketing costs, including:

- Labor costs of sales and marketing department
- Devices that used by sales and marketing staff—computers, phones, and printers
- Software applications and tools (customer relationship management (CRM) and marketing automation)
- Costs of subcontractors—marketing and advertising agencies
- Advertising (online advertising, direct mail, television advertising, and out-of-home advertising)
- Cost of customer service
- Costs of sponsors, events, and conferences [14]

Customer acquisition cost is calculated as the sum of sales and marketing costs divided by the number of customers acquired [15].

Table 1 Key performance indicator definitions by various authors [8, 11–13]

Publication		KPI—key performance indicator	
Title	Author		
12 Business Metrics That Every Company Should Know (2019)	Karola Karslon	Sales revenue	
		Net profit margin	
		Gross margin	
		Sales growth year-to-date	
		Cost of customer acquisition	
		Customer loyalty and retention	
		Net promoter score	
		Qualified leads per month	
		Lead-to-client conversion rate	
		Monthly website traffic	
		Met and overdue milestones	
		Employee happiness	
12 Key Financial Performance Indicators You Should Be Tracking (2018)	Bill Gerber	Operating cash flow	
		Working capital	
		Current ratio	
		Debt-to-equity ratio	
		LOB revenue vs. target	
		LOB expenses vs. budget	
		Accounts payable turnover	
		Accounts receivable turnover	
		Inventory turnover	
		Return on equity	
		Quick ratio	
		Customer satisfaction	
11 Key Business Performance Metrics for Better Operations (2017)	Alicia Wiliams	Financial business performance metrics	Gross profit margin
			Net profit
			Net profit margin
			Debt-to-asset ratio
		Marketing business performance metrics	Return on advertising spend
			Customer acquisition cost
			Time to payback
			Marketing originated customer percentage
		Employee business performance metrics	Employee efficiency
			Quality of work
			Adherence to values

(continued)

Table 1 (continued)

Publication		KPI—key performance indicator	
Title	Author		
18 Key Performance Indicator (KPI) Examples Defined (2018)	Ted Jackson	Financial metrics	Profit
			Cost
			LOB revenue vs. target
			Cost of goods sold
			Day sales outstanding (DSO)
			Sales by region
			LOB expenses vs. budget
		Customer metrics	Customer lifetime value (CLV)
			Customer acquisition cost (CAC)
			Customer satisfaction and retention
			Net promoter score (NPS)
			Number of customers
		Process metrics	Customer support tickets
			Percentage of product defects
			LOB efficiency measure
		People metrics	Employee turnover rate (ETR)
			Percentage of response to open positions
			Employee satisfaction
13 Bonus Key Performance Indicator Examples & Definitions (2018)	Ted Jackson	People metrics	Retirement rate
			Knowledge archived with training
			International promotions vs. external hires
			Salary competitiveness ratio (SCR)
		Customer metrics	Customer churn rate
			Contact volume by channel
			Percentage of customers who are "very" or "extremely" satisfied
			Number of new vs. repeat site visits
		Financial metrics	Cash flow from financial activities
			Average annual expenses to serve one customer
			EBITDA (earning before interest, taxes, depreciation, and amortization)
			Innovation spending
			Customer lifetime value/ customer acquisition cost

Customer acquisition cost consists of product costs and costs related to marketing, research, and sales. This parameter is important because it helps in the calculation of the so-called the importance of the customer [16].

There are many methods to reduce customer acquisition cost, for example:

- Focus on increasing "paying" customers—focus on specific data. The aim of this strategy is to find suitable methods—they bring the desired effect and are affordable. Only 52% of companies are making good use of their conversions. It is important to focus on content's copy, CTA layout, structure, testimonials, and other. Another method is online marketing. One way of online marketing is, for example, e-mail marketing. Choosing the right way to drive conversions depends on the amount of products and services, the demographic situation, and the nature of business.
- Marketing automation—improve lead search, support analysis and reporting, and more. The benefit of automation is the reduction of labor costs and acceleration of processes. Supporting the automation of marketing tools can speed up a process that requires manual effort. Marketing automation has a positive impact on time and cost savings. Companies that automate their marketing methods have seen their revenue increase in 6–9 months.
- Cut down customer churn and use them to bring new customers—the loss of the customer increases the acquisition costs. It's 6–7 times more expensive to acquire a new customer than it is to keep a current one. Dissatisfied customers share negative reviews and can damage the company's reputation. It's important to support customer service and respond promptly to dissatisfied reviews and comments.
- Creating interesting content—the aim is to create content that is directly targeted to the customer. Targeted content can minimization of additional cost [17, 18].

3.2 *Innovation Performance Cost (IPC)*

Innovation performance cost is the amount of costs the organization invests in innovation process. Innovation is a key strategic point for each business unit [8].

Successful implementation of new ideas is an important aspect for improving processes, promoting product and service improvement, increasing efficiency, and increasing profit. The development of new technologies increases competitiveness [3].

Innovation is the key to gaining a competitive advantage and is a major aspect of economic growth. There exist many definitions of the term innovation, such as:

- Innovations are new ways of producing products or processes that contribute to wealth [19].
- Innovation is the application of better solutions that meet new requirements, unspecified needs, or existing market needs [20].

- In an organizational context, innovation can be linked to positive changes in efficiency, productivity, quality, competitiveness, and increased market share. Innovation is a means by which an entrepreneur generates new wealth-generating resources or provides existing resources with increased wealth creation potential [21].

Implementing innovations has a positive impact on:

- Improving productivity.
- Reducing costs.
- Increasing competitiveness.
- Building brand and reputation of company.
- Supporting partnerships and sponsors.
- Increasing turnover and profitability [3].

Before the introduction of new technologies, a detailed market analysis needs to be carried out and the opportunities need to be identified [3].

Innovative technologies are used in the construction sector. Compared to other sectors, their implementation is relatively low. McKinsey & Company points out that construction is one of the least digitized sectors (spending on innovation does not exceed 1%). Support for innovation implementation is growing. Many organizations and associations are being established to promote innovation in the construction sector, such as the initiative Constructech. Constructech has identified three key challenges:

- People—possibility to optimize employee performance through resources and tools
- Intelligent materials—increasing productivity and reducing environmental impact with smart materials and tools
- Data—improving data collection and analysis [22]

Promoting innovation of building is one of the most important ways of planning, realization, and managing the building's life cycle. Use and implementation of innovation have a positive impact on construction industry.

There are a lot of innovative technologies that help for planning, realization, and managing buildings, for example:

- *Embedded Sensors*—bring a lot of opportunities to acquisition and manage safety, material performance, and operational workflow data. Sensors that are fixed on a load-bearing structure are linked to software for building management and provide image of progress in real time.
- *Artificial Intelligence*—are machines that simulate intelligent human behavior. The computer system performs tasks that normally require human intelligence, as well as visual perception, choosing, etc.
- *Machine Learning*—is learning knowledge graphs and expert systems that are able to modify more data (dynamic determination), and they do not require human intervention to carry out certain changes.

- *Virtual Reality*—enables real-time comparison of 3D models with real space; supports location data analysis, virtual tutorials, etc.; increases adaptation, security, and accuracy; and promotes collaboration between building process participants.
- *Building Information Modelling*—is an emerging innovative technology. Building information modelling is a digital tool used throughout the building's life cycle. BIM supports the visualization, planning, communication, and collaboration of project participants [23–25].

4 Building Information Modelling (BIM)

The construction sector is still more making attempts to increase productivity through the use of information and communication technologies (ICT), which certainly include building information modelling. The topic about building information modelling is not new in construction industry. Many participants understand the term building information modelling than software application but construction design process. For some participants, this technology is a completely new method that requires the implementation of new practices, methods, contracts, relationships, etc. [26].

There are many definitions of the term building information modelling (BIM). BIM presents digital, three-dimensional virtual project presentation that helps in designing, planning, building, and cost planning and building project maintenance [27]. The National Building Information Model Standard Project Committee defined the BIM as follows: BIM presents a digital representation of the physical and functional characteristics of the building. It is used as a general source of knowledge, information on equipment, and construction. BIM is a helpful tool to make decisions throughout the building's life cycle [28].

BIM is an intelligent 3D model-based process that gives architecture, engineering, and construction (AEC) professionals the insight and tools to more efficiently plan, design, construct, and manage buildings and infrastructure [29].

The result of the creation process in building information modelling is a model of object. The model contains graphic and non-graphic information. This information can be considered as individual dimensions of building information modelling. 3D model contains graphic information. In addition to the graphical representation of the object, the model also contains non-graphical information that represents each dimension. These include:

- Time parameters (4D time)
- Cost parameters (5D cost)
- Information about sustainability (6D sustainability)
- Information needed to manage buildings (7D facility management)

One of the benefits of implementing BIM at pre-investment phase is to improve coordination and cooperation. BIM is a virtual knowledge database that facilitates easy information exchange. Other benefits include system conflict detection, constructivist assessment and risk elimination, a high level of customization of building systems and building activities, and optimization of time schedule and cost estimation. BIM provides tools that enable early identification and reduction or complete elimination of collisions in building implementation [30].

In the building maintenance and operation phase, BIM delivers all available information that has been collected from design to commissioning. This information is a valuable data that will help to effectively manage the operation and maintenance of buildings. Maintenance and operation efficiency is demonstrated primarily by reducing the time and cost of repair and maintenance [28, 31].

5 Methodology

5.1 Research Aim and Hypotheses

The research is focused on the issue of cost changing and technology in lifecycle building and construction process. We focused on two types of costs, costs for innovations and costs for customers. The respondent's opinion was found out for overall costs too. Based on theoretical background, it was set main research aim to analyze the impact of BIM implementation on overall costs reducing in construction projects. This main aim was supported to part aims:

1. Check the normality a dependence of measured data
2. Bulk comparison of all types of costs in every year and pairwise comparisons of these groups and results arranged in an appropriate form

5.2 Data Collection and Research Sample

All data in research were conducted by online form questionnaire. The research sample was approached by e-mail (1073 respondents) with the request to participate in the research to Slovak construction enterprises which participated in projects whose structure had several phases. One part of questionnaire was about supporting innovative ways in construction.

Based on these requirements, a return of questionnaire was 5.12%. The sample consisted of 55 companies/respondents (architects, contractors, investors in Slovakia), which implemented their projects divided into at least two phases. These companies had information about the costs, and 16 of them supported innovative ways of building, specifically building information modelling, and had required information about various types of costs. Other companies did have no information about costs in this sense, and these respondents had to be excluded from our sample.

To determine the answer was used "Likert scale ranging" from 1 to 5. Based on fixed values had been done the arithmetic average of the values for the selected area under consideration: 1- slightly affected costs, and 5 - very affected costs (growth of costs). The extent and interval values for each stage of the scale were thoroughly explained to the respondents.

5.3 *Research Steps and Methodology*

First of all, it was set a problem statement based on theoretical background. It was set to research questions and hypotheses in this issue. Next step, it was data collection (all information about research sample and data collection process).

In the sense of statistical evaluation, first, the normality of data had to be tested. The normality assumption needs to be considered for validation of data presented in the literature as it shows whether correct statistical tests have been used [32].

For this, it was used very common test called Shapiro-Wilk test. It is a test for normal distribution exhibiting high power, leading to good results even with a small number of observations (respondents). The desired significance level alpha in all used tests was used 5%. The desired hypothesis for the Shapiro-Wilk test tested if (H0) research sample does have a normal distribution (with unspecified mean and variance) against hypothesis (H1) research sample which does not have a normal distribution. Computation of this test was done in MATLAB. Most of research sample did not have a normal distribution. Based on this result, it was chosen nonparametric test called Friedman's test. One of the disadvantages of using of this test is that it is focused on all of groups at once. If one group has different mean rank, the test rejects the null hypothesis, and then further analysis is needed. All results of used tests are shown in "Results" chapter.

To confirm our hypotheses about mean ranks in all of groups at one, Friedman's test was used. It is a nonparametric alternative to ANOVA, and it is appropriate for dependent data. It relies on the rank-ordering of data rather than calculations involving means and variances and allows you to evaluate the differences between three or more repeated (or matched) samples (treatments) [33].

One of the requirements for using this test is that data should be ordinal (e.g., the Likert scale) or continuous.

The null H0 and alternative H1 hypotheses for Friedman's test are as follows:

- H_0: $M_1 = M_2 = \ldots = M_k$ (M_i is the position parameter for sample i).
- H_1: There is at least one pair (i, j) such that $M_i \neq M_j$.

Let n be the size of k paired samples. The Q statistic from Friedman's test is given by:

$$Q = \frac{12}{nk\left(k+1\right)} \sum_{i=1,\ldots,k} \left[R_i^2 - 3n\left(k+1\right)\right]$$

where R_i is the sum of the ranks for sample i. The *p*-value associated with a given value of Q can be approximated by a X^2 distribution with $(k - 1)$ degrees of freedom. This approximation is reliable when kn is greater than 30, the quality also depending on the number of ties. The *p*-values associated with Q have been tabulated [34].

When the *p*-value is such that the $H0$ hypothesis has to be rejected, then at least one sample (or group) is different from another. To identify which samples are responsible for rejecting H0, multiple comparison procedures can be used.

For this multiple comparison procedure, Dunn's test was used. Computation of this test was done in MATLAB. It can be used to pinpoint which specific means are significant from the others. Dunn's multiple comparison test is a post hoc nonparametric test (a "distribution-free" test that doesn't assume your data comes from a particular distribution). The null hypothesis for the test is that there is no difference between groups (groups can be equal or unequal in size). The alternate hypothesis for the test is that there is a difference between groups [35, 36].

Dunn's (1964) insight was to approximate a *z*-test statistic to the exact rank-sum statistic. Dunn's test is the appropriate procedure following a Kruskal-Wallis test. Making multiple pairwise comparisons following an omnibus test redefines the meaning of α, which usually represents the probability of falsely rejecting the null hypothesis for one test, within the inferential framework of the hypothesis test. Dunn (1961) described how to address this issue with a Bonferroni adjustment, which can modify the rejection level for any test by dividing α by the total number of tests and requires a much smaller *p*-value to reject any test. This adjustment leaves α numerically intact but multiplies the *p*-value [37].

6 Results

In the first step, the all types of costs were monitored with the respect of every year. Shapiro-Wilk test was used to determine if the null hypothesis of normality is a reasonable assumption regarding the population distribution of a random sample. The desired hypotheses were that (H_0) research sample does have a normal distribution (with unspecified mean and variance) and (H_1) research sample does not have a normal distribution. Computation of this test was done in MATLAB. In two cases (costs for customers in 2017 and overall costs in 2016), the null hypothesis was not rejected, and the normal distribution was detected (green-colored numbers). But, generally, it could not be said that research data were normally distributed, so then it had been used nonparametric tests. In Table 2, except for normality, the basic characteristics such as an average, a standard deviation, and a confidence interval of our samples were calculated.

These samples were tested by X^2—test ("chi-square test") of Independence to find out whether our samples are dependent or independent and which type of test could then be used. Sums of gained opinions of respondents were set to contingency

Table 2 Descriptive statistics of samples and the normality test

	2016	2017	2018
Innovation performance cost	$\bar{x} = 2.1875$	$\bar{x} = 4.0625$	$\bar{x} = 1.9375$
	$s = 1.1672$	$s = 0.9287$	$s = 0.8539$
	CI = [1.6155, 2.7594]	CI = [3.6074, 4.5175]	CI = [1.5190, 2.3559]
	$p = 0.022$	$p = 0.0126$	$p = 0.009$
Customer acquisition cost	$\bar{x} = 3.125$	$\bar{x} = 2.9375$	$\bar{x} = 2.6875$
	$s = 0.8851$	$s = 1.2894$	$s = 0.8732$
	CI = [2.6913, 3.5586]	CI = [2.3057, 3.5692]	CI = [2.2596, 3.1153]
	$p = 0.0356$	p = 0.1493	$p = 0.0427$
Total costs	$\bar{x} = 3.25$	$\bar{x} = 4.125$	$\bar{x} = 1.5625$
	$s = 1.0646$	$s = 0.9574$	$s = 0.6291$
	CI = [2.7284, 3.7716]	CI = [3.6558, 4.5941]	CI = [1.2542, 1.8707]
	p = 0.1637	$p = 0.0063$	$p = 0.000638$

Table 3 Results of X^2- the test of independence (the contingency table)

	2016	2017	2018	Row totals
Innovation performance cost	35 (43.12) [1.53]	64 (55.71) [1.23]	31 (31.16) [0.00]	130
Customer acquisition cost	50 (46.44) [0.27]	47 (60.00) [2.82]	43 (33.56) [2.66]	140
Total costs	52 (47.44) [0.44]	66 (61.29) [0.36]	25 (34.28) [2.51]	143
Column totals	137	177	99	413

Table 4 Comparison of all types of costs in every year and all years for every type of costs

	2016	2017	2018	Costs for innovations	Costs for customers	Overall costs
X^2	18.5	18.4063	18.8438	23.1563	3.88	28.2188
p-Value	<0.05	<0.05	<0.05	<0.05	*0.1765*	<0.05

table and the results of X^2—test are arranged in Table 3. It provides the observed cell totals, (the expected cell totals) and [the chi-square statistic for each cell].

The X^2 statistic was 11.8219. The *p*-value is 0.018726, so our samples could be considered as dependent variables. The result is significant at $p < 0.05$.

Because of dependent samples, the nonparametric Friedman's test was used. The data were analyzed for all 3 years and for all three types of costs (Table 4). The desired hypotheses were that (H0) research samples do have an equal means of ranks and (H1) research samples do not have an equal means of ranks.

Table 5 Pairwise comparison of all types of costs in every year and all years for every type of costs

	2016	2017	2018
2016	✓		
2017	✗	✓	
2018	✓	✗	✓

Customer Acquisition Cost (in every year)

	2016	2017	2018
2016	✓		
2017	✓	✓	
2018	✓	✓	✓

Total costs (in every year)

	2016	2017	2018
2016	✓		
2017	✓	✓	
2018	✗	✗	✓

Almost in every case, the null hypothesis was rejected, except for customer's costs in every year. In this case, the null hypothesis was not rejected, so respondents in our sample consider these costs in every year as the same. In the other cases, the differences were observed. This test says that somewhere exist a difference mean of rank, but it does not say where the researcher should find difference exactly.

To find out in which year and which types of costs were perceived differently, Dunn's test to do pairwise comparisons was used. Its results are summarized in the next two tables (Tables 5 and 6). Red cross means that there exists statistically difference between groups. Green thick means there is no significantly difference between groups.

The main aim of the research was to show at least a few respondents that implementation of BIM brought a reduction in the costs. The companies which were included into results had their own experiences of planning the construction process in their project without using BIM and with using BIM. Generally, it means recommendation for implementation of BIM technology as a progressive ICT in construction projects, costs for implementation decreased, and implementation of BIM to

Table 6 Pairwise comparison of all types of costs in every year and all years for every type of costs

Year 2016

	Innovation Performance Cost	Innovation Performance Cost	Total costs
Innovation Performance Cost	✓		
Innovation Performance Cost	✗	✓	
Total costs	✗	✓	✓

Year 2017

	Innovation Performance Cost	Innovation Performance Cost	Total costs
Innovation Performance Cost	✓		
Innovation Performance Cost	✗	✓	
Total costs	✓	✗	✓

Year 2018

	Innovation Performance Cost	Innovation Performance Cost	Total costs
Innovation Performance Cost	✓		
Innovation Performance Cost	✓	✓	
Total costs	✓	✗	✓

construction has the positive effect on overall costs. The costs in the first 2 years are probably higher, but in the third year, costs begin to be significantly lower than the previous 2 years. This trend is an important information for construction sector.

7 Conclusion

Science, technology, research, and development are important strategic aim of any advanced society. The application of science and research knowledge is important for the overall development of the economic and social aspects of society. The result of applying the acquired knowledge is the creation of innovations. Innovations include the use of new production techniques, processes, and products, the use of new raw materials, changes in the production process, and others. Managing construction production and projects is complicated. Construction production is characterized by amount of specifics, such as individual subject of delivery, individual place of delivery, individual time of delivery, and individual way of construction. All these specifications affect cost and key performance indicators. In the construction sector, the implementation of innovative methods is increasing, which affects the implementation of construction and project management. There exist many innovative technologies, materials, and software applications that help these activities. Building information modelling is one of the innovative technologies. The contribution analyzed the impact of BIM on reducing total cost, customer acquisition cost, and innovation performance cost. The results showed a positive impact of BIM on the reduction of selected cost parameters. Of course, taking into account the size of the research sample, it has to be stated that this is a trend, but it cannot make clear conclusions for the whole industry. The use of BIM technology in the Slovak construction industry is on the rise, but still represents a relatively low utilization rate. It is important to support the implementation of building information modelling as it has a positive impact on total cost, customer acquisition cost, and innovative performance cost.

Acknowledgments This work was supported by the Slovak Research and Development Agency under the contract no. APVV-17-0549.

The paper presents partial research results of project VEGA 1/0828/17 "Research and application of knowledge-based systems for modeling cost and economic parameters in Building Information Modeling."

References

1. Bhat, A.: What is research – definition, methods, types & examples. https://www.questionpro.com/blog/what-is-research/amp/m. Accessed 22 Oct 2019
2. Frankelius, P.: Questioning two myths in innovation literature, 2009. J. High Technol. Manag. Res. **20**, 40–51 (2009). https://doi.org/10.1016/j.hitech.2009.02.002. https://www.sciencedirect.com/science/article/pii/S1047831009000054?via%3Dihub. Accessed 22 Oct 2019

3. Canada Business Network: Info entrepreneurs, use innovation to grow your business (2009). https://www.infoentrepreneurs.org/en/guides/use-innovation-to-grow-your-business/. Accessed 22 Oct 2019
4. The NBS: What is building information modelling (BIM)? https://www.thenbs.com/knowledge/what-is-building-information-modelling-bim. Accessed 22 Oct 2019
5. Inc.com. Costs (2019). https://www.inc.com/encyclopedia/costs.html. Accessed 23 Oct 2019
6. Business Jargons: Business cost (2019). https://businessjargons.com/business-cost.html. Accessed 23 Oct 2019
7. Reh, F.J.: The Basics of Key Performance Indicators (2019). https://www.thebalancecareers.com/key-performance-indicators-2275156. Accessed 23 Oct 2019
8. Jackson, T.: 18 key performance indicator (KPI) examples defined (2018). https://www.clearpointstrategy.com/18-key-performance-indicators/. Accessed 23 Oct 2019
9. Windermere, A.: Common traits of successful organizations. https://smallbusiness.chron.com/common-traits-successful-organizations-56860.html. Accessed 23 Oct 2019
10. Key Performance Indicators: Establishing the metrics that guide success (2016). https://axiomgroupe.com/research-surveys/2472-key-performance-indicators-establishing-the-metrics-that-guide-success.html. Accessed 23 Oct 2019
11. Karslon K.: 12 Business Metric That Every Company Should Know (2019). https://www.scoro.com/blog/12-business-metrics. Accessed 23 Oct 2019
12. Gerber B.: 12 Key Financial Performance Indicators You Should Be Tracking (2018). https://www.accountingdepartment.com/blog/12-key-performance-indicators-you-should-be-tracking. Accessed 23 Oct 2019
13. Wiliams A.: 11 Key Business Performance Metrics for Metrics Better Operations (2017). https://alistemarketing.com/blog/business-performance-metrics. Accessed 23 Oct 2019
14. Rouse M.: Customer Acquisition Cost. https://searchcustomerexperience.techtarget.com/definition/customer-acquisition-cost. Accessed 24 Oct 2019
15. Kotlyar B.: What is customer acquisition cost and why does it matter? https://www.intercom.com/blog/what-is-customer-acquisition-cost/. Accessed 24 Oct 2019
16. Krishna S.: What is Customer Acquisition Cost and How do You Calculate it (2018). https://landerapp.com/blog/customer-acquisition-cost/. Accessed 24 Oct 2019
17. AppSamurai: 8 Tips To Reduce Customer Acquisition Cost (CAC) (2019). https://appsamurai.com/8-tips-to-reduce-customer-acquisition-cost-cac/. Accessed 24 Oct 2019
18. Olenski S.: How Marketers Can Reduce Customer Acquisition Cost (2017). https://www.forbes.com/sites/steveolenski/2017/01/09/how-marketers-can-reduce-customer-acquisition-cost/#23778afb4063. Accessed 24 Oct 2019
19. Merriam-Webster: Innovation. https://www.merriam-webster.com/dictionary/innovation. Accessed 24 Oct 2019
20. Maranville, S.: Entrepreneurship in the business curriculum. J. Edu. Bus. **68**, 27–31 (1992). https://doi.org/10.1080/08832323.1992.10117582. https://www.tandfonline.com/doi/abs/10.1080/08832323.1992.10117582. Accessed 24 Oct 2019
21. Drucker, P.F.: The discipline of innovation. Harv. Bus. Rev. **76**(6), 149–157 (2002). https://hbr.org/2002/08/the-discipline-of-innovation. Accessed 25 Oct 2019
22. Koutsogiannis A., Construction innovation: how we can help the industry het back in the tech race (2018). https://geniebelt.com/blog/construction-innovation-how-we-can-help-the-industry-get-back-in-the-race. Accessed 25 Oct 2019
23. Eastman, C.M., Eastman, C., Teicholz, P., Sacks, R.: BIM Handbook: A Guide to Building Information Modeling for Owners, Managers, Designers, Engineers and Contractor, p. 2011. Wiley, Hoboken, NJ (2011). ISBN: 9780470541371
24. Higgins, A.: Major Construction Technology Trends to Watch in 2019 (2019). https://connect.bim360.autodesk.com/construction-technology-innovation-2019. Accessed 25 Oct 2019
25. Skymind, Artificial Intelligence (AI) vs. Machine Learning vs. Deep Learning (2019). https://skymind.ai/wiki/ai-vs-machine-learning-vs-deep-learning. Accessed 25 Oct 2019

26. Words & Images, Building Information Modeling: Understanding and Operating in a New Paradigm. The Foundation of Wall and Ceiling Industry, Falls Church, VA (2009)
27. National BIM Standard-United States National Building Information Model Standard Project Committee.: http://www.nationalbimstandard.org/about.php. Accessed 23 Oct 2019.
28. Fadeyi, M.O.: The role of building information modeling (BIM) in delivering the sustainable building value. Int. J. Sustain Built Environ. **6**(2), 711–722. https://www.sciencedirect.com/science/article/pii/S2212609016302047. Accessed 22 Oct 2019
29. Tkáč, M., Mesároš, P.: Potential of 3D laser scanning in construction in connection with building information modelling, In Increasing the Efficiency of Construction Through Modern Methods of Construction, Scientific Paper Proceedings, pp. 103-110, (2015)
30. Richardson, D.: Productivity in the Construction Industry. Australia Institute, Canberra, ACT (2014)
31. Ying, Z.: Significant factors affecting construction productivity (Doctoral dissertation). National University of Singapore, Singapore (2005)
32. Ghasemi, A., Zahedias, S.: Normality tests for statistical analysis: a guide for non-statisticians. J. Endocrinol. Metab. **10**, 486–489 (2012)
33. Stangroom, J.: The Friedman test for repeated measures. Soc. Sci. Stat.. https://www.socscistatistics.com/tests/chisquare2/default2.aspx. Accessed 1 Nov 2019
34. Lehmann, E.L.: Nonparametrics: statistical methods based on ranks, 1975. In: Hollander, M., Wolfe, D.A. (eds.) Nonparametric Statistical Methods, p. 480. Holden-Day, Inc., San Francisco, CA (1999). ISBN 0-471-19045-4
35. Cardillo, G.: Dunn's Test: a procedure for multiple, not parametric, comparisons (2006)
36. Dunn, O.J.: Multiple comparisons among means. JASA. **56**, 54–64 (1961)
37. Dinno, A.: Portland State University. https://pdfs.semanticscholar.org/c381/295bb06ef9df51ebda789e0154821d0d09d4.pdf. Accessed 1 Nov 2019

Employee Suggestion Scheme: Case Study

Marta Szczepaniak and Justyna Trojanowska

1 Introduction

In the age of fierce competition among manufacturers, only those who are continuously growing are able to meet customers' expectations and requirements. In order to flexibly respond to ever-changing external and internal circumstances, an enterprise should act in the spirit of continuous improvement. Continuous improvement is a Japanese enterprise management strategy based on the engagement of all employees in the functioning of the organisation. In order to facilitate the communication and sharing of ideas of low-level staff with senior executives, many companies implement employee suggestion systems. An employee suggestion system is a customised solution for sharing ideas by the staff, aimed at improving processes at the organisation. In this paper, the authors present a kaizen employee suggestion scheme developed for an enterprise operating in the packaging industry and explore the importance of implementation of the continuous improvement approach in an organisation.

2 Analysis of the Literature

Lean manufacturing (LM) is a continuous production improvement concept consisting in the identification and subsequent elimination of waste and raising the value added of activities [1–4]. Melton, Szczepaniak, Womack and Araujo distinguish seven types of wastes: overproduction, waiting, transportation, inventories, excessive processing, motion and defects [5–8]. Liker adds the eighth type of waste,

M. Szczepaniak · J. Trojanowska (✉)
Poznan University of Technology, Poznan, Poland
e-mail: justyna.trojanowska@put.poznan.pl

L. Knapcikova et al. (eds.), *New Approaches in Management of Smart Manufacturing Systems*, EAI/Springer Innovations in Communication and Computing, https://doi.org/10.1007/978-3-030-40176-4_17

which is unused employee creativity [4]. Feld emphasises that understanding, prior to the implementation of LM, why changes are necessary and why they are imperative for the organisation is extremely important, as it ensures that employees are fully engaged in the implementation of changes [9]. Liker, Bhamu and Trapping present a set of mutually complementary LM methods and tools which support enterprises on their path to growth [3, 10, 11]. One of them is kaizen.

Kaizen is a concept of continuous improvement [3, 12] pursued every day, by every employee, at any place [13]. Kaizen is the Japanese word for improvement [4]. It consists in implementing changes in small steps [13–15] while eliminating wastes which do not create value [4]. Topuz and Arasan define kaizen as a process of continuous improvement of a standard way of working [16] and point to the fact that once laid down, standards in a changing production environment must be continuously monitored and adjusted to the prevailing conditions. According to Liker, kaizen is a technique teaching problem-solving, process streamlining, data analysis and productive and independent work [4]. Imai [13] notes that, based on personnel, communication, teamwork, commitment and self-discipline, kaizen is underpinned by common sense and low costs. Bhamu emphasises that in many manufacturing companies, kaizen is treated as an employee suggestion scheme [3] providing for better involvement in the life of the organisation and unhampered communication of ideas for improvement to supervisors. Kejna [17] explores the essence of employee suggestion systems, lying in improvement of own work and processes by employees. People naturally tend to look for ways to facilitate and speed up work and to be more productive [18]. An employee suggestion scheme enables employees to share their observations with the management and colleagues. In [13], he draws attention to the fact that although suggestions submitted by employees may not always bring spectacular economic benefits, the very engagement, self-discipline and willingness to act create great value added for the company. The implementation of an employee suggestion scheme raises employees' satisfaction by giving them a sense of having an impact on the way in which the enterprise operates [18, 19]. This in turn translates into positive attitude and personal commitment [19].

Imai [13] claims that involvement of the executive staff is the driving force of successful implementation of kaizen [13]. By providing support, the management encourages employees to take things in hand [18]. Therefore, Pająk [20] emphasises that all staff should be involved in the implementation of small improvements.

3 Description of the Enterprise

The kaizen employee suggestion system in question has been implemented in a company manufacturing packaging and cardboard labels. The manufacturing processes include cutting, printing, refining and binding. Each stage of the manufacturing process can be carried out on several machines. A general scheme of the manufacturing processes and the machines on which they can be performed are shown in Fig. 1.

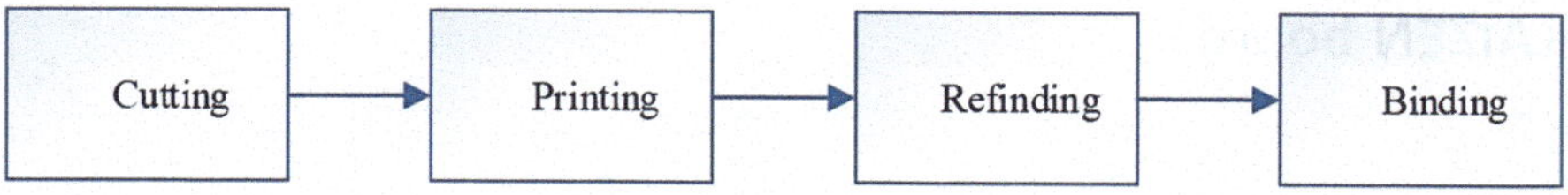

Fig. 1 A general scheme of manufacturing processes and the machines on which they can be performed

Table 1 List of some of the problems faced by the enterprise under analysis

Problems occurring in the enterprise	
1.	Poor communication among employees
2.	Poor motivation for work
3.	Lack of commitment
4.	Employees not willing to share knowledge
5.	Resistance to training new employees
6.	Resistance to changes
7.	Obsolete working time standards in place
8.	Lack of regular updates of manufacturing data

An inherent part of each enterprise are people. The company under analysis has ca. 80 employees. Ca. 60 of them are manufacturing employees; almost 30% of them have a degree of disability (mostly hearing disabled).

The enterprise experienced a number of problems; some of them are listed in Table 1.

The problems were observed by all the employees as well as the company owner; however, no action was taken to eradicate them. Considering that the employees did not feel a need for change, it can be inferred that the underlying problem was a lack of the habit of continuous improvement—a factor of key importance to succeed in the present times.

4 Employee Suggestion Scheme: Case Study

In view of the problem identified in the enterprise—a lack of the habit of continuous improvement—a scheme was developed to collect, assess and implement employees' suggestions. The scheme was aimed to encourage the employees to share ideas with their supervisors and other colleagues. The suggestions submitted were meant to streamline work and optimise processes to reduce costs and time consumption as well as enhance safety and product quality. The scheme was also developed with a view to increase the employees' awareness of the benefits resulting from continuous improvement.

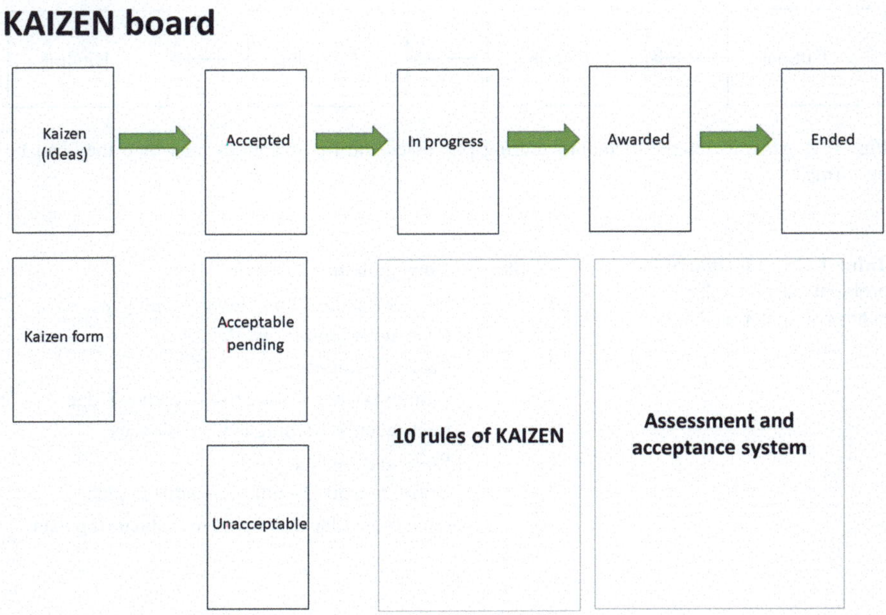

Fig. 2 Kaizen board

As a vehicle of communication, a kaizen board was developed (see Fig. 2). Located in an easily accessible space in the production hall, the kaizen board showed the progress of consideration of the submitted suggestions. Blank forms to submit suggestions were also available.

The kaizen sheet was arranged in two sections. Section one was to be filled in by the employee submitting a suggestion. To make a submission, the employee was supposed to provide the following information:

- The author (authors—max. 2) of the suggestion
- Their position and who they reported to
- Date of submission
- Area of concern (work conditions, safety, quality, efficiency, standardisation, other)
- Location of the area of concern
- Description of the situation/problem and its root cause
- Description of the proposed solution and how it could eradicate the root cause
- The type of waste that the solution would eradicate
- Whether the proposed solution had been implemented and was in place

Section two was to be filled in by the kaizen scheme coordinator. The coordinator was supposed to provide the following data:

- Name and surname of the relevant area manager.
- A number assigned to the submission.

- The deadline for the verification of the submission.
- One of the three statuses assigned to the submission:
  - Accepted—to be implemented—a suggestion assigned this status was accepted and would be implemented. A person responsible for the implementation of the project within a relevant time frame would be appointed.
  - Accepted—pending—a suggestion assigned this status met the requirements for acceptance, but due to high costs, the implementation was postponed and a project team would be appointed responsible for the implementation.
  - Rejected—a suggestion which did not meet the requirements for acceptance.
- If a suggestion was rejected, the reason for the rejection was to be provided in the comment field.
- If a suggestion was accepted, the scheme coordinator would assign it one of the following priorities:
  - Low—implementation time frame up to 6 months
  - Medium—implementation time frame up to 3 months
  - High—implementation time frame up to one month
- If a suggestion was accepted, the coordinator would appoint a person responsible for the implementation and specify the project implementation deadline.
- Once the implementation was completed, the coordinator would fill in the field with the actual date of completion.
- One month after the completion of the implementation, an appointed person would assess the suggestion by assigning a score in each of the following categories:
  - Concept—a score assigned for the degree of precision and development of the submitted suggestion
  - Implementation—a score assigned for meeting the project deadline
  - Evaluation one month from completion—a score assigned for the eradication of the problem and its root cause

Information was also provided on the savings brought about by the implemented change. Finally, the total score for the project was calculated. The final project score and the generated savings were considered at the annual evaluation of all kaizen suggestions aimed to distinguish the so-called super kaizen.

New submissions and statuses of the approved submissions were analysed at monthly meetings of the kaizen committee. The committee consisted of the kaizen coordinator, all department managers and the employee responsible for the OHS. Submissions were considered and results were analysed over time to encourage all employees not only to submit suggestions but also to engage in the implementation. Assessment of the impact of an implemented solution sometime after the completion of the implementation was supposed to push employees to deliver creative solutions which brought real, visible and measureable benefits.

The procedure for filling in the kaizen sheet as well as the rules for the score assignment and awarding the accepted and implemented suggestions were explained

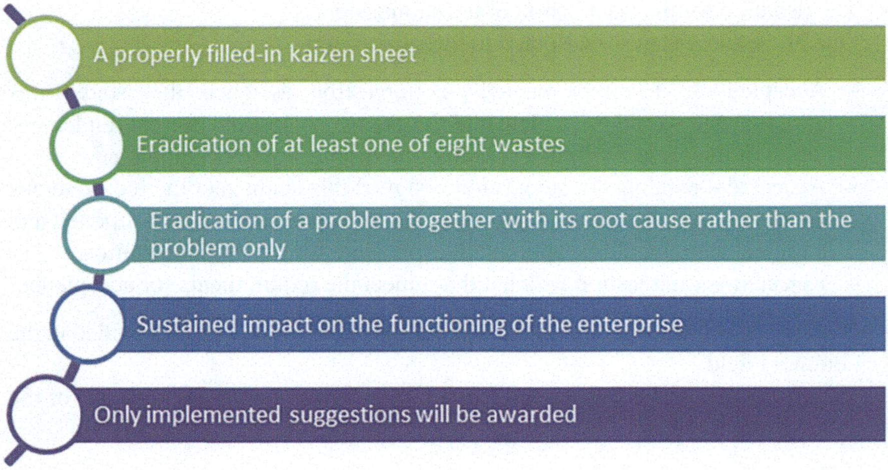

Fig. 3 Requirements for the acceptance of kaizen suggestions

in a document entitled *The System of Evaluation and Acceptance of Kaizen Suggestions*, which was also made available on the kaizen board for everyone to see how the submitted suggestions were evaluated. Another document, entitled *Process Flow*, was developed to identify the procedure from the submission of a suggestion through the implementation to the evaluation of the solution, together with the persons responsible for the completion of particular stages.

The requirements to be met for a suggestion to be accepted were listed in the *Requirements for the Acceptance of Kaizen Suggestions* (see Fig. 3).

In order to familiarise themselves with the kaizen concept and the kaizen employee suggestion scheme, the employees took part in a dedicated training session at which they learnt the following:

- What kaizen is
- The types of wastes
- How important it is for an enterprise to involve employees in the improvement of the value stream
- Ten basic kaizen rules defining how to look at processes and where to look for opportunities to streamline them
- The rules of the kaizen employee suggestion scheme
- The kaizen board and rules of communication
- The system of evaluation and acceptance of suggestions
- The rules for awarding the "super kaizen" prize

Additionally, each employee was provided with a kaizen leaflet listing the most important issues discussed at the training. The last page of the handout featured guidelines on how to look for improvement opportunities.

All the suggestions submitted by the employees were written to a dedicated file. The file contained general information on each suggestion submitted and a link to a scan of the submission.

5 Results of Implementation of the Kaizen Employee Suggestion Scheme

The enterprise has had the employee suggestion scheme in place for more than 3 months. Over the first 3 months, the employees submitted 86 suggestions. The greatest number of submissions concerned the improvement of work conditions and safety. The submissions made during the first 3 months after the implementation of the scheme, broken down by categories, are shown in Fig. 4.

An analysis of the kaizen sheets shows that most of the submissions concerned the eradication of three types of waste: excessive number of operations, redundant traffic and errors. However, it should be noted that the type of waste to be eradicated through the implementation of a given suggestion was indicated by the employees, mostly operators, rather than the evaluating team. Proper assignment of a suggestion to a given type of waste is still confusing for some employees and many of them indicate more than one type of waste. Detailed data on the number of suggestions which should contribute to the mitigation of a given type of waste is shown in the graph below (Fig. 5).

Out of all the submitted suggestions, 71 (i.e. 83%) were accepted. The kaizen team appointed persons responsible for the implementation and most of them have already been completed. Eight suggestions were assigned the status

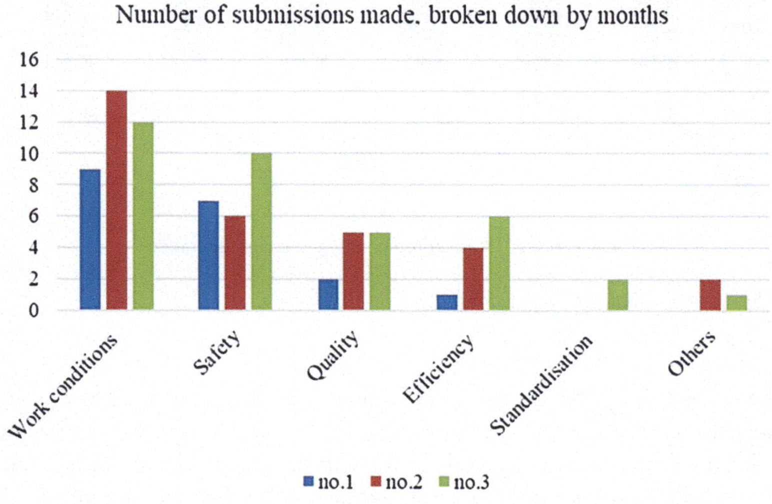

Fig. 4 Number of submissions made, broken down by months

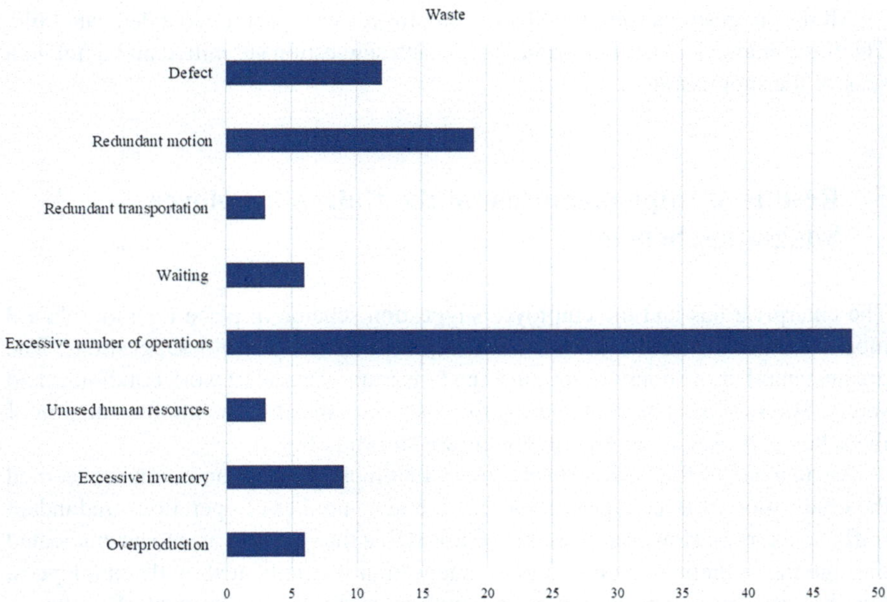

Fig. 5 Number of suggestions to mitigate particular types of waste

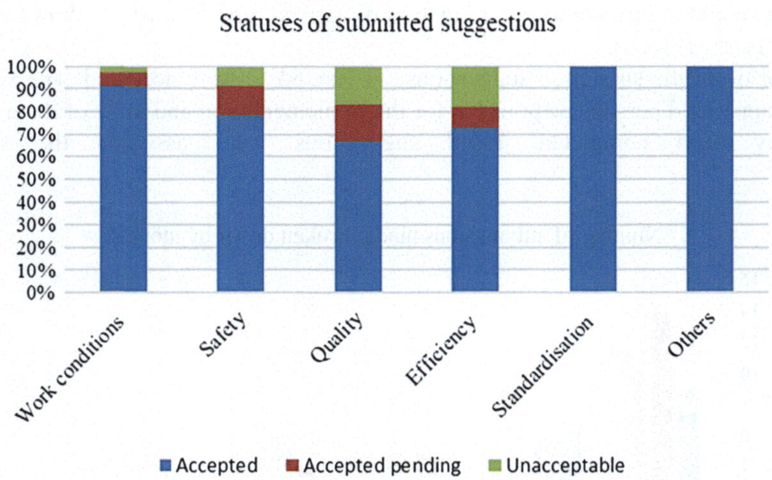

Fig. 6 Statuses of submitted suggestions

"Accepted—pending", due to the cost of implementation. Project teams have been appointed to work on them. Seven suggestions have been rejected. In percentage terms, most of the rejected suggestions were in the categories of quality and efficiency. The statues of the suggestions submitted during the first 3 months of operation of the scheme are shown in Fig. 6.

The number of submissions is growing month over month (in the second month of the scheme's operation, the number of submissions went up by 39% on the first month; in the third month, the increase was 14% on the second month), and most of the submissions are made by the manufacturing employees. Some employees submit more than one suggestion per month. It may be the effect of novelty, as owing to the implementation of the scheme the employees feel that they have a real say in the process of improvement of their workstations and work conditions. The kaizen team anticipates the growth to halt over time and the number of submissions to remain steady thanks to the system of awards.

6 Summary

The implementation of the kaizen employee suggestion scheme has enabled the employees to share their ideas with co-workers. The scheme has facilitated the communication between the management and lower-level employees, including those with disabilities. Employees engaged in the completion of processes have the best insight into their mechanisms, understand causes of waste and know how to eliminate them. Awarding employees for their submissions boosts their creativity and motivation to share ideas with others. An important role in the implementation process is played by the management. The employees must feel that the management believes in them and is ready to afford support and collaboration.

An inherent part of the implementation was popularisation of the concept of continuous improvement and increasing awareness of how small changes can bring about big improvements.

Implementation of the kaizen scheme has solved many problems faced by the enterprise and increased the satisfaction level of its employees.

References

1. Pająk, E.: Production Management. Product, Technology, Organization. PWN, Warsaw (2006)
2. Conner, G.: Lean Manufacturing: Certification Workshop Participant Guide. Lean Enterprose Training, Oregon (2004)
3. Bhamu, J., Sangwan, K.S.: Lean manufacturing: literature review and research issues. Int. J. Oper. Prod. Manag. **34**(7), 876–940 (2014)
4. Likker, J.K.: The Toyota Way. 14 Management Principles from the World's Greatest Manufacturer. McGraw-Hill, New York, NY (2004)
5. Melton, T.: The benefits of lean manufacturing. what lean thinking has to offer the process industries, 2005 institution of chemical engineers, the chemical engineering. Chem. Eng. Res. Design. **83**(A6), 662–673 (2005)
6. Szczepaniak, M., Trojanowska, J.: Methodology of manufacturing process analysis. In: Hamrol, A., Kujawińska, A., Barraza, M. (eds.) Advances in Manufacturing II. Lecture Notes in Mechanical Engineering, pp. 281–294. Springer, Cham (2019)

7. Womack, J.P., Jones, D.T.: Lean Thinking: Banish Waste and Create Wealth in Your Corporation, Revised and Updated, 2nd edn. Free Press, New York, NY (2003)
8. Araújo, A.F., Varela, M.L.R., Gomes, M.S., Barreto, R.C.C., Trojanowska, J.: Development of an intelligent and automated system for lean industrial production adding maximum productivity and efficiency in the production process. In: Hamrol, A., Ciszak, O., Legutko, S., Jurczyk, M. (eds.) Advances in Manufacturing. Lecture Notes in Mechanical Engineering, pp. 131–140. Springer, Cham (2018)
9. Feld, W.M.: Lean Manufacturing: Tools, Techniques, and How to Use Them, The CRC Press Series and Resource Management. CRC Press, Boca Raton, FL (2000)
10. Trapping, D.: The Lean Pocket Guide. MCS Media, Chelsea, MI (2003)
11. Liker, J.K.: The Toyota Way: 14 Management Principles from the World's Greatest Manufacturer. McGraw-Hill Education, New York, NY (2004)
12. Pavnaskar, S.J., Gershenson, J.K., Jambekar, A.B.: Classification scheme for lean manufacturing tools. Int. J. Prod. Res. **41**(13), 3075–3090 (2003)
13. Imai, M., Commonsense, G.K.A.: Commonsense, Low-Cost Approach to Management. McGraw-Hill, New York, NY (1997)
14. Lopes, J.J., Varela, M.L.R., Trojanowska, J., Machado, J.: Production flow improvement in a textile industry. In: Rocha, Á., Guarda, T. (eds.) Proceedings of the International Conference on Information Technology & Systems (ICITS 2018), Advances in Intelligent Systems and Computing, Springer, vol. 721, pp. 224–233 (2018)
15. Gulati, R.: Maintenance and Reliability Best Practices. Industrial Press, New York, NY (2009)
16. Topuz, C., Arasan, Z.: Kaizen-educational: an awareness-raising and motivational-enhancement group counseling model. Proc. Soc. Behav. Sci. **8**(4), 1356–1360 (2013)
17. Kejna, S.: Methodology of Obtaining and Processing of Additional Information for Management Purposes Through the Analysis of Suggestion Systems Based on the Kaizen Philosophy (In Polish). Acta Universitatis Lodziensis. Folia Oeconomica nr 300, Lodz (2014)
18. Verdinejad, F., Mughari, A.M., Ghasemi, M.: Organizational suggestion system in the era of holding by developing an innovative model: the case of bonyade ta avon holding in Iran (an applied model). Iran J. Manag. Stud. **3**, 3 (2010)
19. Adedeji, A.C., Okolie, K.C.: Adopting the Kaizen suggestion system in South African lean automotive components companies. Sci. J. Bus. Manag. (2012)
20. Pająk, E., Klimkiewicz, M., Kosieradzka, A.: Production and Service Management. Polskie Wydawnictwo Ekonomiczne, Warsaw (2014)

Alternative Methods of Using RFID Technology to Identify Transport Units

Jiří Tengler and Jana Pekná

1 Introduction

Radio-frequency identification (i.e., RFID) has been used in various forms and areas for decades. This technology has gradually been used in various industries. This technology has been used mainly in areas where it was necessary to load a large number of identifiers very quickly. Thus, it found application in those areas where the bar code itself was no longer sufficient and it was necessary to extend it with an RFID identifier. The biggest application of RFID technology was recorded especially in the field of logistics. The RFID technology is used to make logistics and customer-supply chain processes visible in these processes. RFID technology enables real-time tracking of relevant data in these processes. These data can be used for practical measures after correct interpretation (increasing efficiency, reducing errors, and improving quality).

One of the aforementioned areas of application is the monitoring of transport units, respectively logistics units within the customer-supply chain. In general, a clear case is the application of RFID technology for both parties involved supply chain. The task of the application is then to provide information about where the transport logistics unit is located, where it is directed or where it came from. This issue may occur unless one of the parties (usually the purchaser) does not agree with the application of RFID technology. Subsequently, customers have no choice but to apply RFID at their own level of a company.

There are plenty of ideas for applications of RFID technology within the supply chain. This may include streamlining logistics, managing transport units or identifying possible defects in shipping or handling at the supplier (goods or shipping units). However, such an application can solve at least some of the above problems. There

J. Tengler (✉) · J. Pekná
University of Žilina, Žilina, Slovakia
e-mail: Jiri.tengler@fpeda.suniza.sk; pekna@stud.uniza.sk

L. Knapcikova et al. (eds.), *New Approaches in Management of Smart Manufacturing Systems*, EAI/Springer Innovations in Communication and Computing, https://doi.org/10.1007/978-3-030-40176-4_18

are many RFID applications operating at own level of a company. These applications are used primarily for monitoring of transport units between individual logistics centers or organizational units of their own company. However, there are no problems with supply chains. However, in some specific nonstandard cases and issues, RFID can still be used. And this issue is the subject of this article.

2 About RFID Technology

The RFID technology is complex, combining a number of different computing and communications technologies to achieve desired objectives. Each object which has to be identified is fitted with a small object called a RFID tag stuck to it. The application at only one side of the supply chain can never bring full value information.

Each RFID tag has a unique identifier that enables additional information about each object to be stored. Devices known as RFID readers wirelessly communicate with RFID tags, with a view to identifying the attached RFID tags, as well as enabling information stored in the RFID label to be read and updated [1].

Every RFID system contains an RF subsystem (see Fig. 1), and most RFID systems also contain an enterprise subsystem. An RFID system supporting a supply chain is a common example of an RFID system with an inter-enterprise. In a supply chain application, a tagged product is tracked throughout its life cycle, from manufacture to final purchase, and sometimes even afterward (e.g., to support service agreements or specialized user applications). Radio-frequency identification is a

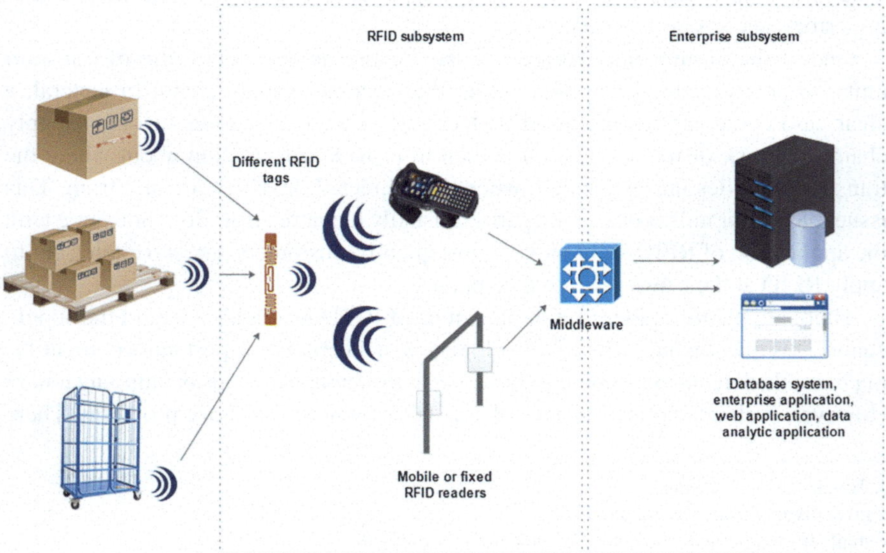

Fig. 1 RFID system

wireless data collection technology that uses electronic tags which store data and tag readers which remotely retrieve data.

It is a method of identifying objects and transferring information about the object's status via radio-frequency waves to a host database. RFID represents a significant technological advancement in AIDC because it offers advantages that are not available in other AIDC systems such as bar codes. RFID offers these advantages because it relies on radio frequencies to transmit information rather than light, which is required for optical AIDC technologies [2].

2.1 *RFID System*

There are three basic components of the RFID system, RFID tags, RFID readers, and middleware, which is responsible for all data transaction in the system [3].

2.1.1 RFID Tag

The RFID tag is a small device that can be attached to an item, case, roll cage, or pallet, so it can be identified and tracked. It is also called a transponder. The tag is composed of microchip and antenna. The label consists of a microchip, antennas, and interconnecting circuits. These elements are attached to a material called a substrate in order to create an inlay. In addition to these components, RFID tags can also have data storage memory, sensors of various types (temperature, humidity, shake, solar radiation), or other security circuits.

Tags are categorized into two types based on the power source for communication and other functionality:

- Active
- Passive

Passive RFID tags are identifiers that do not have their own power supply. To power and communicate with the RFID reader, passive RFID tags use electromagnetic energy absorbed from the RFID reader transmitting antennas. Thus, in order to power passive RFID tags, these must be in the electromagnetic field of the antennas. As a rule, this feature makes them significantly cheaper than active RFID tags.

Active RFID identifiers are identifiers that integrate their own power supply (i.e., batteries). This integrated source then serves either to power the microchip itself or to amplify the radiation signal to the RFID reader antennas.

Active RFID tags can still be divided into semi-passive and semi-active. Semi-active tags use their own power supply for their own integrated circuits. That is, for powering microchips or connected sensors. The semi-active tag uses its own power supply for amplification and control, increasing the range of communication with the RFID reader [4, 5].

2.1.2 RFID Reader

The second component in a basic RFID system is the interrogator or reader. The reader can have an integrated antenna, or the antenna can be separate. The antenna can be an integral part of the reader, or it can be a separate device. Handheld units are a combination reader/antenna, while larger systems usually separate the antennae from the readers. The reader retrieves the information from the RFID tag [2].

2.1.3 RFID Middleware

Middleware is software that controls the reader and the data coming from the tags and moves them to other enterprise systems such as database system. It carries out basic functions, such as filtering, integration, and control of the reader. RFID systems work if the reader antenna transmits radio signals. These signals are captured tag, which corresponds to the corresponding radio signal. Middleware is an important interplay between the physical and software layers and is therefore subject to a large number of requirements [6].

The basic elements that RFID middleware should provide are:

- Simple management and configurability
- Intelligent filtering, routing, and processing of read data by the reader
- High support for RFID readers of various manufacturers
- High support of superior business systems (databases, analytical tools, etc.)
- Wide support of communication protocols, etc. [4, 5]

3 Problem and Solution Design Specification

For the purposes of this article the subject of research may be closer to the example of a fictitious company. Let there be a company manufacturing bearings for rotary engines. This company supplies its products to two customers. For delivery, it uses its means of transport, i.e., containers. The shipping containers are provided with a transport document before the expedition. The company lacks a system that would allow the container movement to be automatically identified. Thus, it lacks the exact information where the particular container was dispatched or where it came from. The transport documents are provided with a bar code and are scanned before the expedition. However, the shipping document is withdrawn upon delivery. Consequently, it is not very clear where the container came from. Application of RFID technology in both the supplier and the customer would to resolve the matter. Unfortunately, customers do not want to implement RFID technology. The following study describes what the supplier has options. Suppose a company wants to know where to go and where they come from. The most likely area implementation RFID technology seems to be an area for loading and unloading. For a better

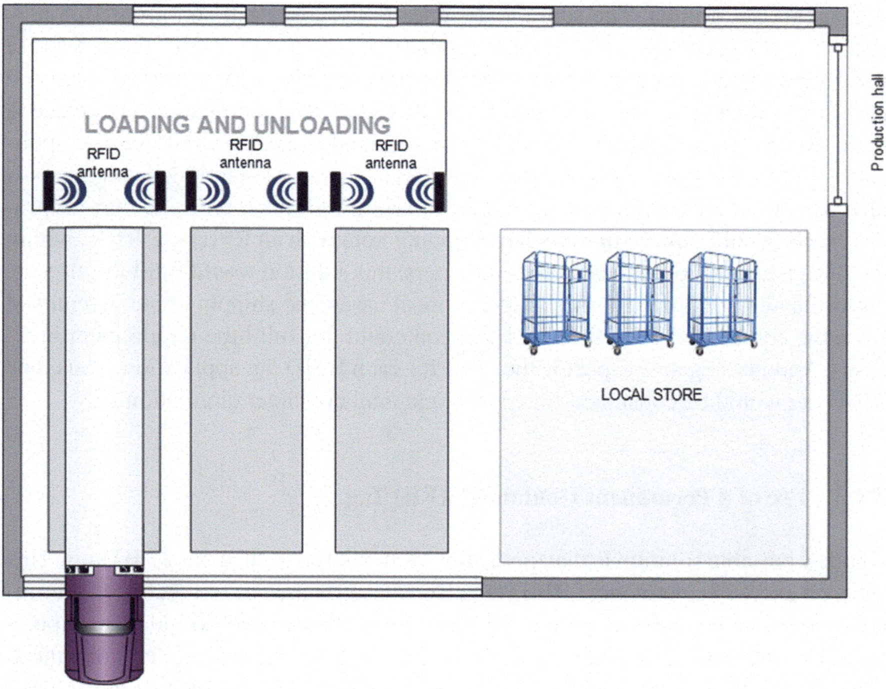

Fig. 2 Area of loading and unloading

understanding, a fictitious company can be illustrated in Fig. 2. Thus, the focal point of retrieving data from RFID technology is the loading and unloading process. Unfortunately, the sample itself does not yet address the provision of the requirement.

3.1 Selected Realized Solutions

There are many solutions to the chosen issue within the scope of RFID technology implementation. Only three solutions have been selected for the purpose of presenting the results of this article.

3.1.1 Use RFID Printer

The first solution is to use an RFID printer. The essence of this solution consists in repeated pushing and programming of the RFID tag and their subsequent application to the container before the expedition. The RFID tag encodes not only the container identifier, but also the customer identification number. The application of the RFID tag could also be possible in the manufacturing area where shipping documents are printed. The solution assumes to use and print and program a new RFID tag for each trip. This solution requires communication with the enterprise database system as a source of input data for printing. Placing the RFID printed and programmed label on the container would be in the production area. This place is the most convenient because the containers are filled here. Also, printing and application of placing a standard bar code dispatch to the abovementioned containers is also performed. This solution would not interfere too much with the current processes but would replace the standard shipping note with an RFID tag. Thus, within the design and in-house location of containers, this solution would fulfil its purpose. Unfortunately, on the customer's side, in most cases, the shipping note is removed from the containers. Therefore, this solution could not fulfil the requirement in all cases. Another negative aspect is the price for each RFID tag application. Thus, one RFID tag would be consumed for each single mail container circulation.

3.1.2 Use of a Permanent Container RFID Tag

Another possible solution is the application of industrial permanent RFID tags. This solution except by permanent RFID tag also envisages the use of touch in the terminal to area loading and unloading. The function of the terminal would be to supplement the retrieved data with the planned direction of the dispatched containers. Before loading the container into a transport vehicle, the employee would select a particular customer. Consequently, within the loading and retrieved container RFID tags, this information would be paired with the customer. This solution has an advantage especially in single use of perennial RFID tags. This means that over time there will be no additional costs for consumables in Frome RFID tags, as was the case with previous solutions. Each container would have a unique and durable identifier that would allow it to be identified both within the reading zones within the enterprise itself and at the output in area of loading and unloading. As mentioned, to record information, where container was heading would be used in small terminal in the area loading and unloading. In this area, the person concerned would determine where the containers would be directed before the loading process. You can see an example of the terminal for selecting the respective customer in Fig. 3.

3.1.3 Use Especially Developer RFID Box

The last solution is RFID box. An RFID box is a switchable RFID tag that allows you to switch between the active statuses of only one of the contained RFID tags. Each RFID should contain a unique container identifier and a customer identifier.

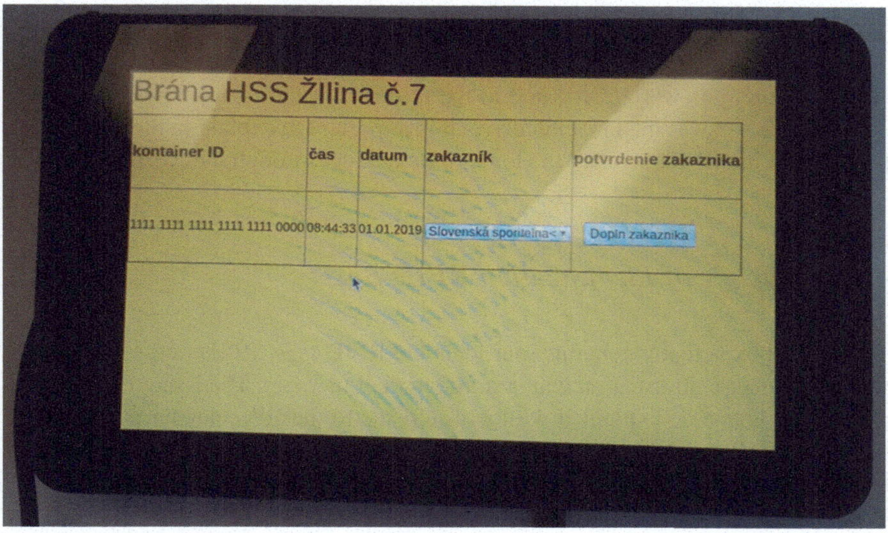

Fig. 3 Touch terminal to select an option

Table 1 Advantage and disadvantage of the proposed solutions

	Solution 1	Solution 2	Solution 3
Description	RFID printer	Permanent RFID tag + touch terminal	Manually switchable RFID box
Advantages	The simplicity of the solution, lower probability of errors	Disposable application of industrial RFID tags for containers	The simplicity of the solution. Lower probability of employee error when switching RFID box
Disadvantages	Large consumption of RFID tags. Possibly increased difficulty in removing an old RFID tag	Necessary to set the direction of the trip in the touch terminal by the unloader. Possible probability of employee error in the terminal	The need to switch the RFID box to the employee in production

3.2 Advantage and Disadvantage of the Proposed Solutions

Based on the description of individual solutions it is possible to specify the advantages and disadvantages of individual solutions. The advantages and disadvantages of the individual solutions are shown in Table 1.

4 RFID Box/Multi-tag

All three solutions were elaborated in the research of the abovementioned issues. The last solution was the most intensive. It is important to explain in more detail the principle of the RFID box and the logic of its reading through the RFID reader.

4.1 Principle of Identification RFID Box

The RFID box actually contains four identical RFID tags. RFID tags vary slightly with each other's information that is stored in their memory. The issue for the needs of this article can be explained without reference to the EPC standard. The developed RFID box uses 96-bit RFID tags. When you load them, we get 24 characters in the hexadecimal system. This string of characters (hexadecimal numbers) can be arbitrarily divided into two or more parts. One of these may serve as an identifier for a particular container (the unique container identifier) and the other may identify the direction the container is shipped to. In the proposed solution, four basic tags were used in the basic version, even with a larger cylinder diameter and a suitably selected RFID tag, and more RFID tags can be achieved. Thus, the container identifying portion for the tag is the same, but the status is different for each of the four RFID tags. A clearer understanding of data storage is shown in Fig. 4. In this picture you can see the individual information stored on each of the four RFID tags. In general, the format may vary considerably in real conditions, but is sufficient for presentation. Information with a green background labeled "id status" is an abbreviated form of the customer company designation. This data is then paired with other identification information of the customer company within the database. In practice, this is the only data that is displayed when switching the multi-tag and then reading the RFID reader. Indication of the blue background, i.e., Id container, then means a unique

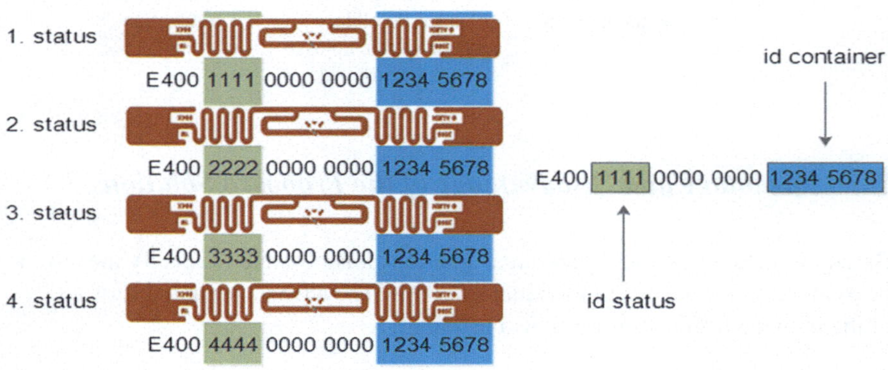

Fig. 4 Simplified data storage principle in the RFID tag

identifier of a particular container and it must be the same even if the position of the multi-tag is switched.

Of course, it is also possible to use a standard format for the same purpose, e.g., EPC standard. However, for a company that will only identify the transport units at its own enterprise level, the use of the EPC standard is irrelevant.

4.2 *Principle of Operation and Description of RFID Box*

The design of the RFID box was based on the idea that RFID tags would be switched by means of a rotating cylinder. RFID tags would then be mounted on this cylinder around the perimeter. For this reason, it was necessary to shield all the cylinders outside the areas on the front of the multi-tag and to reflect the electromagnetic signal. During the design it was assumed that the antireflective measure could have a negative impact on the readability of the active RFID tag. Therefore, two versions of the RFID box were initially designed. These two versions were designed based on the two RFID tags mentioned above. Based on the size of these RFID tags, 3D objects were modeled and then printed on a 3D printer. Both versions of the RFID box differ in size and RFID tag used.

4.2.1 Small RFID Box (Fig. 5)

- Box size 110 × 27 × 27 mm
- Four pieces of AVERY DENNISON AD-226IM UHF RFID WET INLAY (NXP G2IM)

4.2.2 Large RFID Box (Fig. 6)

- Box size 90 × 45 × 45 mm
- Four pieces of AVERY DENNISON AD-236U7 UHF RFID WET INLAY (NXP UCODE 7)

Fig. 5 Small RFID box/multi-tag

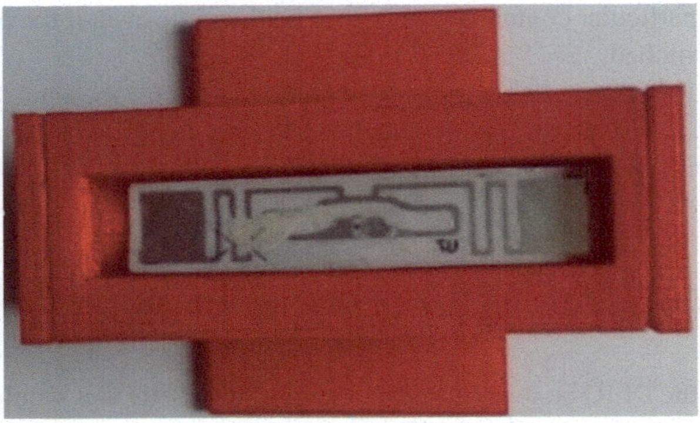

Fig. 6 Large RFID box/multi-tag

Fig. 7 Antireflective treatment of both RFID boxes

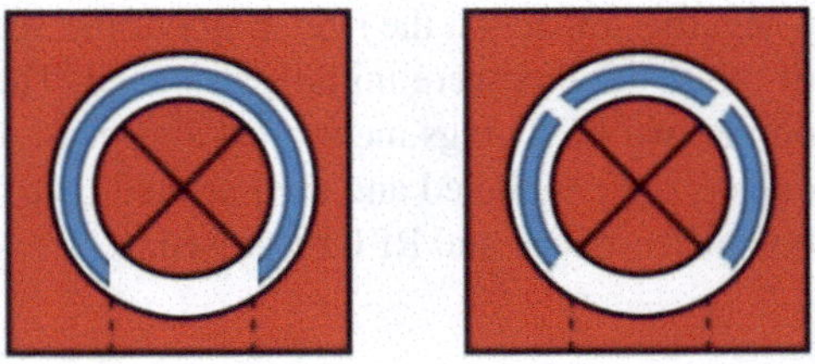

Important spectra for the proper operation of the RFID box are antireflective finishes. Thus, avoid the reading of those tags that are not rotated in the direction of the RFID box opening. Antireflective aluminum foils were used for this purpose. These films were applied both inside and outside the outer cylinder section; see Fig. 7.

5 Testing of RFID Boxes

Several tests were performed to verify the RFID box solution we designed. The aim of the testing was to find out if the solution is usable for real traffic and to what extent the quality of reading RFID tags has changed. The tests were carried out on a real container called "Kontel." The RFID box was placed in different container positions. More precisely, the RFID box was place on the top and side of the cage, as can be seen in Fig. 8.

5.1 Monitored Parameters When Reading a Tag

– Dimensionless value RSSI (the strength of the reflected signal)

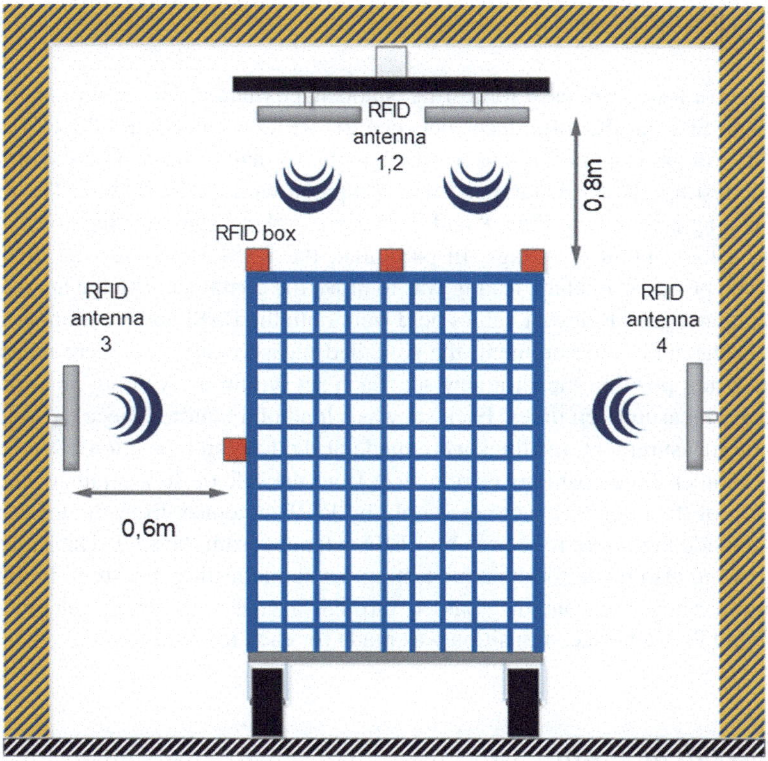

Fig. 8 View on optionality of measurement

5.1.1 Used Tags

- Small RFID box with tags AVERY DENNISON AD-226IM
- RFID tag itself AVERY DENNISON AD-226IM
- Large RFID box with tags AVERY DENNISON AD-236U7
- RFID tag itself AVERY DENNISON AD-236U7

5.1.2 Used Equipment and Material

- Motorola FX7400
- Antenna AN480
- Middleware Aton onID
- Database MariaDB
- Container KONTEL

5.2 Types of Tests

Two types of tests were used for testing: static and dynamic. The goal of static testing was to find out the differences between the RFID tag itself and the RFID box. Usability for real operation was verified within dynamic tests. Even within the dynamic testing, the RFID tag itself was compared with the RFID box. The principle of testing is shown in Figs. 8 and 9. The principle of static testing consisted of reading a fixed number of tags. In particular, the RSSI value was the dominant value. The principle of static testing was to move the container with the tag over the RFID antenna gate. Relevant values here were both the RSSI and the number of tag reads. In the static measurement, the so-called measurement sets were measured. Each new tag position meant a new set. Each set within a given tag position was repeatedly measured ten times. Each set was a load of a hundred piece tag value. In the static measurement, readings were made at the top three positions; see Figs. 7 and 8. Another three positions were realize from the side of the container. The tags at the top of the container were read only by RFID antennas from the top and the container side tags were read only by RFID antennas from the side. The same load settings were also made for dynamic testing. Dynamic testing measured RFID tags in motion. Thus, the container passed through a gate with RFID antennas; see Figs. 7 and 8. A ten-piece transition was made for each tag location.

5.3 Result of Testing

Within the evaluated graphs you can see the minimum, maximum, and average values of the tag itself in comparison with the RFID box. In the results, you can compare an individual tag location, either from the top or side of the container.

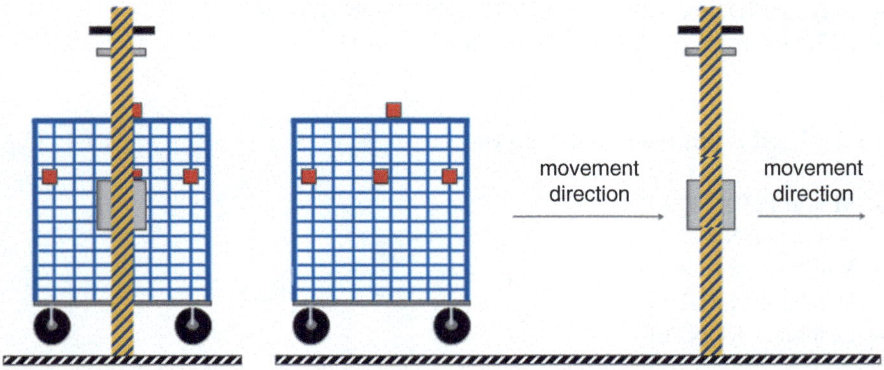

Fig. 9 Static testing (left) and dynamic testing (right)

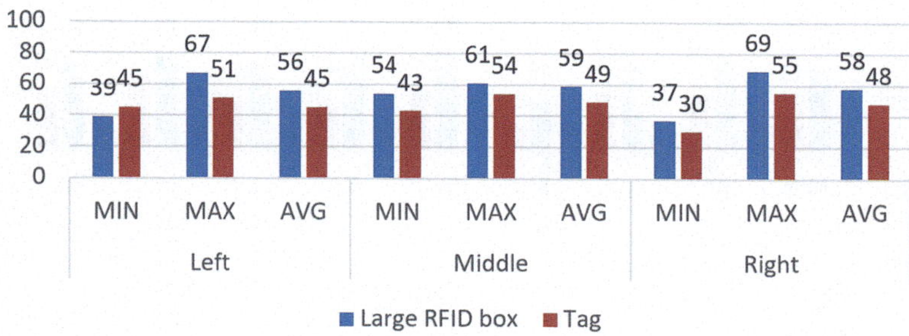

Fig. 10 Result of static testing, large RFID box, two antennas top, all position

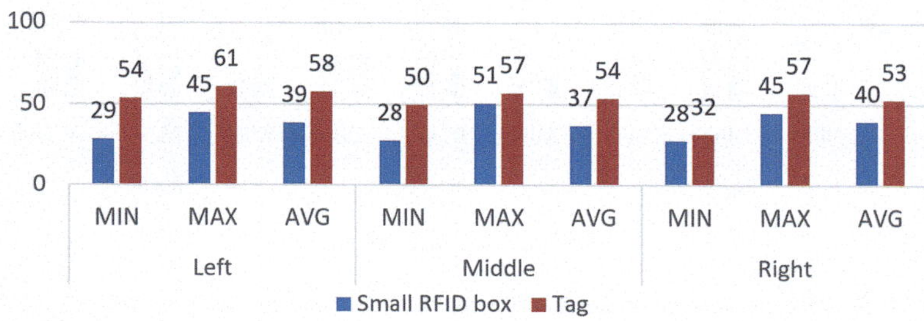

Fig. 11 Result of static testing, small RFID box, two antennas top, all position

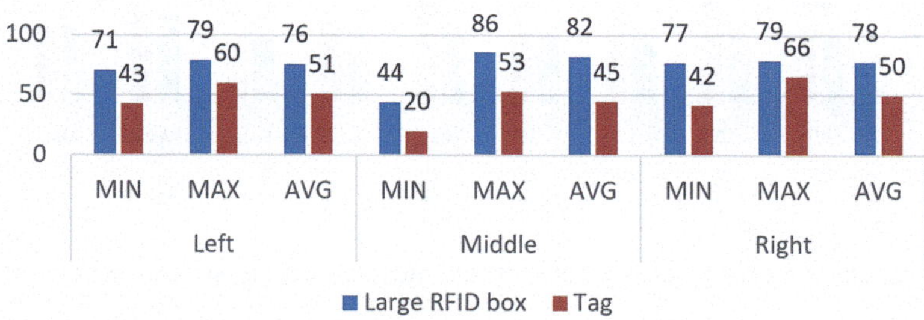

Fig. 12 Result of static testing, large RFID box, two antennas on side, all position

Sets 1–3: large RFID box, antennas top, tags at the top of the container (Fig. 10)
Sets 4–6: small RFID box, antennas top, tags at the top of the container (Fig. 11)
Sets 7–9: large RFID box, antennas on side, tags on the side of the container (Fig. 12)
Sets 10–12: small RFID box, antennas on side, tags on the side of the container (Fig. 13)

It is clear from the results of the static tests that the large RFID box is significantly better than the standard RFID tag. The reason for this phenomenon can be mainly due to sufficient space around the box tag and sufficient distance from the bovine parts of the containers. While the values of a large RFID box are better

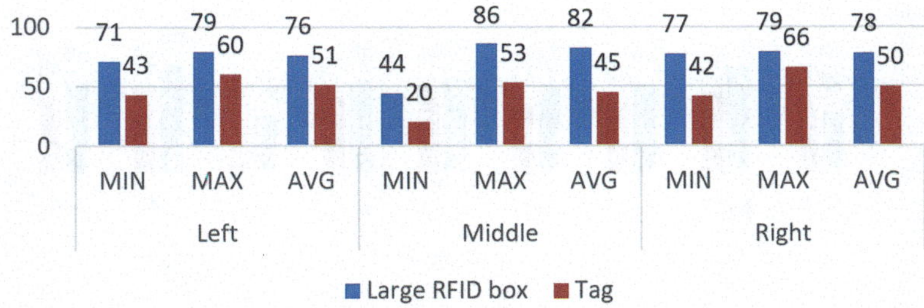

Fig. 13 Result of static testing, large RFID box, two antennas on side, all position

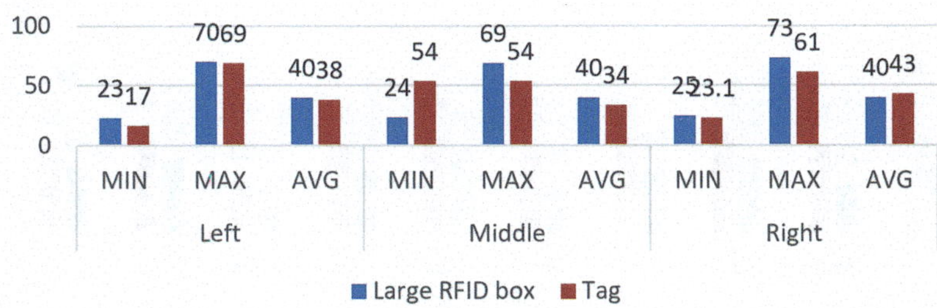

Fig. 14 Result of dynamic testing, large RFID box, two antennas top, all position

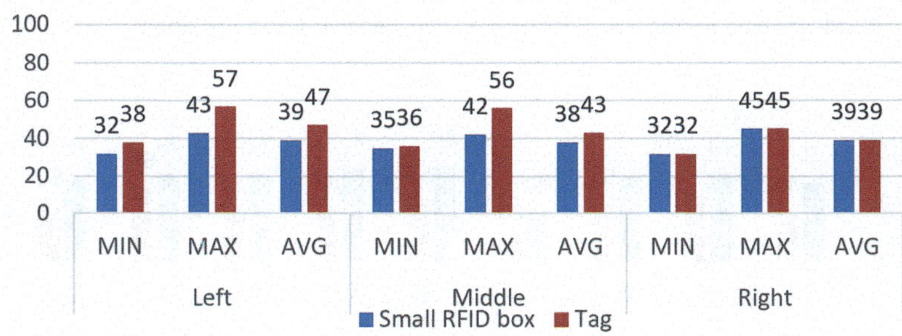

Fig. 15 Result of dynamic testing, small RFID box, two antennas top, all position

compared to a classic tag in a small RFID box, the scenario is the opposite. The small dimensions of the box do not allow the ideal electromagnetic signal propagation. That's why the standard tag in some sets is significantly better. When it is loaded by antennas from the top or bottom, a large RFID box gives the best results.

Transitions 1–3: large RFID box, antennas top, tags at the top of the container (Fig. 14)

Transitions 4–6: small RFID box, antennas top, tags at the top of the container (Fig. 15)

Transitions 7–9: large RFID box, antennas top, tags on the side of the container (Fig. 16)

Transitions 10–12: small RFID box, antennas top, tags on the side of the container (Fig. 17)

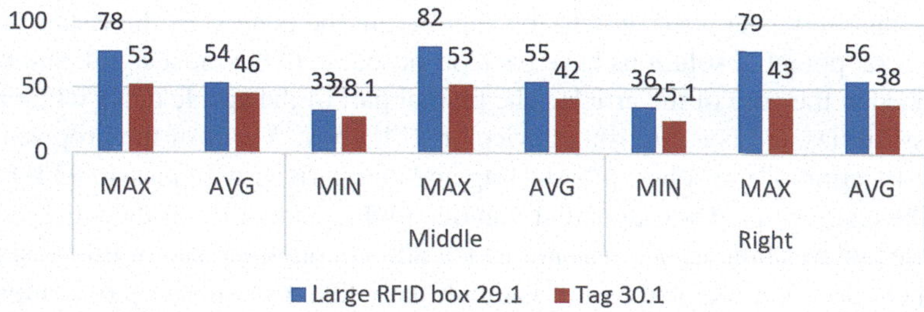

Fig. 16 Result of dynamic testing, large RFID box, two antennas on side, all position

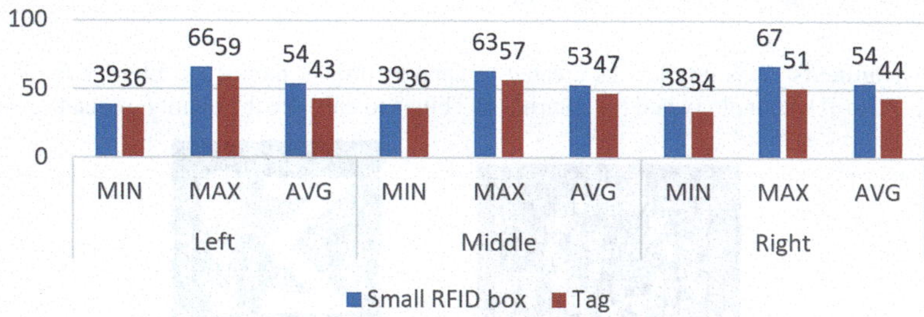

Fig. 17 Result of dynamic testing, small RFID box, two antennas on side, all position

It is clear from the results of the static tests that the large RFID box is significantly better than the standard RFID tag. The reason for this phenomenon can be mainly due to sufficient space around the box tag and sufficient distance from the bovine parts of the containers. While the values of a large RFID box are better compared to a classic tag in a small RFID box, the scenario is the opposite. The small dimensions of the box do not allow the ideal electromagnetic signal propagation. That's why the standard tag in some sets is significantly better. Although the antennas are read from the top or bottom, the big RFID box gives the best results.

The dynamic test results show that the difference in the size of a large RFID box is called when reading in motion. The same is true when loading a small RFID box, where the difference with the classic tag becomes even stronger. Another situation is when reading tags from the side. Here at both RFID boxes, we notice a significant improvement over the classic RFID tag. Although the quality of reading of both boxes varies considerably, in both cases the RFID box was not unread. From these laboratory tests, it is difficult to conclude that the same results could be obtained by RFID boxes even when deployed in real operation.

6 Conclusion

The subject of our research was the possibility of using RFID technology for monitoring of transport units at the level of own company. More specifically, the main aim was to analyze possibilities of using RFID technology within the supply chain, if the technology is applied only to the supplier. In the context of the issues identified, several potential solutions have been proposed, with the most significant being mentioned at the core of the article. The pivotal part of the article deals with one of the possibilities and is a multi-tag device or RFID box. A part of the proposal was also verification of the solution by testing and comparison with classic RFID tags. From the test results, it was apparent that due to the readability of the RFID box in the static and dynamic tests it was not bad at all. Although the use of this or similar equipment does not seem so real, it is quite likely that a small group of companies that want to use RFID technology could reach in the future. The RFID technology has existed for decades and, despite its relatively unchanged form, it can offer a lot of nonstandard applications in the future.

Acknowledgments This article was created to support project named as: EUREKA-E! 11158 U-health-Auto-ID technology and the Internet of Things to enhance the quality of health services.

References

1. Vojtěch, L., Neruda, M.: Application of shielding textiles for increasing safety airborne systems – limitation of gsm interference. In: The Ninth International Conference on Networks (ICN 2010), pp. 157–161. IEEE Computer Society, Los Alamitos (2019)
2. Beneš, F., Kubáč, L., Staša, P., Kebo, V.: RFID and augmented reality. In: Proceedings of the 14th International Carpathian Control Conference (ICCC). Rytro, Hotel Perła Południa, Poland, May 2013, pp. 186–191. IEEE, Vienna (2013)
3. Madleňák, R., Madleňáková, L., Štefunko, J.: The variant approach to the optimization of the postal transportation network in the conditions of the Slovak Republic. Transport Telecommun. **16**, 237–245 (2015)
4. Thornton, F., Lanthem, C.: RFID Security. Syngres Publishing, Inc, Rockland, MA (2006)
5. Hunt, V., Puglia, A., Puglia, M.: RFID: A Guide to Radio Frequency Identification. John Wiley & Sons, Inc., Hoboken, NJ (2007)
6. Mojský, V., Kolarovszki, P.: Simulation model of logistic chain in GS1 Slovakia laboratory. Transcom. **40**, 389–396 (2019)

Inventory Replenishment Strategy Proposals Using a Software Tool Tecnomatix Plant Simulation

Peter Trebuňa, Miriam Pekarčíková, and Marek Kliment

1 Introduction

The intelligent logistics network of the production process will create added value through communication between relatively autonomous entities, thereby achieving speed, flexibility and quality of value flow. This is conditioned by the implementation of key elements:

- Implementation of sensors for continuous automatic data collection from production equipment, production lines, transport equipment, storage facilities and warehouses, materials and parts.
- Creation of an information-communication network between all objects and subjects of the enterprise's value flow.
- Application of intelligent data generation software, i.e. collecting, processing, analysing and evaluating the collected data.

The convergence of the physical world with the digital world is now a new paradigm of the autonomous and decentralized world of production.

Process of inventory replenishment in dynamic models is usually deterministic (inputs are always delivered in ordering quantity and time), but it can also be random in time as well as in quantity. In these models it is not a determination of time and the size of one order and it also cannot be planned a series of orders (an order cycle) [1, 2]. The aim is to find a way of managing the stock size by orders. The random dynamic models include P-system and Q-system, whose specification is given below. On the other side, static models are characteristics with the determination of one unique delivery that do not repeat. It is also possible to include here

P. Trebuňa · M. Pekarčíková (✉) · M. Kliment
Faculty of Mechanical Engineering, Institute of Management, Industrial and Digital Engineering, Technical University in Kosice, Košice, Slovakia
e-mail: peter.trebuna@tuke.sk; miriam.pekarcikova@tuke.sk; marek.kliment@tuke.sk

L. Knapcikova et al. (eds.), *New Approaches in Management of Smart Manufacturing Systems*, EAI/Springer Innovations in Communication and Computing, https://doi.org/10.1007/978-3-030-40176-4_19

models determining the amount of inventory of spare parts, inventory to cover the one-time sale [3, 4].

Computer simulation is an attempt to model a real hypothetical situation, process or thing using a computer with the goal to study it and see how it works by different and changing conditions. The model has to represent key characteristics and own behaviour of system to find the answers on tasked questions or verify a result that is otherwise costlier or impossible to achieve. Tecnomatix Plant Simulation performs a specific type of simulation, defined by NASA as "a tool to model organizational activities and system responses to discrete events in the operational flow, supports Monte Carlo analysis and can be used for manufacturing flows, operational processing flows, supply chain flows, and flows of information through an organization" [5]. Implementation and use of simulation tools has potential in connection with decision-making competencies of managers in logistics greatly strengthened.

2 Demand-Driven Inventory Replenishment Models

The models showing the dynamic inventory management problems with determined movement probabilistically incorporate all the types of costs: costs associated with the acquisition of supplies, maintaining inventory costs, costs of lack of safety stock and cost of excess inventory. Some of them have new features: for example, the cost of excess inventory does not have such a costly impact as in the case of the static models. This is because of the long unlimited time period of stock maintenance and so inventory excess generated in one period can be used in the next period by a reduction in orders. This means that the cost of inventory excess is shown in the form of the high cost of maintaining inventory [6–8]. The probabilistic nature of demand and consumption of inventory accept fluctuation in the consumption of supplies. In the event that causes a lack of prompt supply and significant losses in the supply system, the company secures itself by increasing inventory corresponding to the expected consumption of so-called safety stock to ensure a supply system to random fluctuation in consumption. This safety stock reduces the shortage risk of prompt supplies, but it binds capital and also requires the storage, keeping, and thus causes additional costs of maintaining inventory. The maintaining costs of safety stock stand against the costs from the shortage of prompt supplies in the dynamic probabilistic models [9–11]. Therefore it has created a new agent that acts in these models which is the stock replenishment interval. Managing of inventory state in the dynamic models can be influenced by two types of interventions [12–14]:

- Orders' frequency—Q-system.
- Orders' sizes—P-system.

The number of orders "N" and their size "x" are bound together by precise quantitative relation: $x = Q/N$. This relation is valid next only when it comes to middle values of these variables. Particular sub-periods lead to random deviations from the true consumption of its middle value and thus inventory fluctuations around its expected size. These effects need to balance which is possible, for example, by changing the frequency of orders at a constant order size—it is "Q"-systems—or by changing the amount of orders at a fixed interval between particular orders—then it is "P"-systems. These are two basic approaches in inventory management used for the absorption of probabilistic fluctuations in consumption [15–17].

2.1 Q-System of Inventory Replenishment

This system operates with a fixed order size, which is indicated by the symbol "Q" (from the word "quantity"). Fluctuation in the stock consumption is balanced by changes in the frequency of orders. Q-system works on the basis of determining the amount of inventory that is used to cover consumption during the particular period, so-called signal level of stocks. If the actual stock reaches that level, then the new order is executed. Random fluctuations in consumption and cost of the lack of emergency stock are taken into account when determining the signal level. Q-system is also known as "a system of two stores". The name comes from the use of the mentioned system in practice, where each item is stored in two separate stores. The major part of inventory is kept in the first one and in the second store inventory is maintained and it also serves to cover consumption for the time period of replenishing. The new order is realized when the stock in the first store is used. By the time a new delivery arrives to the warehouse the consumption is covered from the second warehouse. After the arrival of new delivery the second warehouse is firstly supplied for the signal level and the rest of supply is placed into the first store. It is essential to monitor whether the remaining stock does not drop under the signal level which is in case of several stock items quite difficult [18, 19].

Application of the Q-system in practice means determining the constant size of the order and inventory signal level for each item. Since this is the system where consumption fluctuation is expressed by changes in the ordering cycle, it is not necessary to maintain safety stocks for the random increased consumption during the ordering cycle—in this case, the actual supply drops to the signal level and thus there is a realization of a new order. This however cannot be applied during the time of order realization, so to cover possible consumption fluctuation exactly during this time, the safety stock is required (dependency exists between the size of the safety stock and length of ordering cycle); see Fig. 1 [20, 21].

2.2 P-System of Inventory Replenishment

Stock replenishment is realized by orders in fixed delivery times, usually with constant intervals between orders. The system is referred as "P" (from the word "period") inventory management system. It is the second basic inventory management system with probabilistically determined movement. It is based on the fact that

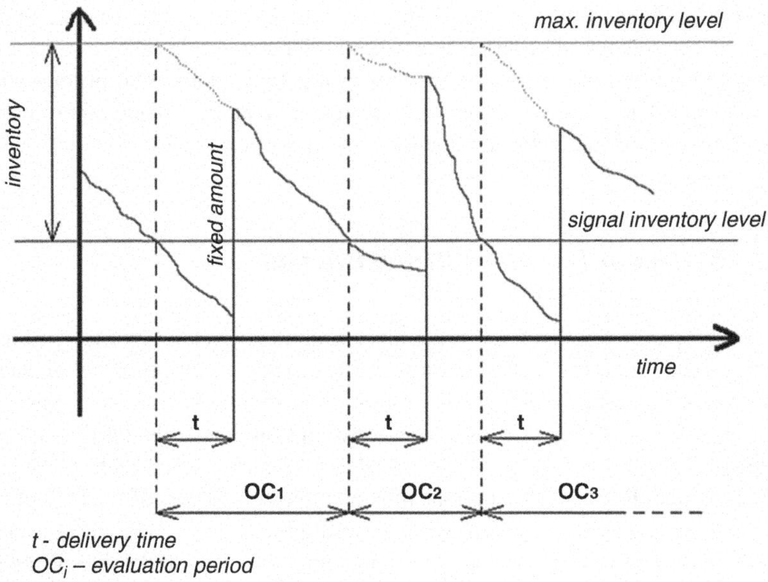

Fig. 1 Graphic's schema of the Q-system of inventory replenishment [20]

orders of variable size are carried in advance fixed ordering times. The size of particular orders is set so that the sum of the actual stock size at the time of the order was equal to a predetermined variable of which size takes into account the consumption fluctuation. Fluctuation in actual consumption around its middle value is balanced by the size of particular orders. It is possible by using this model to find the optimal length of ordering cycle and the optimal ordering level at which the actual stock is replenished. The size of a particular order is determined as the difference between the stock ordering level and actual stock at the time of order. This system does not require a permanent inventory control which is advantage; it only needs a periodic control [18, 19, 21].

It is possible to find out by comparing the mechanism of the Q and P systems that the main difference between them is in the way how they receive sales fluctuation. While in the Q-system shortening of the ordering cycle induces higher demand and the safety stock is formed only to cover the higher consumption during the time of stock acquisition. In the P-system, the safety stock must compensate fluctuation in sales during the entire ordering cycle. During which time there is an interaction between the size of the order of one period and the shortage risk of prompt stock in next periods. The P-system uses a simplified assumption which is to determine the safety stock only with respect to one ordering cycle plus the next interval of stock acquisition. Q- and P-systems form the main inventory management systems where the practical use is relatively simple; see Fig. 2, [20, 21].

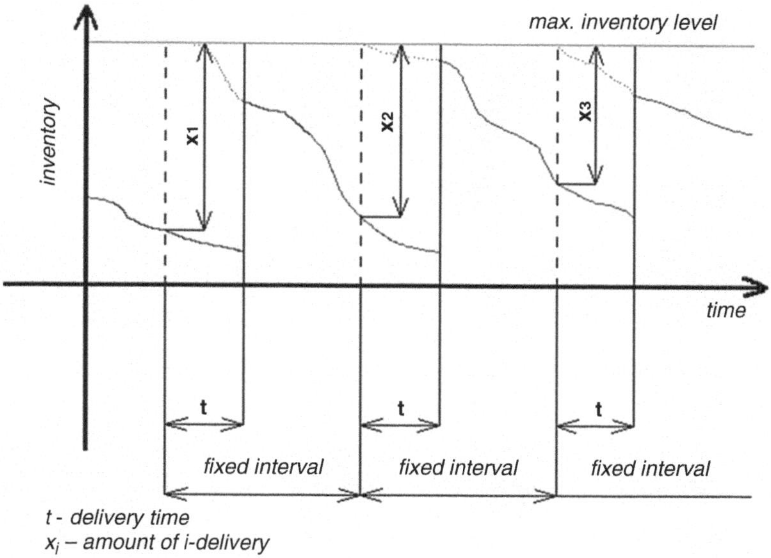

Fig. 2 Graphic's schema of the P-system of inventory replenishment [20]

3 Testing Replenishment Strategy by Using Tecnomatix Plant Simulation

Simulation model that is processed below is oriented on testing the replenishment strategy. Case study was made in the area of simulation software Tecnomatix Plant Simulation (Siemens). Specifically, it is the simulation model of procurement for demand-driven consumption, i.e. customer order simulation and consumption, which consumes the inventory and initiates the supply process from the supplier when it reaches buffer minimum of inventory level. The way of representation of inventory consumption is viewed as a silo.

Input data for case study: maximum inventory level = 5000 pcs, delivery size = 2000 pcs and delivery cycle = 2 days.

The storage strategy is part of the process of deciding on the quantity and time of the input acquisition, i.e. the process of initializing the purchase and supply of material, depending on the length of the delivery time. The supply method can be implemented in two variants in terms of time [20]:

- Order time will depend on the stock; i.e. ordering will be realized upon acquisition of the order point; the warehouse will be filled at regular intervals (P-model—with relatively constant demand or relatively high insurance stocks).
- As far as the quantity is concerned, the filling of the warehouse can be carried out in fixed quantities (Q-model); the amount will vary depending on actual consumption. The essence of the supply is to reduce the storage in the storage tank to a minimum level, which necessitates the implementation of the supply.

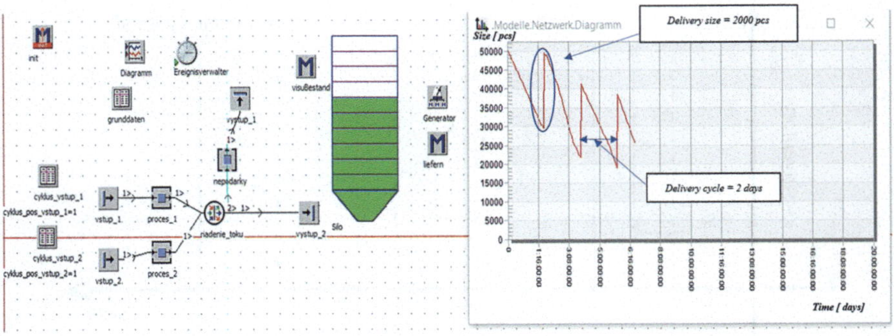

Fig. 3 Simulation model—maximum inventory level (processed according to [22])

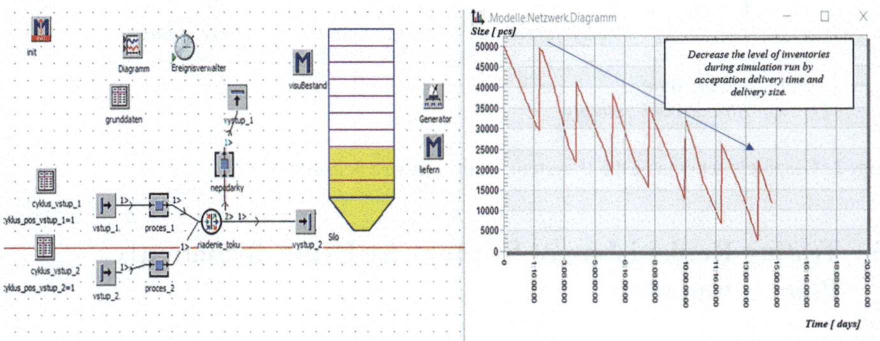

Fig. 4 Simulation model—average inventory level (processed according to [22])

The way of showing the consumption of supplies is in the form of silos. It is an illustrative depiction of the process of replenishment and consuming of inventory by simultaneously displaying the maximum, average and minimum stock levels by using colours, green (maximum), yellow (average) and red (minimum). The simulation ends when the silo is empty. This is due to the set replenishment method and to the amount of consumption, delivery time and order size that is constant. According to the results of the simulation, it is possible to set the optimum size of the order, which would be closer to the needs of customer requirements. Figures 3, 4, and 5 show simulation models for the above-mentioned consumption and replenishment phases. There is also an inventory sawtooth diagram showing the consumption progress in each phase of the simulation run [22–24].

However, efforts are being made to keep stocks to a minimum; i.e. it is necessary to know when the level of inventory is approaching the minimum level and to enter a higher-order requirement in order to cope with future consumption and avoid stock shortages. In the case study, there is a supply option whose output is as follows: decrease in the level of inventories in a way of supply and consumption and decrease to 0 level in about 37 days. Therefore, it can be stated that in the case of

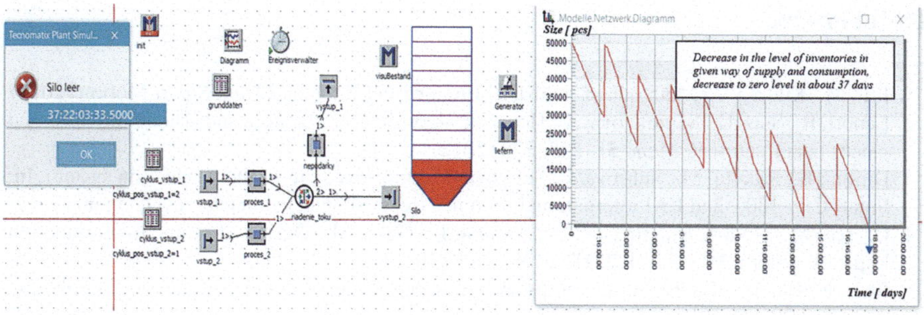

Fig. 5 Simulation model—empty buffer (processed according to [22])

demand fluctuations considered and with respect to the delivery period of 2 days, this order will have to be increased.

4 Conclusion

The introduction of new ICT in the context of Industry 4.0 (CPS, IoT (Internet of Things), cloud computing, big data) into business processes and systems creates scope for forecasting applications that significantly improve the quality of forecasted values based on the processing of daily data available online, which helps to make order size more efficient.

The main tasks of integrated supply logistics are as follows:

- Integrated planning, optimization, visualization and supply process management:
  - Ensuring optimal delivery (JIT/JIS).
  - Ensuring material inputs in the right quantity, right quality, in the right place.
  - Provision of capacity requirements – personnel, transport, storage.
- Determination of logistics costs for different variants of the supply process:
  - Determination of logistics costs and expenses.
- Standardization of the planning process, which contributes to the optimization of the business system as a whole:
  - Communication with the supplier regarding the correct method of delivery.
  - Communication with other operators regarding the optimal way of storage.

Acknowledgments This article was created by implementation of the grant project APVV-17-0258 Digital engineering elements application in innovation and optimization of production flows.

References

1. Edl, M., Kudrna, J.: Meth ods of Industrial Engineering, 1st edn. Plzen, Smart Motion (2013)
2. Lenort, R., Stas, D., Samolejova, A.: Heuristic algorithm for planning and scheduling of forged pieces heat treatment. Meta. **51**, 225–228 (2012)
3. Dulina, L., Rakyta, M., Sulirova, I., Šeligova, M.: Improvement of the production system. In: Smart City 360°. 2nd EAI international Summit: Revised Selected Papers. EAI, Ghent (2017)
4. Gregor, M., Hodon, R., Biňasová, V., Dulina, Ľ., Gašo, M.: Design of simulation-emulation logistics system. MM Sci. J. **2018**, 2498–2502 (2018)
5. SE Research Consortium. https://www.nasa.gov/consortium/DiscreteEventSimulation. Accessed 20 Apr 2019
6. Vavrík, V., Gregor, M., Grznár, P.: Computer simulation as a tool for the optimization of logistics using automated guided vehicles. Procedia Eng. **192**, 923–928 (2017)
7. Grznár, P., Gregor, M., Mozol, Š., Krajčovič, M., Dulina, Ľ., Gašo, M., Major, M.A.: System to determine the optimal work-in-progress inventory stored in interoperation manufacturing buffers. Sustainability. **11**(14), 2071–1050 (2019)
8. Straka, M., Malindzakova, M., Trebuna, P., Rosova, A., Pekarcikova, M., Fill, M.: Application of EXTENDSIM for improvement of production logistics' efficiency. Int. J. Simul. Model. **16**, 422–434 (2017)
9. Laciak, M., Sofranko, M.: Designing of the technological line in the SCADA system PROMOTIC. Proceedings of the 2013 14th International Carpathian Control Conference, Rytro, Poland (2013)
10. Malindzakova, M., Rosova, A., Baranova, V., Futo, J.: Modelling of outbursts and ejections occurrences during steel production. Meta. **54**, 247–250 (2015)
11. Saniuk, S., Saniuk, A., Lenort, R., Samolejova, A.: Formation and planning of virtual production networks in metallurgical clusters. Meta. **53**, 725–727 (2014)
12. Straka, M., Bindzar, P., Kadukova, A.: Utilization of the multicriteria decision-making methods for the needs of mining industry. Acta Montan. Slovaca. **19**, 199–206 (2014)
13. Straka, M.: Simulation of Discrete Systems and Simulation Languages, 1st edn. Kosice, TU Kosice (2005)
14. Badida, M., Gombar, M., Maslejova, A., Sobotova, L., Kmec, J., Vagaska, A.: Evaluation of zinc coating quality by statistical methods. Przemysl Chem. **94**, 2146–2149 (2015)
15. Rosova, A., Malindzakova, M.: Material flow – starting point for recovery of inputs in the production company, International Multidisciplinary Scientific GeoConference Surveying Geology and Mining Ecology Management, 14th International Multidisciplinary Scientific Geoconference and EXPO, SGEM 2014; Albena, Bulgaria SGEM Vol. 3, no. 5 (2014)
16. Lerher, T. , Edl, M., Rosi, B.: Energy efficiency model for the mini-load automated storage and retrieval systems, Int. J. Adv. Manuf. Technol., Vol. 70, no. 1–4, pp. 97–115 (2014)
17. Wicher, P., Staš, D., Karkula, M., Lenort, R., Besta, P.: A computer simulation-based analysis of supply chains resilience in industrial environment. Meta. **54**(4), 703–706 (2015)
18. Pekarčíková, M., Trebuňa, P., Markovič, J.: Case study of modelling the logistics chain in production, Conference: 6th Conference on Modelling of Mechanical and Mechatronic Systems (MMaMS), Vysoke Tatry, Slovakia, Modelling of mechanical and mechatronic systems book series. Procedia Eng. **96**, 355–361 (2014)
19. Klos, S., Ptalas-Maliszewska, J.: The impact of ERP on maintenance management. Manag. Prod. Eng. Rev. **4**, 15–25 (2013)
20. Trebuňa, P., Fiľo, M., Pekarčíková, M.: Supply and Distribution Logistics, 1st edn. Amos, Ostrava (2013)

21. Gregor, T., Krajčovič, M., Wiecek, D.: Smart connected logistics. Procedia Eng. **192**, 265–270 (2017). Transcom 2017 12th International Scientific Conference of Young Scientists on Sustainable, Modern and Safe Transport. High Tatras, Grand Hotel Bellevue, Slovakia
22. Bangsow, S.: Tecnomatix Plant Simulation: Modeling and Programming by Means of Examples. Springer, Berlin (2016)
23. Siemens. Process simulate: Manufacturing process verification in powerful 3D environment (2018). https://www.plm.automation.siemens.com/en/products/tecnomatix/manufacturing-simulation/robotics/process-simulate.shtml#lightview%26url=/en_us/Images/7457_tcm1023-80351.pdf%26title=ProcessSimulate%26description=ProcessSimulateFactSheet%26docType=pdf. Accessed 11 Apr 2019
24. Siemens. Tecnomatix (2017). http://www.sova.sk/sk/produkty/tecnomatix. Accessed 16 Apr 2019

Innovation of the Production Line in the Enterprise with the Help of Module TX Process Simulate

Jozef Trojan, Peter Trebuňa, and Marek Mizerák

1 Introduction

PLM is an abbreviation of the English term product lifecycle management, which refers to the process of managing the complete lifecycle of a product, from its first concept through detailed design, manufacturing, and after-sales service to disposal.

Sometimes the term is mistakenly understood only as so-called PDM software, which is used in the subphases of the PLM process for managing engineering design data and communication between designers. Product lifecycle management includes standardized workflows, business systems, key data as well as adequately trained staff in addition to the necessary computing applications. Such a complex entity can be used as a core system of a manufacturing enterprise or any other company, and its business results in a physical product [1].

The definition of PLM, as a software solution, includes a smooth collaborative infrastructure of computer applications used to work with product data throughout its "life," its "conception" being understood as recording the first idea of the form or function of a product in that system. As such, the PLM system covers all product data management and electronic communications between all stakeholders, including customers (CRM), all necessary resources (ERP), and the supply chain [2].

Each institution may have a different perspective on the product lifecycle since each focuses on a different industry or different sectors of different importance. A variety of tools are used to manage the product lifecycle. The most comprehensive of these are digital packages from large companies such as Siemens PLM Software. These packages consist of several tools from CAD/CAM to planning tools to simulation and innovation tools. The advantage of these complex packages is the data

J. Trojan (✉) · P. Trebuňa · M. Mizerák
Faculty of Mechanical Engineering, Institute of Management, Industrial and Digital Engineering, Technical University of Košice, Košice, Slovakia
e-mail: jozef.trojan@tuke.sk; peter.trebuna@tuke.sk; marek.mizerak@student.tuke.sk

L. Knapcikova et al. (eds.), *New Approaches in Management of Smart Manufacturing Systems*, EAI/Springer Innovations in Communication and Computing, https://doi.org/10.1007/978-3-030-40176-4_20

link between the various tools and the overall management of the entire "virtual enterprise." The JT format ensures the data continuity of the individual sub-modules, while the tonto format is supported by all the tools in the package, eliminating the problem of communication between sites (technology, construction, logistics, planning, etc.) [3].

The lifecycle is thus captured from the initial stage of product design to the final stage of its production simulation. Of course, the entire lifecycle also includes service, shipping, and final disposal of the product, but these phases are not yet sufficiently underpinned by software, and so we are not working with them yet [4].

2 Create 3D Models Using SolidWorks Software

After opening the SolidWorks software and then creating a new project, we draw the desired shape in the selected plane in Sketch. We need to pull the given part or machine shape into space, leaving Sketch on, switching to the Features option next to Sketch, and using Extruded Boss/Base. Enter the specified length in the following table. The length is preset in millimeters, but if we have dimensions in larger scales, just write a scale abbreviation after the number, e.g., cm, m, etc. Also, you can use the arrows next to Blind to change the pull direction [5].

Conversely, if we have a rotating component, it is clear that it is rotated about its axis. That's why we have to think about it when drawing. We proceed as follows: first, we enter the axis around which the shape of the part will rotate later, proceed as in the case of the nonrotating part, and draw the shape of the part with all the curves and roundings. We draw the entire shape of the part above or below the axis of rotation. If there is a hole in the part, then we will also adjust the distance of the created drawing from the axis accordingly. After drawing the desired shape and correctly taking the distance from the axis, we reopen the Features, still labeled Sketch, and choose Revolved Boss/Base. In the window then set the axis of rotation, by clicking on the drawn axis, and we can also specify how many degrees to rotate around the axis (Fig. 1).

The next step is to cut out the holes with the Extruded Cut tool found in Features. The procedure is the same as in the previous step, so we draw a shape that we need to cut to the object's surface in Sketch. Then, through Extruded Cut, we set the cut length as well as the cutting direction [6].

Using other tools, we will gradually complete all the necessary 3D models for the work (Fig. 2). For models that consist of multiple parts, they need to be folded somewhere. This is the Assembly interface. First, we have to choose a base part or object that will serve as a matrix for the other parts. The other parts are put in the same way or just by right-clicking from the storage place directly into the Assembly Workspace. If we have a matrix and all the necessary parts, we will start using the Mate option, on the main tooling menu, to gradually associate the parts with the matrix or parts together. Mate determines the bonds between the edges, between the surfaces, between the surface of the bar and the wall of the hole, and so on [7].

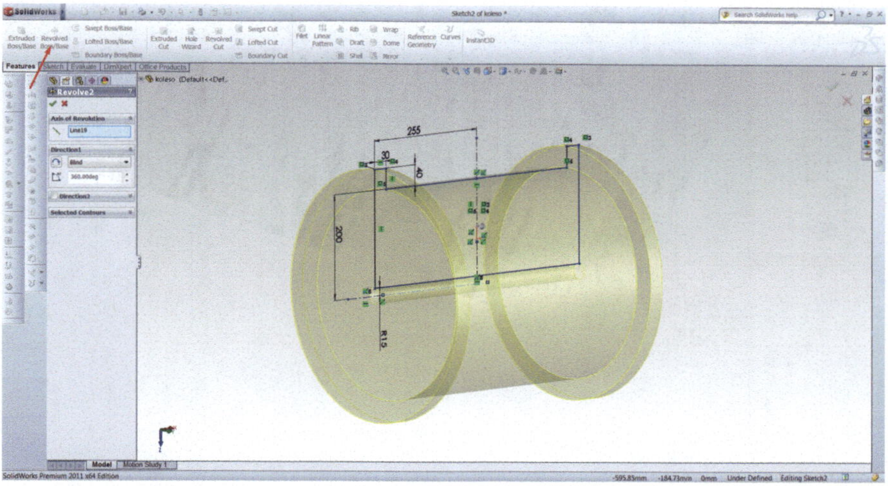

Fig. 1 Rotate the drawing around the axis through Revolved Boss/Base

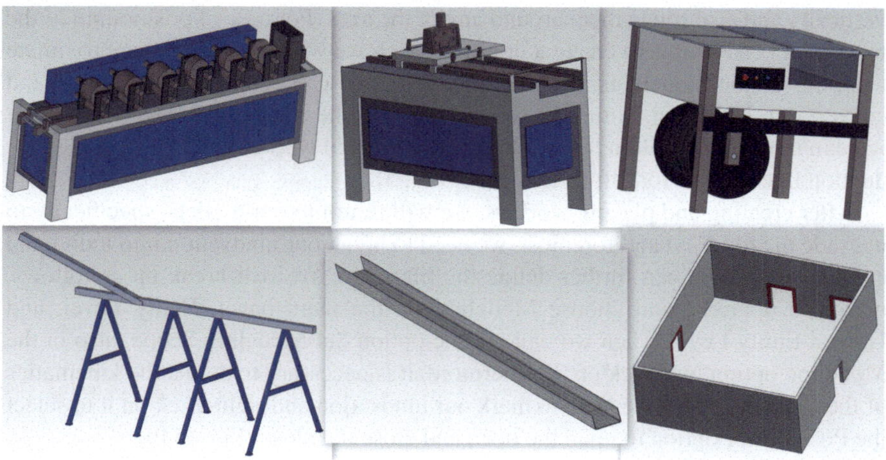

Fig. 2 3D models modeled in SolidWorks

3 Creating Simulation Using Drawn Models

Insert the modeled parts and machines into the Process Simulate library of Tecnomatix software. Then we put them into the project from the library. If we have put all the 3D models into the project, then we have to place these models correctly according to the plan of deploying machines and warehouses in the factory production hall. So we start by placing the wall, through the lines, to placing the individual stores.

Click on the displayed model and select Placement Manipulator on the top toolbar. We will show the axes with which we can move the models horizontally and

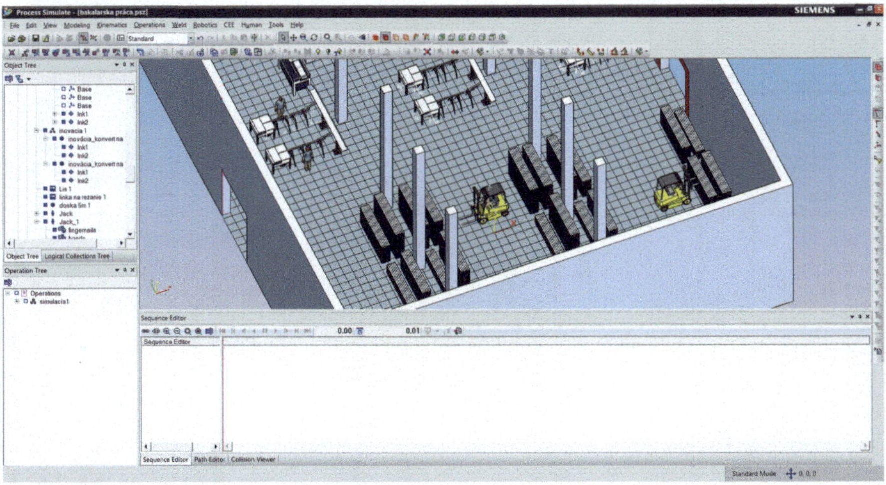

Fig. 3 Location of 3D models, forklift, and factory workers

vertically and also rotate them around any of the axes. For ease of positioning of the models, it is advisable to create a floor on which we will visualize the approximate locations of the models using fine projections. We can create it in any CAD and convert it to JT format. Similarly, we will place all the converted 3D models as they were in the factory building. We also create or embed workers who operate the production line and the forklift operator [8] (Fig. 3).

After creating and placing workers, we will return to our models, specifically to upgrade the tilt. First and foremost, we need to break our innovation into individual entities so that we can further define the tilt poses. We will break up as follows, mark the upgrade, and choose Modeling on the main panel, Entity Level, and Unload Entity Level. Then we will get the option Set Modeling Scope, also in the Modeling option, we click on. Furthermore, it is necessary to define the kinematics of the tilt valve. First, we need to mark our innovation and right-click on it to select the Pose Editor option to open the open and close the close (Fig. 4) [9].

After placing and configuring all variables, proceed to the simulation itself. Set the tracks for the U-profile up to the tilt. We create a new operation via the New Device Operation. We will enter the operation name, Device, in our case upgrade_ convert to 1part_1; Scope will be the workplace_1. Next, we define From pose (starting pose) and for the To pose, enter "open" that we created earlier in the Pose Editor. With the lift-up lifted, reset the U-profile path through the Path Editor so that the U-profile comes out on the raised lift. Thus, the first U-profile remains on the hinge until another is encountered and the first profile is inverted. After joining the profiles, the shutter is lowered and the U-profile connected by inertia is lowered to a defined path on the take-off table (Fig. 5) [10].

At the take-off table, there is one worker on each side, who is ready to grab and transfer the mentioned U-profiles to the wrapper. After placing the U-profiles in the wrapper, it is necessary that we also give the workers a way back through Human,

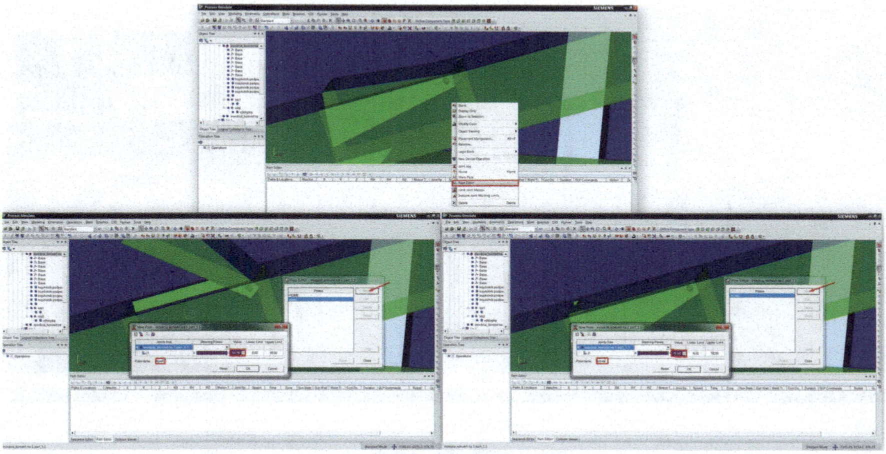

Fig. 4 Create poses through Pose Editor

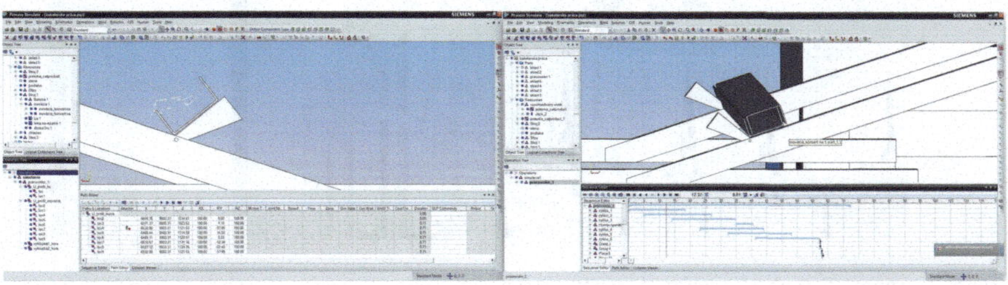

Fig. 5 Innovation—tilt

Walk Creator. This cycle can be repeated. In Walk Creator, click on Path Creator and then Select a Location and use the axes to return the workers to the specified location. After tying the entire stack in the wrapper, you still need to define the path for the forklift that takes it and takes it to the warehouse [11].

4 Comparison of Actual Production with Simulation

In the simulation, we focused on two shortcomings in production and storage. We have found in the production that said U-profile bending and cutting machine has been unused. According to the calculations, we found out that the machine only worked at 50%. This has been caused by the human factor when stackers of stacks at a faster machine run did not catch up. To solve this problem, we have designed an innovation in the form of a hinge located on a drop table. With this innovation, we have allowed the machine to accelerate up to 70%, while workers have been able to collect the aforementioned U-profile stacks (Fig. 6).

Fig. 6 Compare actual take-off table with innovation

In storage, we found the warehouse deployed incorrectly, because the forklift worker had great difficulty in storing the stacks he had made and had to dial and rotate many times to store them correctly and also to find the right stacks for shipment. This problem was solved by better warehouse orientation, which resulted in batch removal and faster shipment handling (Fig. 7). More specifically, the time and capacity evaluations are given at the conclusion.

5 Conclusion

When analyzing the production line, it was found that the line was used only for 50% of its capacity, which marked the times for packing one stack of U-profiles averaging 35.4 s. These times meant that to increase the orders of profiles produced on the line (because it was not possible to produce faster), it was necessary to introduce overtime and work changes on weekends. Therefore, it was necessary to propose some innovation that would speed up production. After modeling the drop-up innovation on the take-off table and creating a production process simulation,

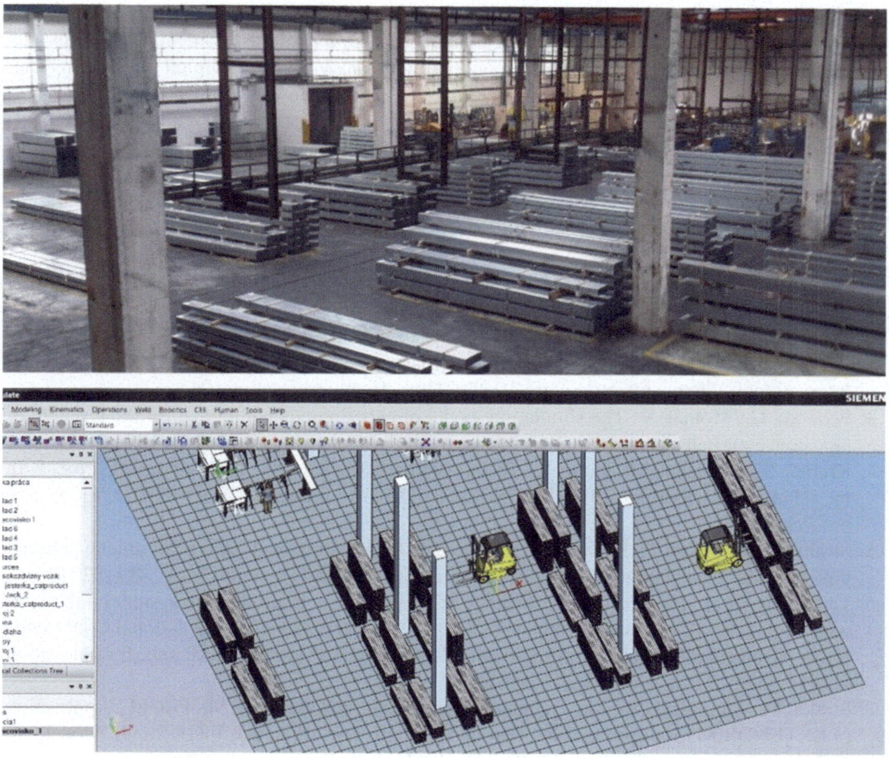

Fig. 7 Comparison of stock management in normal operation and simulation

time savings of 5.7 s per stack were found. By converting to a single production dose, time savings of up to 1:54 min were found.

One production line produces 4320 profiles for one work change (8 h). By introducing the proposed innovation, the production line is capable of producing the same number of profiles in 6 h and 30 min. This insight allows an increase in production capacity of 20% (in the case of a larger number of orders) without increasing the number of employees or purchasing new machines.

Another problem solved was the distribution of warehouses. When observing the orientation of the workers in the warehouse orientation, the forklift worker's great downtime was found when looking for orders when they were shipped to customers. The dispatch of one production batch took 3:20 min. After upgrading the stores as a change in storage orientation, this time was reduced by half a minute to 2:50.

Currently, most large companies are already using different simulation software to streamline and innovate production. In the future, simulation software such as Tecnomatix will be an indispensable part of any business, as the simulation will save the company money, reduce production times, provide flexible supplies, eliminate downtime, and detect and eliminate problems [12].

Acknowledgement This article was created by implementation of the grant project VEGA 1/0708/16 Development of a new research methods for simulation, assessment, evaluation, and quantification of advanced methods of production and APVV-17-0258 Digital engineering elements application in innovation and optimization of production flows.

References

1. Bureš, M., Görner, T., Šimon, M., Sekulová, K.: Využití digitálních nástrojů ergonomie v praxi. J. Saf. Res. Appl. **4**(1), 1–7 (2011)
2. Poor, P., Basl, J.: Czech Republic and processes of industry 4.0 implementation. In: Katalinic, B. (ed.) Proceedings of the 29th DAAAM International Symposium, pp. 0454–0459. DAAAM International, Vienna. https://doi.org/10.2507/29th.daaam.proceedings.067
3. Fedorko, G., Molnár, V.: The simulation model of experimental equipment for the research of pipe conveyor belts using ABAQUS software. Appl. Mech. Mater. **611**, 183–187 (2014)
4. Krehel, R., Kocisko, M., Poor, P.: Technical diagnostics in the paper industry. Proc 27th Int. DAAAM Symp. **2016**, 0775–0784 (2016). https://doi.org/10.2507/27th.daaam.proceedings.112
5. Straka, M., Kacmary, P., Rosova, A., Yakimovich, B., Korshunov, A.: Model of unique material flow in context with layout of manufacturing facilities. Manuf. Technol. **16**(4), 814–820 (2016)
6. Trebuňa P., et al.: Increasing the efficiency of logistics flow in production system by the software tool tecnomatix process simulate – 2018. MMS 2018: 3rd EAI International Conference on Management of Manufacturing Systems. Gent (Belgicko): European Alliance for Innovation s. 1–11. ISBN 978-1-63190-167-6. ISSN 2593-7642,
7. Straka, M., Malindzakova, M., Trebuna, P., Rosova, A., Pekarcikova, M., Fill, M.: Application of EXTENDSIM for improvement of production logistics' efficiency. Int. J. Simul. Model. **16**, 422–434 (2017). https://doi.org/10.2507/IJSIMM16(3)5.38
8. Poór, P., Šimon, M., Karková, M.: CMMS as an effective solution for company maintenance costs reduction. Prod. Manag. Eng. Sci. **2015**, 241–246 (2015)
9. Dulina, L., Rakyta, M., Sulirova, I., Šeligova, M.: Improvement of the Production System. Smart City 360°. 2nd EAI International Summit: Revised Selected Papers. EAI, Ghent (2017)
10. Edl, M., Kudrna, J.: Methods of Industrial Engineering, 1st edn. Smart Motion, Plzen (2013)
11. Lenort, R., Stas, D., Samolejova, A.: Heuristic algorithm for planning and scheduling of forged pieces heat treatment. Meta. **51**, 225–228 (2012)
12. Lerher, T., Edl, M., Rosi, B.: Energy efficiency model for the mini-load automated storage and retrieval systems. Int. J. Adv. Manuf. Technol. **70**(1–4), 97–115 (2014)

IoT System with Switchable GSM, LoRaWAN, and Sigfox Communication Technology for Reliable Data Collection to Open-Source or Industrial Cloud Platforms

Kamil Židek, Ján Piteľ, and Peter Lazorík

1 Introduction to Data Collection

Data can be collected from the production process by two types of technologies: contact and contactless.

The contact sensor approach can measure some technological parameters of materials (for example, it can assess roughness or hardness by using a mechanical contact prong) or it can perform a simple task such as assessing on/off status using contact switches for part presence in a fixed position. A typical example is measurement of the temperature of air or liquid in construction nodes using water or oil in cooling equipment.

The contactless approach usually uses contactless sensors for measuring physical values. Typical example are MEMS (microelectromechanical system) sensors, which indirectly measure values by measuring a physical phenomenon with transfer of its value to a measured value. The next most widely used contactless technologies for part measuring are based on two-dimensional or three-dimensional vision systems or laser sensors based on triangulation/interferometry or the confocal principle.

The main aim of the Industry 4.0 concept is full data digitalization from machines, products, and production processes. This can be achieved by contactless sensors because they can monitor every product in high-batch production.

For suitable product monitoring, IoT (internet of things) devices equipped with adequate sensors provide wireless connectivity, ideally independent of the standard wireless network (Wi-Fi or Bluetooth), with automatized data collection to some cloud platform (commercial or open-source). These devices can be battery powered

K. Židek (✉) · J. Piteľ · P. Lazorík
Faculty of Manufacturing Technologies with a Seat in Prešov, Department of Industrial Engineering and Informatics, Technical University of Kosice, Prešov, Slovakia
e-mail: kamil.zidek@tuke.sk; jan.pitel@tuke.sk; peter.lazorik@tuke.sk

L. Knapcikova et al. (eds.), *New Approaches in Management of Smart Manufacturing Systems*, EAI/Springer Innovations in Communication and Computing, https://doi.org/10.1007/978-3-030-40176-4_21

for a long time and can be mounted in fixtures or used for monitoring the production environment too.

These digital data can be accumulated in PLC (programmable logic controller) or SCADA (Supervisory Control and Data Acquisition) systems, but they cannot be stored long term, because industrial control and monitoring systems are quite limited in terms of storage space. A possible solution is use of a cloud platform combined with a PLC system used as an IoT device. Cloud platforms are used for data analysis and visualization (data mining).

The main advantage of using IoT devices with wireless communication is extensive support for data-mining functions. These cloud platforms include preprogrammed functions for simplification of data mining and knowledge extraction from big data collected from production processes, products, or production systems.

Below, we list the main six technologies of IoT communication on the market, which are usable in connection with industrial processes:

- *GSM* (Global System for Mobile Communications): The first technology used for IoT devices with wireless communication, but with many problems, very high power consumption, and low penetration of buildings underground, like parking environments.
- *NB-IoT* (Narrowband Internet of Things): Mobile operators' IoT networks with half-duplex data transfer and a low bandwidth.
- *GSM* (*LTE-M* (Long-Term Evolution for Machines)): Competitor to NB-IoT with a comparatively higher data rate and full-duplex with voice over the network, but it requires more bandwidth.
- *LoRa/LoRaWAN*: Open standard for simple data transfer between two nodes. LoRaWAN extends communication to star topology with internet gateways to store data in some clouds.
- *Sigfox*: Reliable closed technology with a low bandwidth and roaming features.

2 Related Works

Industry 4.0 (referred to as the fourth industrial revolution) is a current trend of control, automation, monitoring, and data exchange in manufacturing technologies. It includes IoT, CPS (cyber-physical systems), cloud platforms, cognitive computing, and other related disciplines [1]. The main task is digitization with data transfer, analysis, and cumulated knowledge extraction. There are many papers describing research results concerning CPS in the Industry 4.0 concept, big data processing [2], and combination of CPS with IoT systems [3]. Industry 4.0 is aimed at technologies that are usable in product customization, because customer needs are directed toward unique products. This requirement must satisfy the need for low-cost products with maximized customization.

Industrial parts are partially suitable only for digitalization of production process, machines, and environments. We need to specify technologies that are suitable for permanent installation in products and that are low cost. The promising technologies for these tasks are:

- RFID (radiofrequency identification) labels for wireless part identification.
- MEMS sensors for contactless data measurement.
- IoT devices with independent wireless communication.

A modern approach to product identification is use of RFID UHF (ultrahigh-frequency) technology. Some research on RFID systems is described in papers about security of tags [4], detection of missing tags [5], and new searching protocols [6]. MEMS sensors have minimal power consumption and can be powered by a battery during product operation. We can monitor product overheating, vibration, and the status of the environment continuously.

Research on monitoring devices using MEMS sensors has been published in papers discussing monitoring of mechatronic systems by MEMS [7], vibration analyses by MEMS [8], and structural monitoring by MEMS [9].

Research on cloud systems and data mining has been described in papers discussing clouds and big data connections [10], cloud robotics data [11], clouds in industrial automation [12], and combination of chaos theory with cloud systems [13]. There are also research papers discussing industrial process simulation and data digitalization in Tecnomatix software [14, 15].

3 Example of Data Acquisition from Industrial Production

Our task was specification of technologies for data acquisition from the control process. We selected three technologies for data collection:

- A vision system (which is usually an industrial IoT system).
- An RFID reader and tags (a combination of an industrial system with IoT technology).
- MEMS sensors (battery-powered IoT technology).

Industrial control systems (for example, PLC) do not have direct support for connection to IoT devices and data collection. Bridge devices are needed to transfer data from IoT to industrial networks. The components for bidirectional data transfer are called gateways and provide security support by encryption and isolation of industrial networks from IoT networks.

Our solution consisted of an OPC (Open Platform Communications) server using Python language for data collection from IoT devices, which communicated with a MindConnect IoT2040 gateway.

A scheme of the components used in an experimental conveyor system for collection and data transformation from IoT to a cloud platform is shown in Fig. 1.

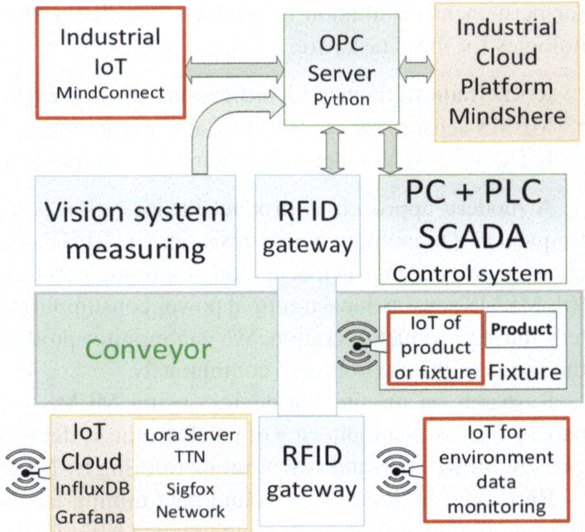

Fig. 1 Concept of an experimental system with integrated IoT (internet of things) devices and a cloud platform for data collection

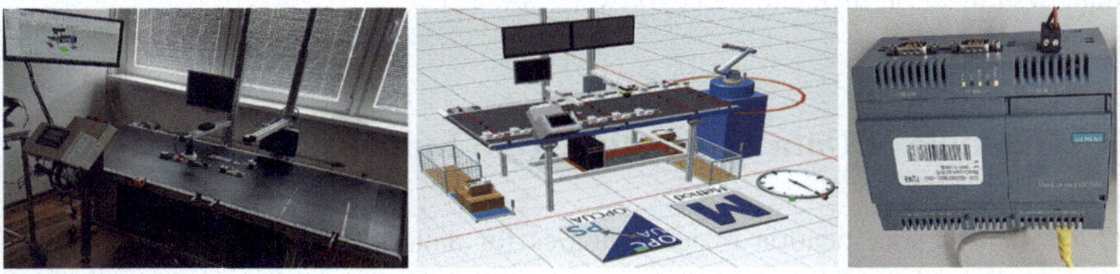

Fig. 2 Experimental conveyor system (*left*), three-dimensional model (*middle*), and an industrial IoT (internet of things) gateway (*right*)

3.1 Device Sources for IoT Implementation and Data Acquisition

An experimental system for IoT implementation and data acquisition is shown in Fig. 2.

Industrial data collection technologies are described in the following sections.

3.1.1 Industrial Vision System

An integrated vision system provides measurement data from parts positioned on fixtures. An example of data acquisition by measurement of a product in a production system is shown in Fig. 3.

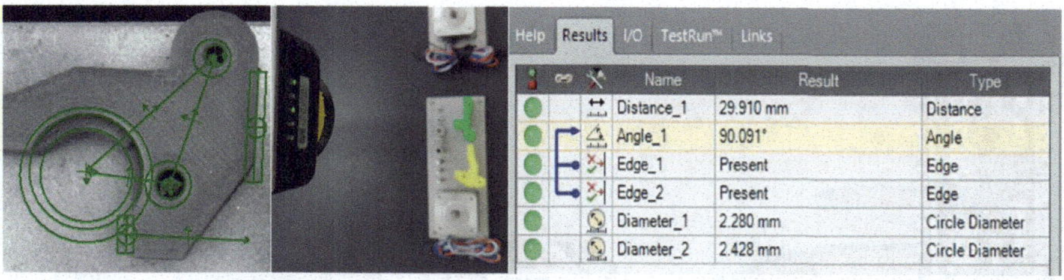

Fig. 3 Dimension check (*left*), Cognex vision system (*middle*), and measured data (*right*)

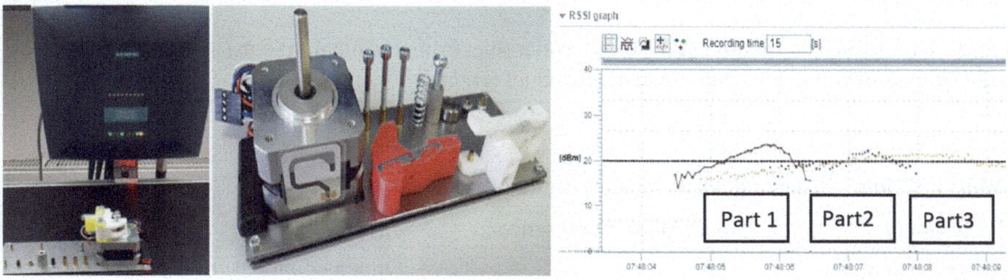

Fig. 4 RFID (radiofrequency identification) reader (*left*), assembly parts and fixture with tags (*middle*), and signal from tags (*right*)

3.1.2 RFID System

RFID technology can provide other data for cloud systems. Our experimental RFID system consisted of one RFID reader and three tags fixed to parts and fixtures. The RFID signal intensity (RSSI (received signal strength indicator)) from the transponder or tags can help to acquire the exact position of the tags according to the signal intensity value.

An example of a subsystem for data acquisition from an RFID system based on a reader and tags is shown in Fig. 4.

3.1.3 IoT Devices with MEMS Sensors

MEMS sensors are suitable for integration into the final product as IoT devices and can be used for long-term monitoring/diagnostics by customers. Our proposal for the final product data collection (lifetime monitoring) was integration of MEMS sensors combined with an MCU (microcontroller unit) controller and an IoT communication module. Connection with the manufacturer can be realized at fixed intervals to send accumulated data to a cloud platform. We selected two variants of testing boards—an ESP32 and a Pycom MCU—and connected them to external MEMS (an accelerometer, gyroscope, magnetometer, and temperature gauge) with Sigfox, LoRa/LoRaWAN, and a GSM communications module. The wireless IoT

Fig. 5 Pycom Sigfox module and MEMS (microelectromechanical system) sensors (*left*), and an IoT (internet of things) module fixed to a product (*right*)

Table 1 Comparisons of devices for data digitalization

Technology	Device/product presence	Tasks
Vision systems	–/yes	Dimension measuring , object or error presence
RFID	–/yes	Position detection, product presence, or personalization
IoT with MEMS	yes/yes	Product or device monitoring

IoT internet of things, *MEMS* microelectromechanical system, *RFID* radiofrequency identification

devices used for our experiments with data collection from MEMS sensors are shown in Fig. 5.

The main problem of wireless connectivity for the customer is additional configuration for an IoT connection. This problem can be solved by specialized IoT device networks (Sigfox, LoRaWAN, GSM), which can be set up and managed by the IoT manufacturer. A summary and comparison of data digitalization devices in use is shown in Table 1.

4 Problem Formulation

Standard IoT devices have some disadvantages that are not acceptable by Industry 4.0 standards. The main problem is their low reliability for wireless data transfer to a cloud platform. Our idea was combination of more IoT wireless technologies with different parameters to increase reliability and maintain continuous data flow to a cloud platform.

For this task, we selected a combination of three basic IoT communication technologies in one device to ensure data collection when signal quality degradation increased in currently used technology. The main idea was continuous measurement of signal quality via the selected IoT device communication channels and switching between them.

For reliable switching between IoT communication technologies, some theoretical knowledge about signal quality, noise, and prediction trends is needed. Signal quality can be measured using the LQI (link quality index), RSSI, SNR (signal-to-noise ratio), and BER (bit error rate).

SNR is a ratio-based value that evaluates a signal based on the noise being seen. SNR comprises two values and is measured as a positive value between 0 and 120 db; the closer it is to 120 db, the better. The signal value and noise value are typically expressed in decibels.

RSSI is measured in decibels from 0 to −120. When looking at this value, the closer to 0 it is, the stronger the signal is, which is better. The signal is affected by the access points (APs) transmission power and antenna, as well as the client's antenna.

$$\text{Actual RSSI} + \text{antenna gain} = \text{displayed RSSI}$$
$$-68\text{db} + 2\text{db} = -66\text{db}$$

IoT devices work with lower frequencies like standard voice and video data transfer, and can have significantly lower signal quality with stable communication.

FSPL (free space path loss) is a measure of how much signal power is lost over a given distance. Typically, about 0.020 db per foot is lost in an outdoor or large open office space; doors, walls, glass, etc. affect this. This is why, as you walk away from an AP, your signal gets weaker.

LQI information is often useful for quality-of-service measurements or location approximation algorithms. The LQI value is an unsigned 8-bit integer (int8u) ranging from 0 to 255. The relationship between the LQI and the link "cost" for purposes of route/neighbor maintenance in the stack is such that LQI values of 200 map to the lowest costs of 1, 3, and 5, while LQIs below 200 represent links with high error rates, so the worst-case cost of 7 is assigned. A LQI value of 200 represents approximately 80% reliability of receiving the packet intact. LQI measurement is based on the chip error rate of the current packet being received from the previous hop of the inbound route, so it provides information specific to the link–layer connection to the neighboring device relaying the current packet to the local device.

BER is the number of bit errors divided by the total number of transferred bits during a selected time interval. The bit error ratio is a unitless value, often expressed as a percentage.

We now can define the switching level by these values according to published quality tables of selected technologies.

4.1 GSM Signal Quality Evaluation

The GSM signal quality can be evaluated using the AT command AT+CSQ (a network signal quality query) and is rated from 0 to 31. We can divide the resulting values into different levels of quality, which are explained in Table 2.

Table 2 GSM (Global System for Mobile Communications) signal quality qualification

Measured value	RSSI (dBm)	Condition
2 to 9	−109 to −95	Marginal
10 to 14	−93 to −85	OK
15 to 19	−83 to −75	Good
20 to 30	−73 to −53	Excellent

RSSI received signal strength indicator

Table 3 Sigfox signal quality quantification in RSSI (received signal strength indicator) for the RC1 region (Europe)

RSSI	Number of base stations	Link quality indicator
−122 dBm < RSSI	3	Excellent
−135 dBm < RSSI ≤ −122 dBm	3	Good
−122 dBm < RSSI	1 or 2	Good
−135 dBm < RSSI ≤ −122 dBm	1 or 2	Average
RSSI ≤ −135 dBm	Any	Limited

4.2 Sigfox IoT Signal Evaluation

Sigfox signal quality gives much better results with low signals like standard Wi-Fi connections because lower frequencies are used. Sigfox provides four basic signal indicators: LQI, RSSI, SNR, and receiver redundancy (the number of stations that receive a message). Signal evaluation for Sigfox communication is divided into five groups, which are defined in Table 3.

4.3 LoRaWAN IoT Signal Evaluation

LoRa point-to-point communication modules do not provide information about signal quality. LoRaWAN modules provide basic two-signal quality evaluation (RSSI and SNR), as shown in Table 4.

4.4 Limitations of IoT Technologies

The main problem of IoT devices are the number of messages (downlinks, uplinks) allowed per day, the size of messages, and the need for minimal delay. These limitations of IoT communication technologies currently in use are summarized in Table 5.

Table 4 LoRaWAN signal quality qualification by RSSI

SNR value	RSSI (dBm)	Condition
2 to 9	−130 to −120	Marginal
10 to 20	−119 to −100	OK
21 to 30	−100 to −90	Good

SNR signal-to-noise ratio

Table 5 Comparison of IoT (internet of things) technologies

	Number of messages per day			
Technology	Uplink	Downlink	Bytes	Delay (s)
LoRaWAN (TTN)	1440	Maximum 10	10	30
Sigfox	140	Maximum 4	12	30
GSM	Not defined	Not defined	UDP/TCP	–

GSM Global System for Mobile Communications, *TCP* transmission control protocol, *TTN* The Things Network, *UDP* user datagram protocol

5 Main Principle of IoT Hybrid Device Combination

We selected three technologies that could provide different features in the IoT sector: GSM technology, Sigfox, and LoRaWAN. The main principle of combination of IoT technologies into one module with data collection to an open-source system is shown in Fig. 6.

For simple testing of communication, we used only a temperature sensor for output data (uplink) and relay on/off (downlink).

We defined the cascade leveling of switching according to the signal quality degradation of the signal. The IoT communication technologies were LoRaWAN and Sigfox, and the last backup was GSM technology.

The LoRaWAN condition was an RSSI value of less than −119 dBm and an SNR value of less than 10. The Sigfox condition for switching was the LQI limit and an RSSI value of less than −134 dBm. At a minimal RSSI of −109 dBm, the GSM tried to switch to the other two technologies. If some IoT communication technology at a higher level reached a stable signal, the backup technologies switched back. The proposed principle for switching between these IoT communication technologies is displayed in Fig. 7a. We tested the signal degradation by moving the device between two towns. The trajectory for signal monitoring is shown in Fig. 7b.

The first experiments with signal quality monitoring were performed using each of the separate technologies.

5.1 GSM Module as an IoT Device

The principle scheme of an IoT solution based on a GSM module with input and output devices is shown in Fig. 8.

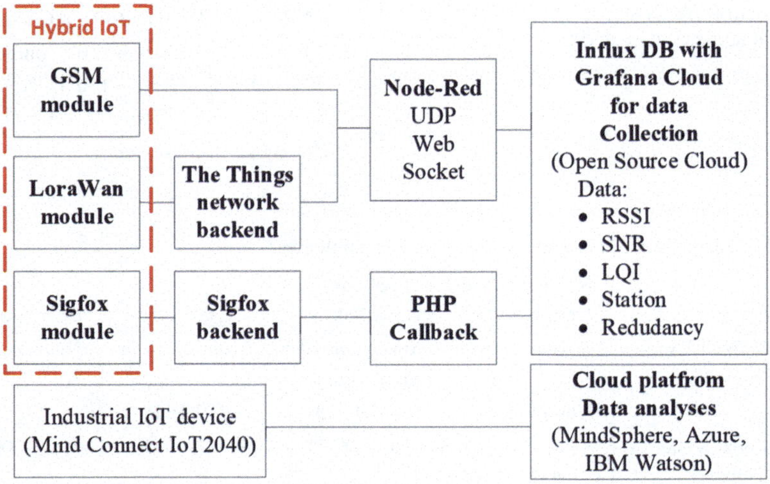

Fig. 6 Principle scheme of a hybrid experimental IoT (internet of things) device

The experimental GSM device used for measuring signal quality with a temperature sensor and relay is shown in Fig. 9.

The visual software Node-RED provided transfer of data from the GSM module controlled by Arduino Nano V3 to an InfluxDB database with visualization by the Grafana UI (user interface), as shown in Fig. 10.

5.1.1 GSM Signal Acquisition

The signal quality measurements along the 30 km trajectory between the two towns are shown in Fig. 11.

The Grafana application received the actual temperature, a signal represented by the CSQ (cell signal quality) value, and the actual credit associated with a SIM (subscriber identification module) card.

5.2 Sigfox IoT Device

The main principle of Sigfox data acquisition and transfer to the cloud platform is shown in Fig. 12.

We tested two devices. The first was a Pycom module with an integrated Sigfox module, and the second was our design with a Wisol Sigfox module combined with an Arduino Nano V3 MCU. The experimental devices are shown in Fig. 13.

The graphic control application for IoT device control was based on a PHP web page and Grafana data collection, as shown in Fig. 14.

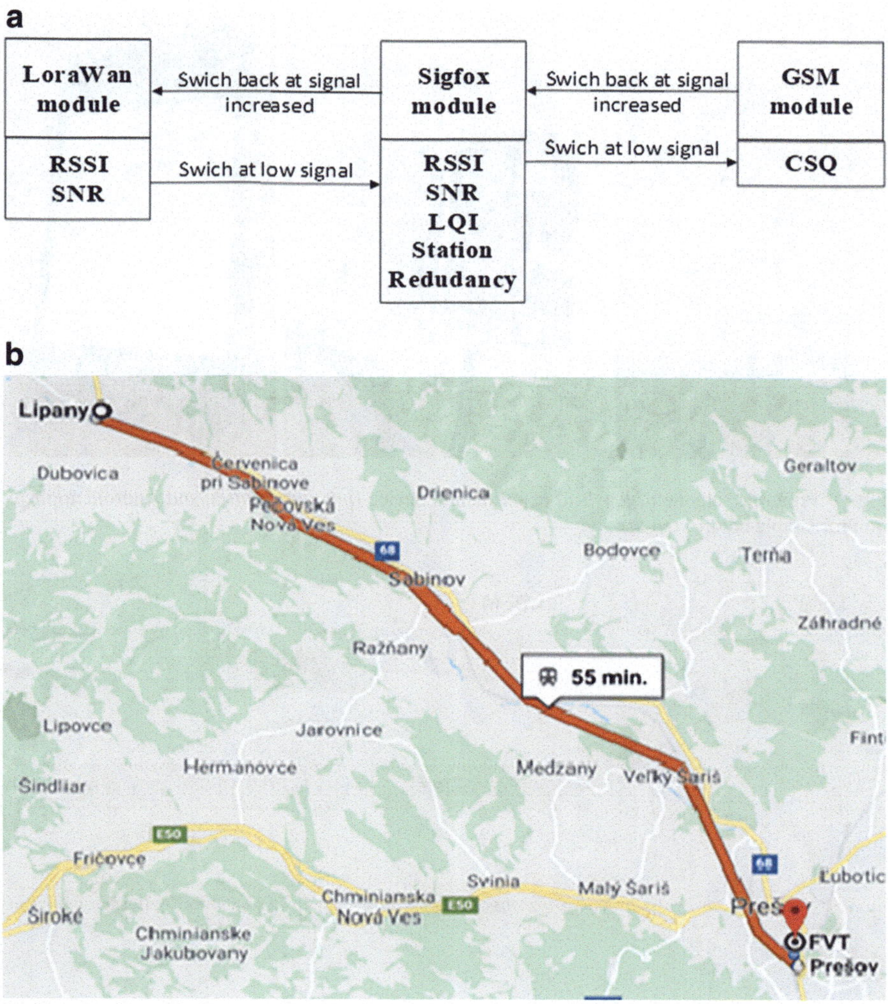

Fig. 7 (**a**) Methodology for switching between IoT (internet of things) technologies according to signal quality. (**b**) Testing trajectory for all devices

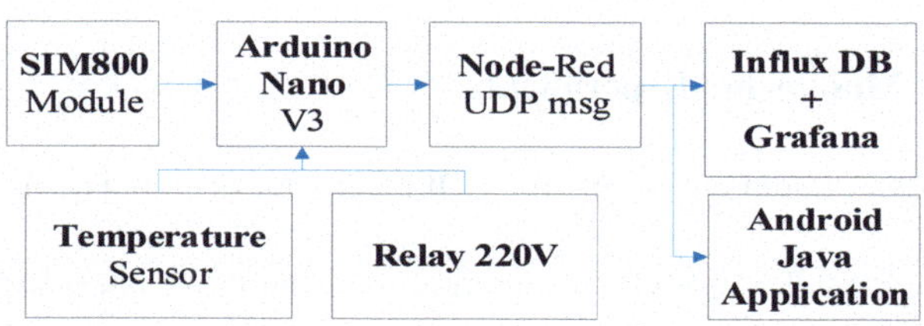

Fig. 8 Principle scheme of a hybrid experimental device

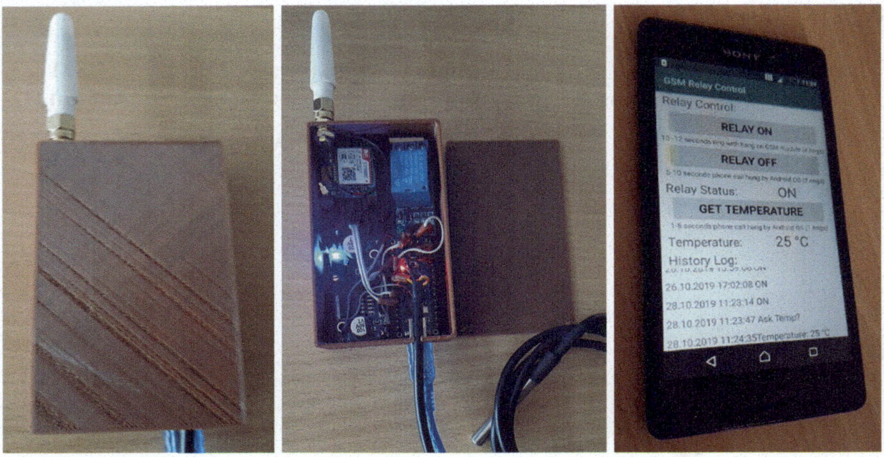

Fig. 9 Principle scheme of a hybrid experimental device (*left* and *middle*), and Android application for simple control (*right*)

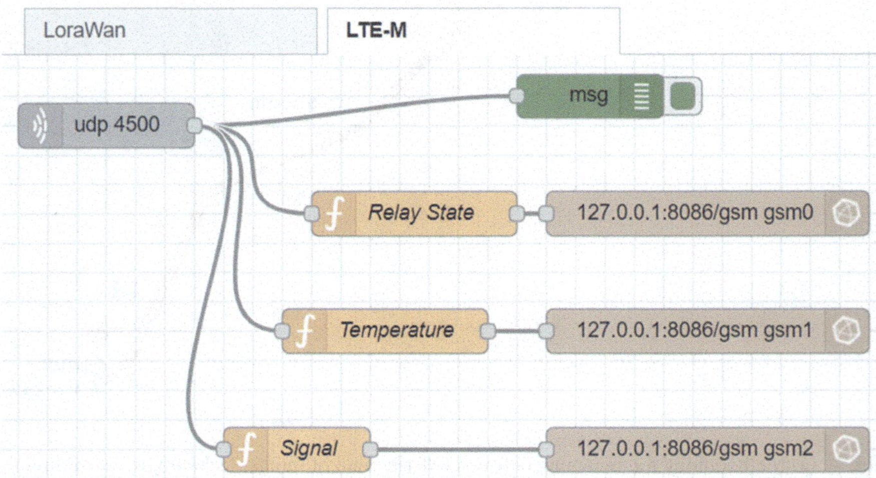

Fig. 10 Node-RED application for data transformation to InfluxDB and Grafana

5.2.1 Signal-Monitoring Experiments

The two-way experiment measuring the LQI parameter of signal quality is shown in Fig. 15.

The signal quality RSSI/SNR measurements during movements between the two towns are shown in the Grafana UI in Fig. 16.

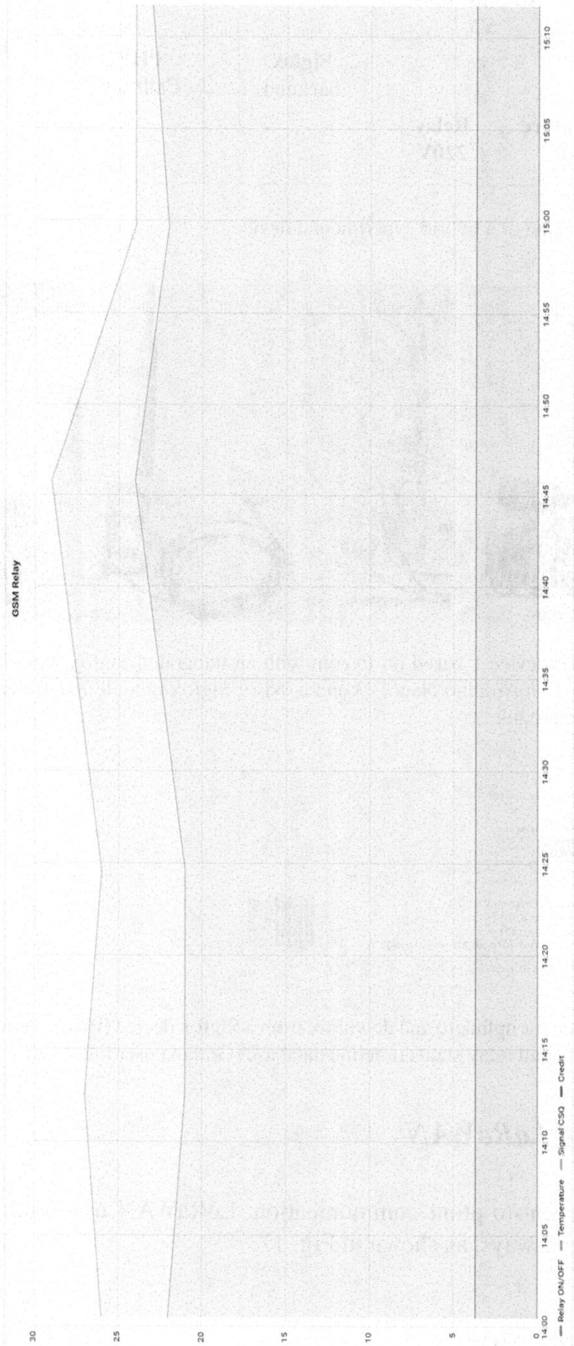

Fig. 11 Node-RED application for data transformation to InfluxDB and Grafana

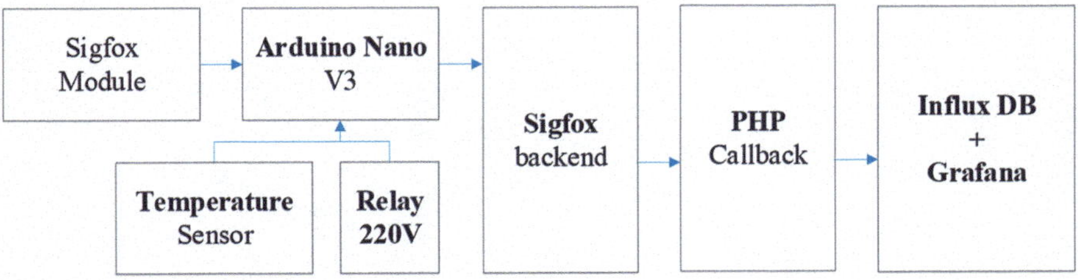

Fig. 12 Principle scheme of a hybrid experimental device

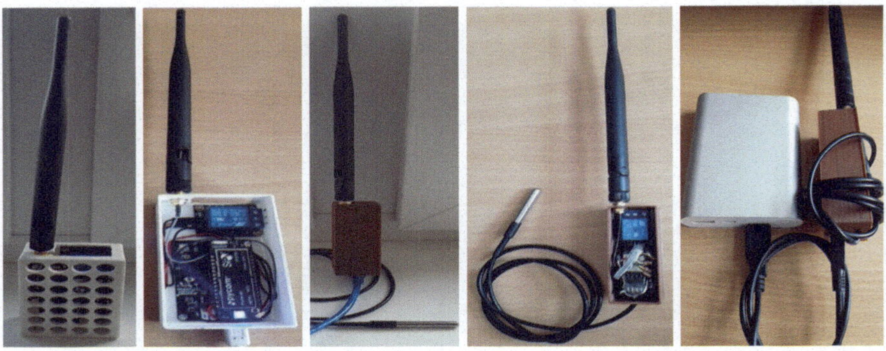

Fig. 13 Experimental device 1 based on Pycom with an integrated Sigfox module, and experimental device 2 based on Arduino Nano V3 and a Wisol Sigfox module IoT (internet of things) testing set with a power bank

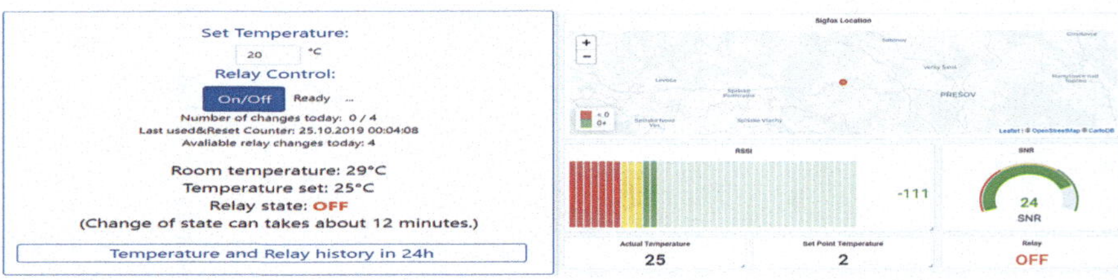

Fig. 14 User interface for uplink to and downlink from a Sigfox device (PHP), showing the actual temperature, set point, and relay status timeline database (Grafana user interface)

5.3 *LoRa and LoRaWAN*

LoRa uses only point-to-point communication. LoRaWAN uses multipoint communication with gateways, as shown in Fig. 17.

5.3.1 LoRa Device Nodes

The experimental devices, based on an Ebyte module combined with an MCU, are shown in Fig. 18.

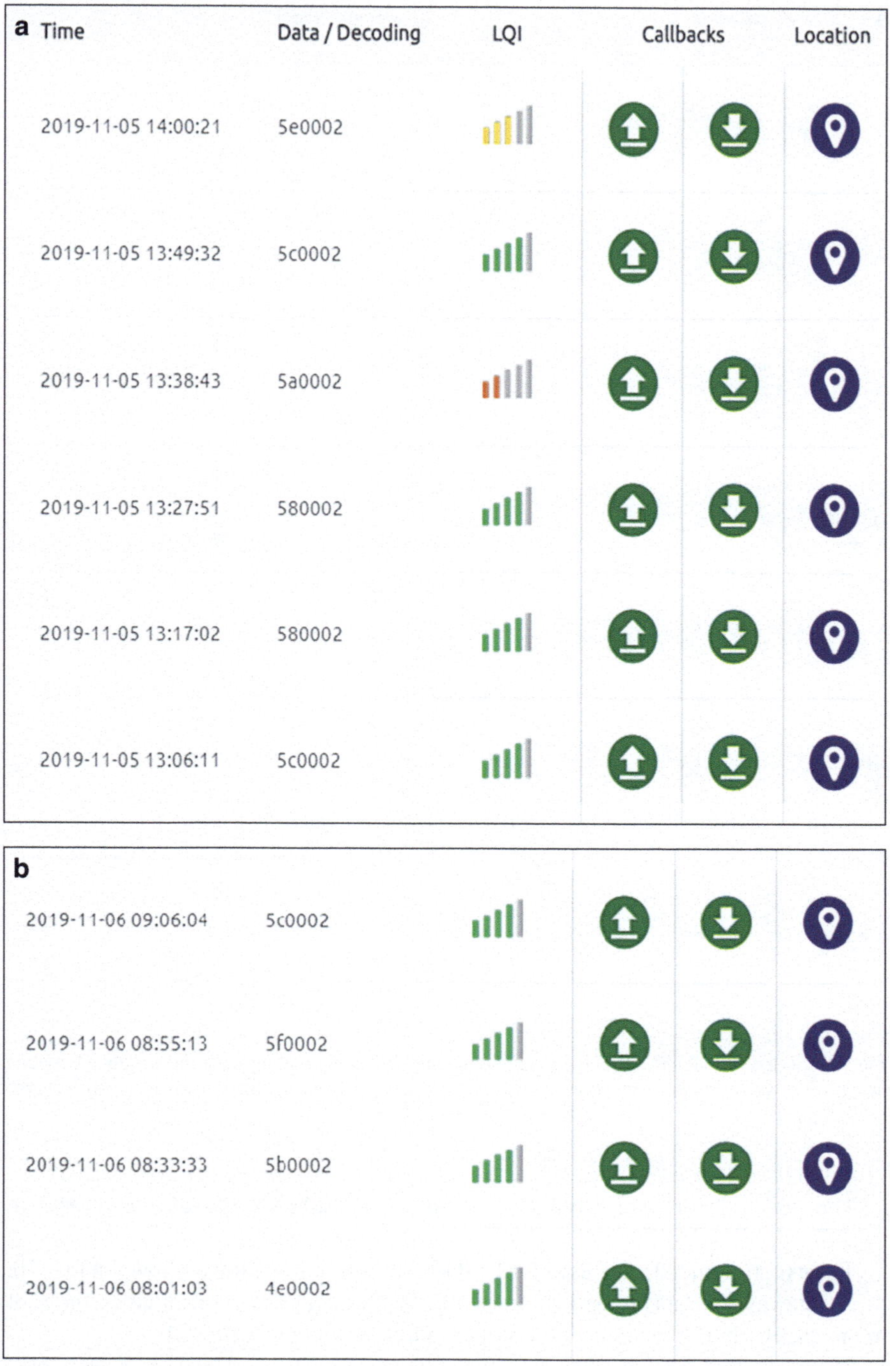

Fig. 15 Signal test (LQI (link quality index)) between two towns over a distance of about 30 km

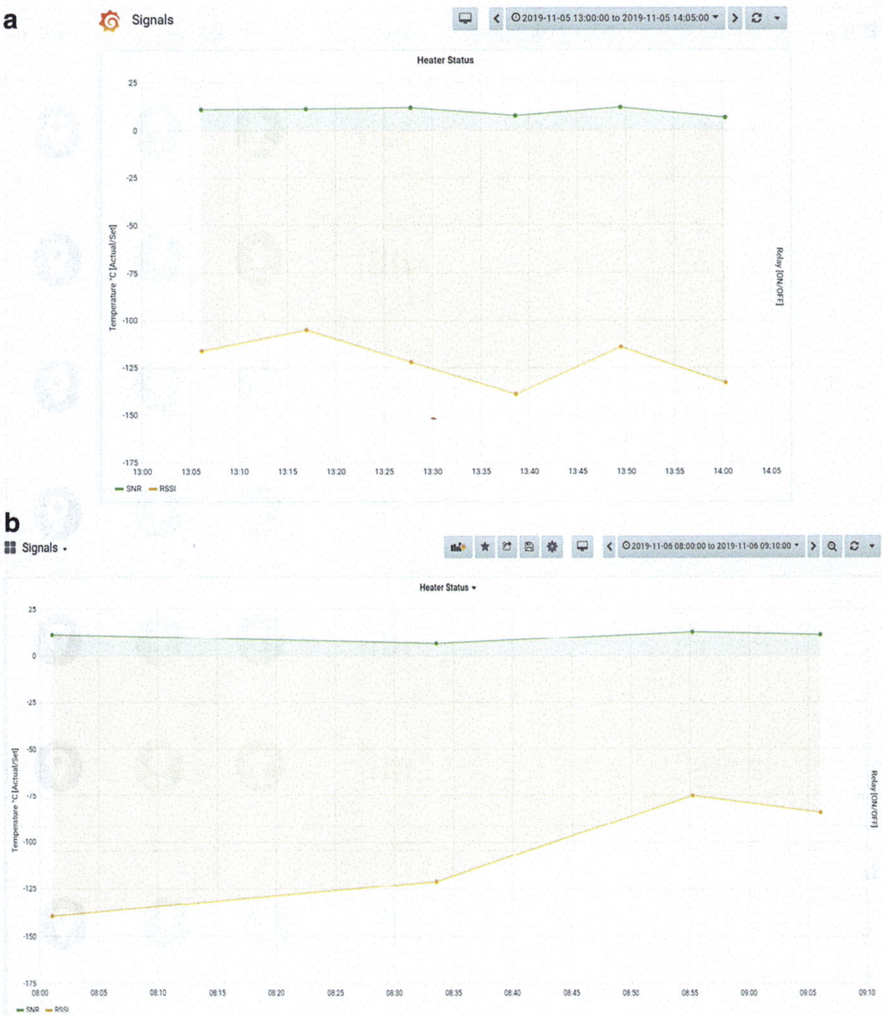

Fig. 16 Signal test (RSSI/SNR) (received signal strength indicator/signal-to-noise ratio) between two towns over a distance of about 30 km, with a two-way check. (**a**) A-to-B test. (**b**) Reverse B-to-A test

The principle of data acquisition from the LoRaWAN modules is shown in Fig. 19.

The experimental device used for LoRaWAN communication is shown in Fig. 20.

A gateway for data collection from LoRaWAN devices was installed on the roof in our faculty buildings to cover a bigger radius, as shown in Fig. 21.

The algorithm used for data transformation from a TTN (The Things Network) backend to our department server cloud, based on InfluxDB and the Grafana visualization interface, is shown in Fig. 22.

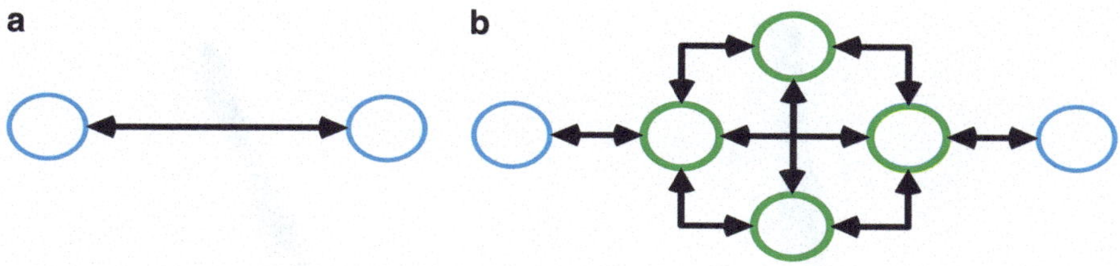

Fig. 17 (**a**) LoRa point-to-point communication. (**b**) LoRaWAN multipoint communication with gateways (*green circles*)

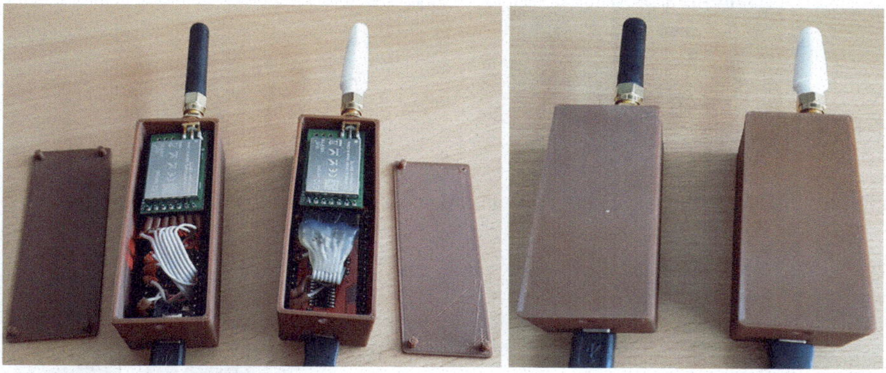

Fig. 18 Experiments using E32 Ebyte LoRa server/client devices

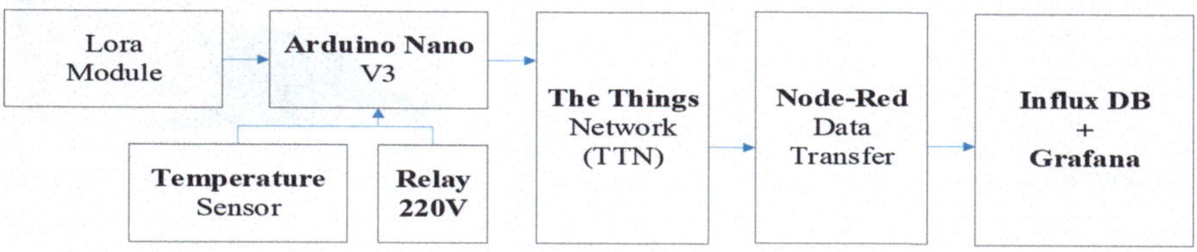

Fig. 19 Basic principle of LoRaWAN data acquisition

5.3.2 Signal-Monitoring Experiments

Uplink and downlink messages from LoRaWAN communication with the JSON (JavaScript Object Notation) data example frame used for signal quality extraction are shown in Fig. 23, and an actual signal status is shown in Fig. 24.

5.4 Industrial Data Processing by a Cloud Platform

Industrial data that do not require wireless communication can use the industrial IoT gateway MindConnect. These data are provided by a vision system and an RFID reader, and can be stored in some kind of commercial cloud platform. An example of data collection to MindSphere Cloud, IBM Watson, and Azure is shown in Fig. 25.

Fig. 20 LoRaWAN Dragino module with an Arduino Uno microcontroller unit board

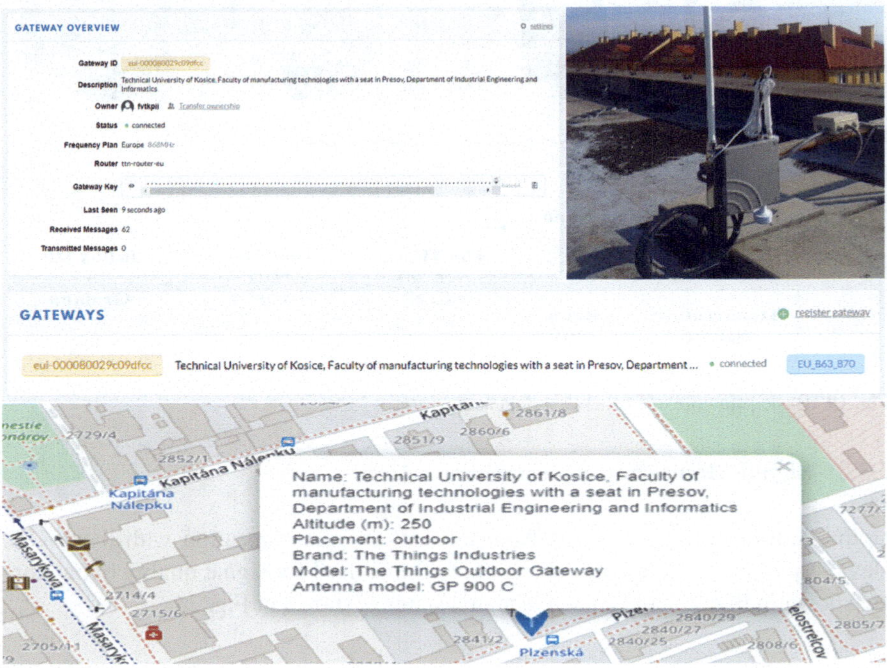

Fig. 21 Gateway implementation on the roof of our department

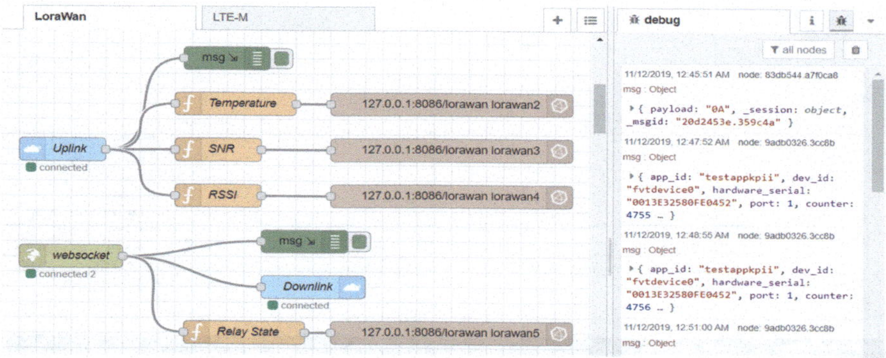

Fig. 22 Data transfer to InfluxDB by Node-RED

Fig. 23 Example of one downlink and two uplink communications between a node and gateway, and JSON (JavaScript Object Notation) data for signal data extraction RSSI/SNR (received signal strength indicator/signal-to-noise ratio)

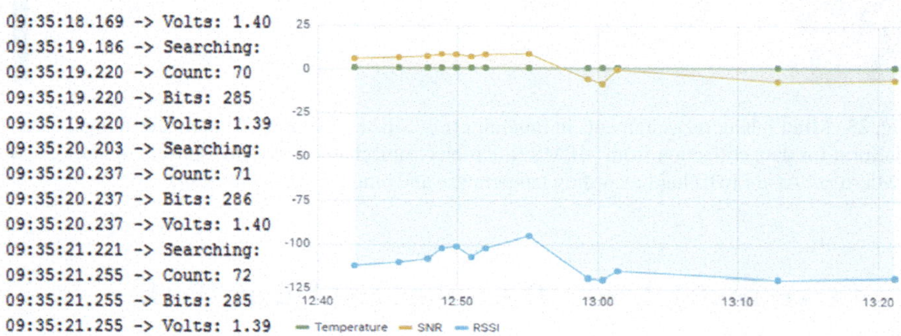

Fig. 24 (**a**) LoRa messages from a receiver. (**b**) LoRaWAN signal experimental test

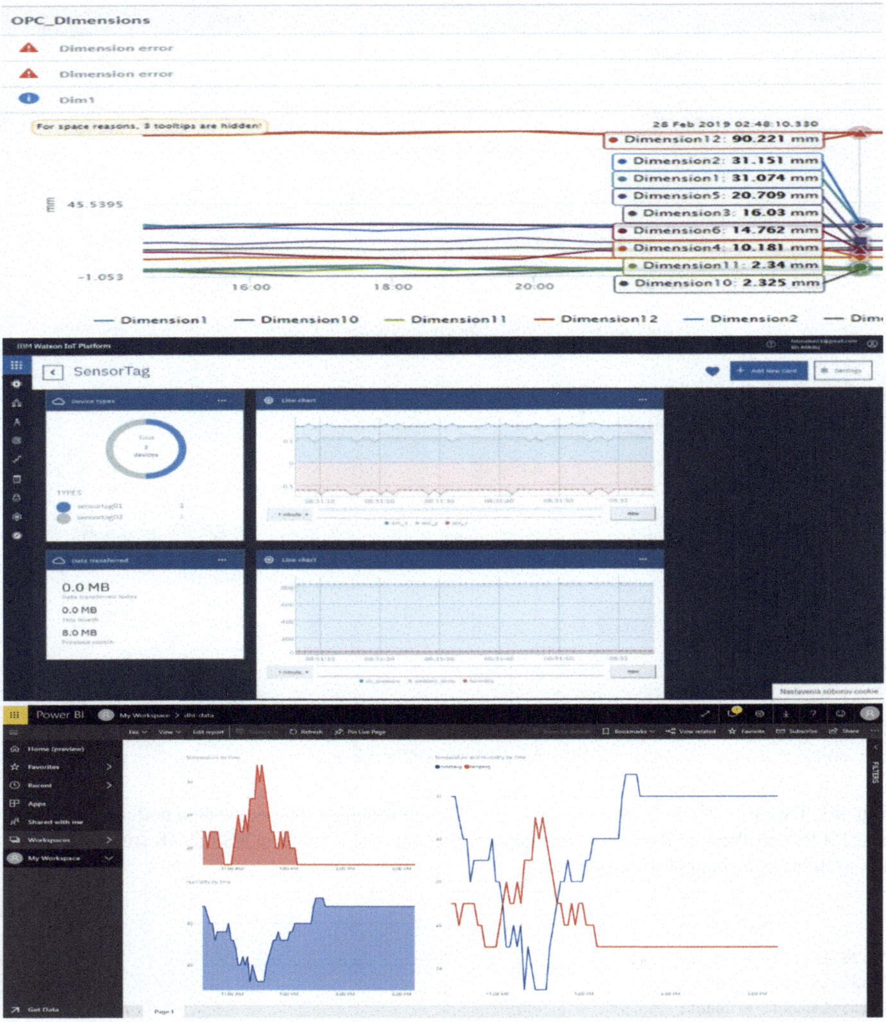

Fig. 25 MindSphere measurements in timelines (*top*), IBM Watson IoT (internet of things) cloud platform for data collection from MEMS (microelectromechanical system) sensors (*middle*), and a Microsoft Azure IoT cloud recording temperature and humidity data (*bottom*)

6 Experimental Results, Discussion, and Future Work

In accordance with the limitations of IoT communication technologies in terms of bandwidth and delay between messages, use of switching technologies cannot be very frequent. The results from our experiments on the maximum and minimum signal quality according to the evaluation values for all three tested devices are shown in Table 6.

Table 6 Signal evaluation results

	Signal evaluation parameter			
	SNR	RSSI	LQI	CSQ
LoRaWAN	9 to −9	−95 to −124	–	–
Sigfox	23 to 10	−75 to −135	Poor/average	–
GSM	–	–	–	23 to 28

CSQ cell signal quality, *GSM* Global System for Mobile Communications, *LQI* link quality index, *RSSI* received signal strength indicator, *SNR* signal-to-noise ratio

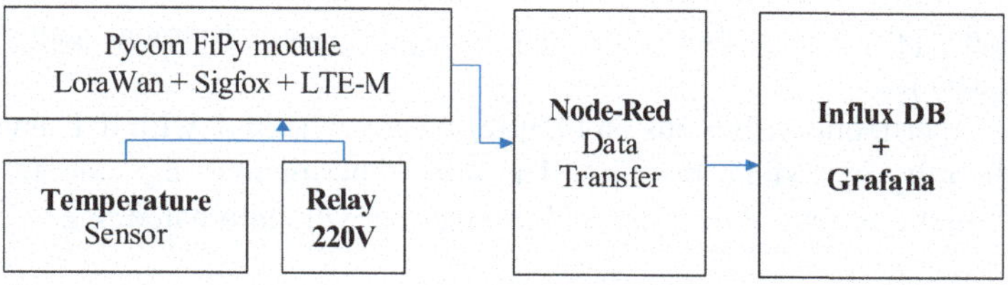

Fig. 26 Principle of a hybrid IoT (internet of things) device integrated into one module

The best technology for industrial use is LoRaWAN because device data can be collected by the company gateway and stored in a local cloud with minimum cost. The first IoT backup technology is a Sigfox IoT device, but it requires payment of a yearly fee. The last backup technology is a GSM (LTE-M/NB-IoT)–based IoT device with higher power consumption, which is not suitable to be powered from a battery for a long time but can be used as a reliable backup for a short time.

Our next experiments will be aimed at minimizing the dimensions and increasing the compactness of the solution. It is possible to integrate all of these technologies into one control system. GSM (LTE-M), Sigfox, and LoRa/LoRaWAN communication technology can be continuously tested for signal quality by one control system. A cloud system with history data mining can predict degradation of signals in time. An experimental hybrid IoT device using a Pycom FiPy module is shown in Fig. 26.

7 Conclusion

Data from industrial RFID and vision systems can be transferred to a cloud platform by wire transfer and an IoT gateway with synchronization to an OPC server. This communication is very stable in comparison with wireless data transfer, which is the main feature of IoT devices. IoT wireless communication is influenced by many factors such as noise, distance, and the number of walls or buildings between the real device and the cloud platform.

This paper has described the basic ideas, methodology, and experiments involved in combination of three IoT communication technologies (LoRaWAN, Sigfox, and GSM) with one hybrid device, with increased reliability.

Experiments with these IoT communication technologies provided reliability during transfer between two towns. We want to test their usability in a highly urbanized environment and an area with a lower density of buildings.

According to the results of the experiments, the most suitable technology for the main data collection is LoRaWAN communication because it provides a local star topology managed in an acceptable range of about 3 km, which is acceptable for small and medium-sized factories. International companies can extend this range by use of interconnected gateways. Sigfox can be used as a first-level backup data device because it provides roaming features and can be used with battery power over a long period. IoT GSM modules have limited roaming connections, can operate on battery power for a short time, and in very noisy environments can switch to other frequencies.

Future experiments will focus on long-term data collection with IoT devices and signal reliability in a static position. The next experimental implementation will take place in the setting of existing real production with automated lines.

Acknowledgements This work was supported by the Slovak Research and Development Agency under contract no. APVV-15-0602, VEGA 1/0700/20 and also by the Project of Structural Funds of the EU, ITMS code 26220220103.

References

1. Xu, L.D., Duan, L.: Big data for cyber physical systems in industry 4.0: a survey. Enterp. Inf. Syst. **13**(2), 148–169 (2019). https://doi.org/10.1080/17517575.2018.1442934
2. Li, G., Tan, J., Chaudhry, S.S.: Industry 4.0 and big data innovations. Enterp. Inf. Syst. **13**(2), 145–147 (2019). https://doi.org/10.1080/17517575.2018.1554190
3. Sanin, C., Haoxi, Z., Shafiq, I., Waris, M.M., Silva de Oliveira, C., Szczerbicki, E.: Experience based knowledge representation for internet of things and cyber physical systems with case studies. Futur. Gener. Comput. Syst. **92**, 604–616 (2019). https://doi.org/10.1016/j.future.2018.01.062
4. Han, W., Liu, W., Zhang, K., Li, Z., Liu, Z.: A protocol for detecting missing target tags in RFID systems. J. Netw. Comput. Appl. **132**, 40–48 (2019). https://doi.org/10.1016/j.jnca.2019.01.027
5. Lee, C.-C., Chen, S.-D., Li, C.-T., Cheng, C.-L., Lai, Y.-M.: Security enhancement on an RFID ownership transfer protocol based on cloud. Futur. Gener. Comput. Syst. **93**, 266–277 (2019). https://doi.org/10.1016/j.future.2018.10.040
6. Liu, C.-G., Liu, I.-H., Lin, C.-D., Li, J.-S.: A novel tag searching protocol with time efficiency and searching accuracy in RFID systems. Comput. Netw. **150**, 201–216 (2019). https://doi.org/10.1016/j.comnet.2019.01.011
7. Židek, K., Hošovský, A.: Wireless device based on MEMS sensors and Bluetooth low energy (LE/smart) technology for diagnostics of mechatronic systems. Appl. Mech. Mater. **460**, 13–21 (2013). https://doi.org/10.4028/www.scientific.net/AMM.460.13
8. Varanis, M., Silva, A., Mereles, A., Pederiva, R.: MEMS accelerometers for mechanical vibrations analysis: a comprehensive review with applications. J. Braz. Soc. Mech. Sci. Eng. **40**(11), 527 (2018). https://doi.org/10.1007/s40430-018-1445-5
9. Dumont, M., Wolf, D.: Usage of MEMS capacitive acceleration sensors for structural monitoring. Dyn. Civil Struct. **2**, 77–89 (2019). https://doi.org/10.1007/978-3-319-74421-6_11

10. Lu, Y., Xu, X.: Cloud-based manufacturing equipment and big data analytics to enable on-demand manufacturing services. Robot. Comput. Integr. Manuf. **57**, 92–102 (2019). https://doi.org/10.1016/j.rcim.2018.11.006
11. Aissam, M., Benbrahim, M., Kabbaj, M.N.: Cloud robotic: opening a new road to the Industry 4.0. In: Derbel, N., Ghommam, J., Zhu, Q. (eds.) New developments and advances in robot control. Studies in systems, decision and control, vol. 175, pp. 1–20. Springer, Singapore (2019). https://doi.org/10.1007/978-981-13-2212-9_1
12. Mahmoud, M.S.: Architecture for cloud-based industrial automation. In: Yang, X.S., Sherratt, S., Dey, N., Joshi, A. (eds.) Third international congress on information and communication technology. Advances in intelligent systems and computing, vol. 797, pp. 51–62. Springer, Singapore (2019). https://doi.org/10.1007/978-981-13-1165-9_6
13. Hu, Y., Zhu, F., Zhang, L., Lui, Y., Wang, Z.: Scheduling of manufacturers based on chaos optimization algorithm in cloud manufacturing. Robot. Comput. Integr. Manuf. **58**, 13–20 (2019). https://doi.org/10.1016/j.rcim.2019.01.010
14. Hrehova, S.: Predictive model to evaluation quality of the manufacturing process using Matlab tools. Procedia Eng. **149**, 149–154 (2016). https://doi.org/10.1016/j.proeng.2016.06.649
15. Lazár, I., Husár, J.: Validation of the serviceability of the manufacturing system using simulation. ERISE J. **5**(4), 252–261 (2012). https://doi.org/10.7160/eriesj.2012.050407

Index

L. Knapcikova et al. (eds.), *New Approaches in Management of Smart Manufacturing Systems*, EAI/Springer Innovations in Communication and Computing, https://doi.org/10.1007/978-3-030-40176-4

E

The manufacturer's authorised representative in the EU is Springer
Nature Customer Service Centre GmbH, Europaplatz 3, 69115 Heidelberg,
Germany. If you have any concerns regarding our products, please
contact ProductSafety@springernature.com

Printed and bound by CPI Group (UK) Ltd, Croydon, CR0 4YY
05/10/2026
02228126-0011